I0064447

Textbook of Animal Genetics and Breeding

Textbook of Animal Genetics and Breeding

Edited by **Dominic Fasso**

SYRAWOOD
PUBLISHING HOUSE

New York

Published by Syrawood Publishing House,
750 Third Avenue, 9th Floor,
New York, NY 10017, USA
www.syrawoodpublishinghouse.com

Textbook of Animal Genetics and Breeding
Edited by Dominic Fasso

© 2016 Syrawood Publishing House

International Standard Book Number: 978-1-68286-059-5 (Hardback)

This book contains information obtained from authentic and highly regarded sources. Copyright for all individual chapters remain with the respective authors as indicated. All chapters are published with permission under the Creative Commons Attribution License or equivalent. A wide variety of references are listed. Permission and sources are indicated; for detailed attributions, please refer to the permissions page and list of contributors. Reasonable efforts have been made to publish reliable data and information, but the authors, editors and publisher cannot assume any responsibility for the validity of all materials or the consequences of their use.

The publisher's policy is to use permanent paper from mills that operate a sustainable forestry policy. Furthermore, the publisher ensures that the text paper and cover boards used have met acceptable environmental accreditation standards.

Trademark Notice: Registered trademark of products or corporate names are used only for explanation and identification without intent to infringe.

Printed in the United States of America.

Contents

Preface

The study of animal genetics and breeding are essential for practices like animal husbandry, etc. This book contains some path-breaking studies in the field of animal genetics which will enable the reader to gain a comprehensive insight into this discipline. Also included in this book are detailed discussions on genomics, DNA structure and modeling, chromosomes, etc. These topics are of utmost significance, especially for students and research scholars of zoology, veterinary sciences and related fields. This book is a complete source of knowledge on the latest advances in the field of animal genetics and breeding.

Significant researches are present in this book. Intensive efforts have been employed by authors to make this book an outstanding discourse. This book contains the enlightening chapters which have been written on the basis of significant researches done by the experts.

Finally, I would also like to thank all the members involved in this book for being a team and meeting all the deadlines for the submission of their respective works. I would also like to thank my friends and family for being supportive in my efforts.

Editor

Development and epithelial organisation of muscle cells in the sea anemone *Nematostella vectensis*

Stefan M Jahnel[1], Manfred Walzl[2] and Ulrich Technau[1*]

Abstract

Introduction: *Nematostella vectensis*, a member of the cnidarian class Anthozoa, has been established as a promising model system in developmental biology, but while information about the genetic regulation of embryonic development is rapidly increasing, little is known about the cellular organization of the various cell types in the adult. Here, we studied the anatomy and development of the muscular system of *N. vectensis* to obtain further insights into the evolution of muscle cells.

Results: The muscular system of *N. vectensis* is comprised of five distinct muscle groups, which are differentiated into a tentacle and a body column system. Both systems house longitudinal as well as circular portions. With the exception of the ectodermal tentacle longitudinal muscle, all muscle groups are of endodermal origin. The shape and epithelial organization of muscle cells vary considerably between different muscle groups. Ring muscle cells are formed as epitheliomuscular cells in which the myofilaments are housed in the basal part of the cell, while the apical part is connected to neighboring cells by apical cell-cell junctions. In the longitudinal muscles of the column, the muscular part at the basal side is connected to the apical part by a long and narrow cytoplasmic bridge. The organization of these cells, however, remains epitheliomuscular. A third type of muscle cell is represented in the longitudinal muscle of the tentacle. Using transgenic animals we show that the apical cell-cell junctions are lost during differentiation, resulting in a detachment of the muscle cells to a basiepithelial position. These muscle cells are still located within the epithelium and outside of the basal matrix, therefore constituting basiepithelial myocytes. We demonstrate that all muscle cells, including the longitudinal basiepithelial muscle cells of the tentacle, initially differentiate from regular epithelial cells before they alter their epithelial organisation.

Conclusions: A wide range of different muscle cell morphologies can already be found in a single animal. This suggests how a transition from an epithelially organized muscle system to a mesenchymal could have occurred. Our study on *N. vectensis* provides new insights into the organisation of a muscle system in a non-bilaterian organism.

Introduction

Muscles are present in all metazoans except sponges and placozoans. Their emergence marks an important step in evolution because it allows organisms to disperse, escape, hunt and explore new habitats. Muscle cells are a major derivative of the mesoderm in Bilateria, but can also be found in two non-bilaterian phyla, the Ctenophora and the Cnidaria. The diploblastic Cnidaria are of particular

interest for understanding the evolution of key bilaterian traits because, they are considered to be the sister group of the Bilateria [1,2] and therefore occupy a crucial phylogenetic position. Cnidarian polyps generally have smooth muscles, yet medusae also have striated muscles [3]. The striking structural similarity of striated muscles in Cnidaria and Bilateria has led to the suggestion that striated muscles of Cnidaria and Bilateria are homologous [4]. These authors extended their arguments by proposing that cnidarians are reduced Mesodermata [4]. However, a recent phylogenomic study tracing the evolutionary origin of all muscle components known from model bilaterians revealed the absence of several crucial muscle proteins from the genome of non-

* Correspondence: ulrich.technau@univie.ac.at
[1]Department of Molecular Evolution and Development, Centre for Organismal Biology, Faculty of Life Sciences, University of Vienna, Althanstrasse 14, 1090 Wien, Austria
Full list of author information is available at the end of the article

bilaterian organisms as well as the bilaterian lineage-specific innovations of other crucial muscle proteins [5]. These phylogenetic and expression analyses led to the conclusion that striated muscles evolved convergently in cnidarians and bilaterians, on the basis of ancestral proteins, which predate the divergence of animals [5]. Furthermore, several key myogenic transcription factors such as MyoD and MRFs (myogenic regulatory factors) have not been identified in cnidarians. This raises questions of how muscles in cnidarians develop and how they are structured.

In recent years, *N. vectensis*, a representative of the Anthozoa, has been established as an important model for studying embryology, phylogenetic relationships, comparative genomics and the origin of triploblasty [6-8]. This makes it a promising addition to the existing group of cnidarian model systems. Gene expression and functional studies have led to a much better understanding of the molecular regulation of embryogenesis and larval development during the last decade. Nonetheless, our knowledge about the anatomy and cellular composition of this new model organism remains poor. Often, not even the cell types that account for a highly specific gene expression pattern are known. In particular, the adult stage is poorly understood. Hence, as *N. vectensis* continues to develop into a major cnidarian model organism, we need to reach a deeper understanding of the composition, connections and differentiation kinetics of the different cell types at various developmental stages.

Frank and Bleakney [9] investigated the general anatomy of *N. vectensis* at a histological level, yet the level of resolution and the detail of analysis did not enable conclusions to be drawn about the development and precise cellular composition of the various cell types.

Here, we present a detailed anatomical description of the muscular system of *N. vectensis* using histology, electron microscopy, confocal microscopy and transgenic lines, specifically expressing reporter genes in retractor muscles of the column and tentacles. We show that muscle cells display different levels of epithelial organization, dependent on their position in the organism. They vary from an epitheliomuscular organisation to a basiepithelial muscle cell, which has lost all apical cell-cell junctions and subsequently is positioned at the base of the epithelium. Our data suggest that epitheliomuscular cells can be highly modified to comply with spatial constraints. Based on these results we interpret the different modes of epithelialization to represent intermediate steps of detachment from the epithelium, which finally would enable muscle cells to become located between the ectoderm and endoderm.

Methods
Animal culture
Animals were cultured and spawning was induced as described elsewhere [10,11]. In brief, animals were kept at

18°C, predominantly in the dark, and spawning was induced by raising the temperature to 24°C in the presence of light for 10 h.

Histology
Adult animals were anesthetized with a few drops of 7% $MgCl_2$ for 30 min and fixed in 4% PFA (paraformaldehyde) in PBT (phosphate buffered saline, 0.2% Triton) overnight, washed in PBT, dehydrated in ethanol and methyl benzoate and embedded in paraffin. Sections (7 μm) were stained with Azan.

Transmission electron microscopy
Samples were put on ice and anesthetized as described above with $MgCl_2$ (primary polyps, adults) and subsequently fixed with 2.5% glutaraldehyde in 0.1 M cacodylate buffer (pH 7.2) for 1 h (planulae, primary polyps) or overnight (adults) at 4°C. After fixation, samples were either stored in 1.25% glutaraldehyde in 0.1 M cacodylate buffer (pH 7.2) at 4°C or processed immediately. Then they were washed in the same buffer used for fixation. Samples were postfixed in 1% OsO_4 in 0.1 M cacodylate buffer (pH 7.2) for 30 min and washed with 0.1 M cacodylate buffer (pH 7.2). Thereafter they were dehydrated through a graded series of ethanol and acetone and embedded into the Low Viscosity Resin (Agar) following sectioning using routine techniques. After staining with watery solutions of uranylacetate (20 min) and lead citrate (10 min), sections were examined with a Zeiss Libra 120 transmission electron microscope.

Phalloidin stainings
Animals were anesthetized as described above and fixed in 4% PFA in PBT overnight at 4°C and subsequently washed in the same buffer. Permeabilization was enhanced by putting embryos into ice-cold (–20°C) acetone for 7 min followed by thoroughly washing in PBT. Subsequently samples were incubated in Phalloidin-Alexa 488 (Invitrogen) (3 μl/100 μl PBT) for at least overnight at 4°C in the dark, washed in PBT and mounted on glass slides in Vectashield.

Cryosectioning
Stained animals were cut into smaller pieces and incubated in OCT infiltration solution (20% OCT compound (Sakura), 25% sucrose in PBS) overnight, followed by putting them in a drop of 80% OCT (80% OCT, 25% sucrose in PBS) on a glass slide. A plastic mold was placed over them and filled with two additional drops of 80% OCT. After orientation the slide was placed on a metal block (precooled in liquid nitrogen) and immediately filled with 100% OCT. The solid frozen samples were stored at –20°C or processed immediately. Sections (14 μm) were cut in a Leica CM3050S cryostat at –24°C

and collected on warm (RT) Superfrost Ultra Plus slides (Thermo Scientific). After a drying step (RT, overnight in the dark) the slides were placed in PBS for 5 min to wash away the embedding medium and mounted immediately in Vectashield.

Antibody staining

Samples were anesthetized (primary polyps) and fixed in 4% PFA in PBT for 6 h, washed in PBT and blocked in blocking buffer (80% PBT, 20% sheep serum, 1% bovine serum albuminum) for 2 h at room temperature. They were then incubated in primary antibody (rat, anti-RFP; Chromotek) 1:400 in blocking buffer overnight at 4°C, subsequently washed in PBT and incubated in secondary antibody (goat, anti-rat, DyLight 549; Jackson) 1:500 in blocking buffer containing phalloidin (3 µl/100 µl) for 3 h at room temperature. After an additional washing step, samples were mounted on glass slides in Vectashield.

Results

The muscle system of the sea anemone *N. vectensis* can be roughly divided into a column and a tentacle system (Figure 1). The body column contains three morphologically and functionally distinct muscle groups. The parietal and a retractor muscle are longitudinal muscles. They are orientated along the oral-aboral axis of the polyp body column and located in different regions of the mesentery. The columnar ring muscle cells span the whole body column (Figure 1). Similarly, in the tentacle, ring and longitudinal muscles are present (Figure 1). With the exception of the tentacle longitudinal muscle, all muscles are of endodermal origin. The following subchapter describes in detail the two muscle systems in adults (Figures 2, 3, 4 and 5), as well as the development of the longitudinal muscles of the column (Figures 6, 7 and 8) and tentacle muscles (Figure 9).

The body column and its muscles

Cross-sections of adult polyps revealed the presence of eight uniformly constructed mesenteries, each of which extends from the body wall to the actinopharynx. The typical architecture of a fully differentiated mesentery below the pharynx is shown (Figure 2A, Figure 3A). A characteristic bilateral rosette-like folding of the mesoglea is present at the base of each mesentery. These foldings are aligned by the myonemes of the parietal muscle cells (Figure 2B, Figure 3B). These myonemes connect directly to the neighboring ring muscle at the base of the mesentery (Figure 2D, Figure 3B) of the body column at both sides of the mesentery. Towards the distal part of the mesentery, the parietal muscle is separated from the retractor muscle by the stalk. In the retractor muscle the extensive folding of the mesoglea is even more pronounced and branched than in the parietal muscle, but it

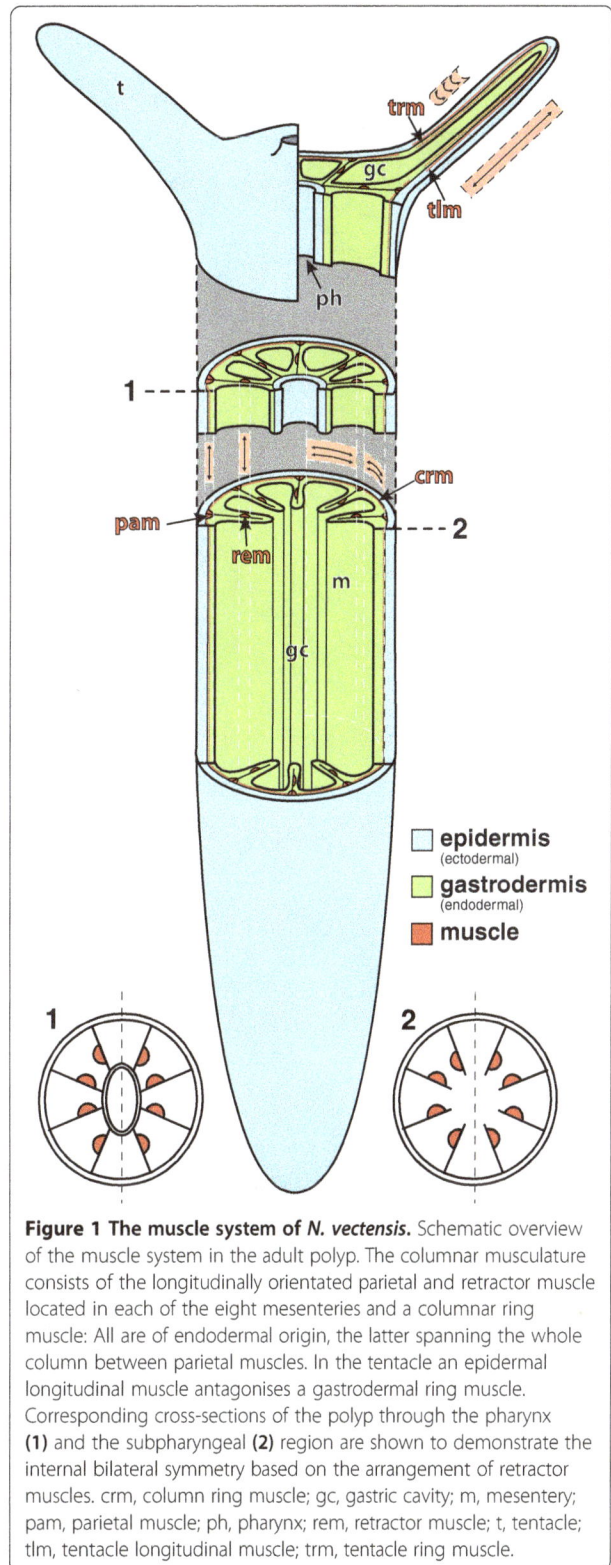

Figure 1 The muscle system of *N. vectensis*. Schematic overview of the muscle system in the adult polyp. The columnar musculature consists of the longitudinally orientated parietal and retractor muscle located in each of the eight mesenteries and a columnar ring muscle: All are of endodermal origin, the latter spanning the whole column between parietal muscles. In the tentacle an epidermal longitudinal muscle antagonises a gastrodermal ring muscle. Corresponding cross-sections of the polyp through the pharynx **(1)** and the subpharyngeal **(2)** region are shown to demonstrate the internal bilateral symmetry based on the arrangement of retractor muscles. crm, column ring muscle; gc, gastric cavity; m, mesentery; pam, parietal muscle; ph, pharynx; rem, retractor muscle; t, tentacle; tlm, tentacle longitudinal muscle; trm, tentacle ring muscle.

is formed only on one side (Figure 2C, Figure 3C). The side of the retractor muscle within the mesentery is non-random and defines the directive axis. Between the mesenterial filament (also termed septal filament) at the

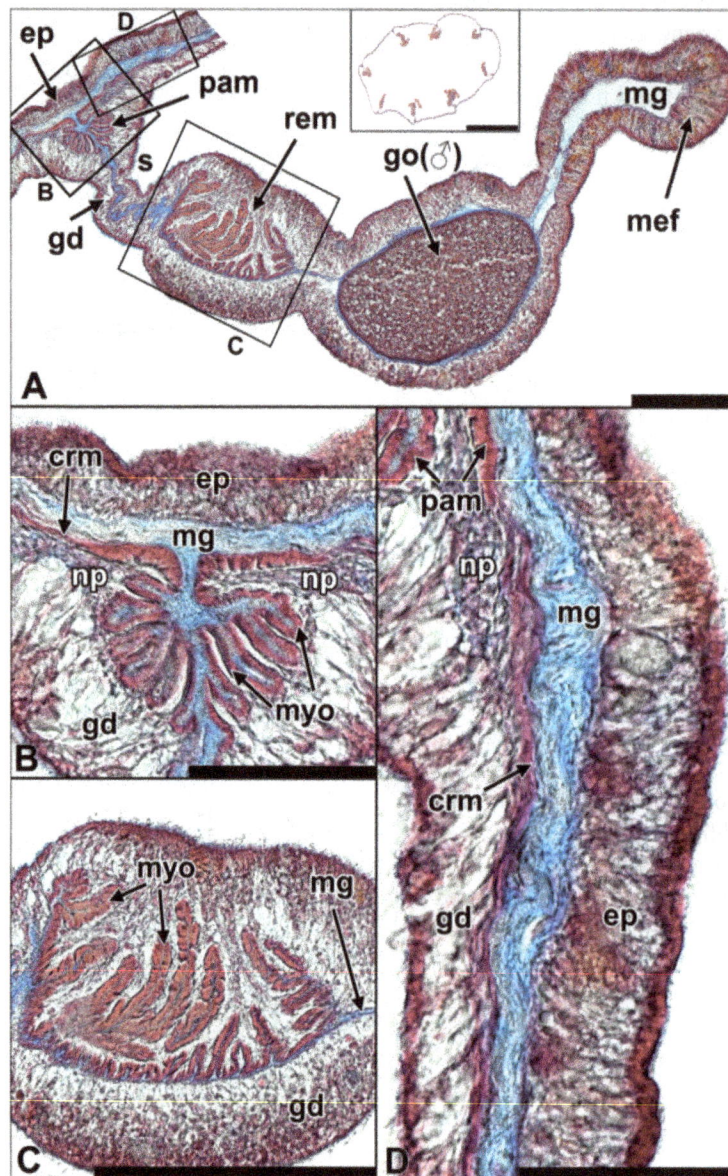

Figure 2 Histology of the muscle system of *N. vectensis*. All sections are cross-sections. **A** Architecture of a single mesentery. The parietal muscle is located at the base of the mesentery, followed by a more distal retractor. Between the retractor muscle and the mesenterial filament at the tip of the mesentery, a gonad, embedded into the mesoglea, is formed. Inlet: Cross-section through the subpharyngeal region showing all eight mesenteries. **B** Detail of **A**. The parietal muscle shows a typical bilateral folding of the mesoglea, on which the myonemes of the muscle cells attach. The muscle consists of two contiguous sheets of myonemes, which merge into the ring muscle laterally. Two distinct neural plexi are visible on either side next to the parietal muscle. **C** Detail of **A**. In the retractor muscle the mesoglea folds and branches only to one side. This folding is generally more pronounced than in the parietal muscle. **D** Detail of **A**. The ring muscle lies between the parietal muscle of each mesentery. Similarly, the myonemes are located at the base of the cells. ep, epidermis; gd, gastrodermis; go, gonad; mef, mesenterial filament; mg, mesoglea; myo, myoneme; np, neural plexus; pam, parietal muscle; rem, retractor muscle; crm, column ring muscle; s, stalk. Scale: **A, C**: 100 µm; **B, D**: 50 µm; inlet: 1 mm. All sections are stained with Azan.

tip and the retractor muscle, a gametogenic region containing either eggs or sperm – directly embedded within the mesoglea – is present in sexually mature animals.

Both parietal and retractor muscle are longitudinal muscles, with the former showing a more complex branching (secondary branching) of the mesoglea (Figure 2C,

Figure 3C, Figure 4A); this increases the surface and enables more muscle cells to attach. As a result the overall surface area of myonemes is generally larger in the retractor than in the parietal muscle. In the ring muscle of the body wall the myonemes are arranged perpendicular to the oral-aboral axis. Instead of being a contiguous ring

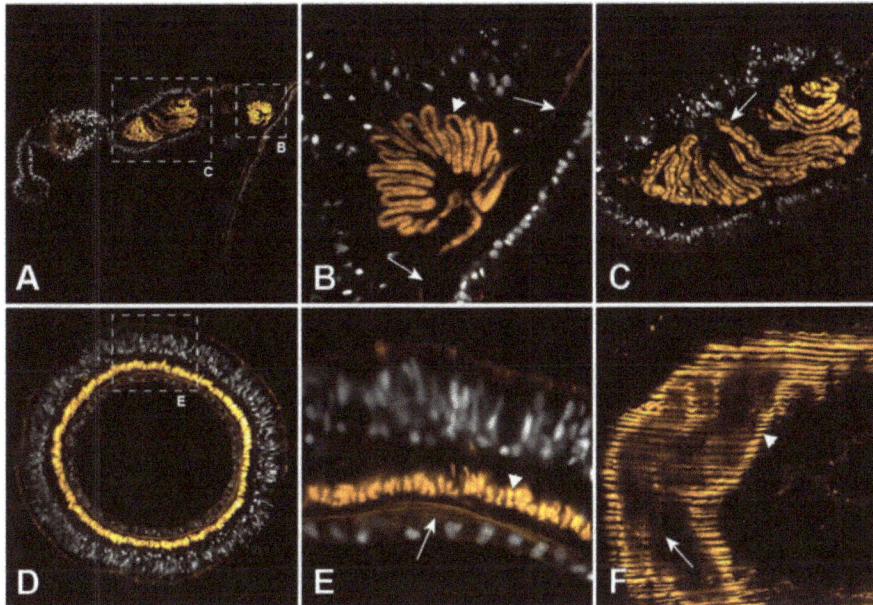

Figure 3 Muscles of the adult polyp revealed by F-actin staining with phalloidin. A Cross-section of a whole mesentery with proximal parietal muscle and distal retractor muscle. **B** Detail of **A**. Parietal muscle. Note the transition of the columnar circular muscle (arrows) into the longitudinally orientated parietal muscle (arrowhead). **C** Detail of **A**. Retractor muscle (arrow). **D** Cross-section of a tentacle. **E** Detail of **D**. The well-formed longitudinal muscle is located basiepithelially in the epidermis (arrowhead), whereas the circular muscle is weakly developed and located in the gastrodermis (arrow). **F** Single longitudinal optical section of a tentacle showing the strong longitudinal muscle filaments (arrowhead) overlying those of the circular muscle (arrow). white: DAPI, orange: phalloidin.

of muscle cells, it consists of eight portions situated between the mesenteries and merging directly into the parietal muscle to each side (Figure 2D, Figure 3B).

Parietal and retractor muscles retain an epithelial organization

In cross-sections of the adult polyp, the cells of the retractor muscle are strongly entangled, making it difficult to follow each cell border from the base to the apex. At the proximal (towards the parietal muscle) and the distal border (towards the mesenterial filament) of the retractor muscle, the myonemes are closer to the apex of the epithelium. Here, a clear continuity can be observed, and the apical part of the cell contributes in building up the epithelium. Via thin cytoplasmic bridges, the apical part is connected to the robust basal, myofilament-containing part, which lies adjacent to the mesoglea (Figure 4E, F). The myonemes are large and well pronounced in the central region of the retractor muscle sheet (Figure 4G). Although not visible in single sections, these myonemes probably also still exhibit a connection to the epithelium. This interpretation is based on the fact that the space between the folded sheet of myonemes is filled with numerous cytoplasmic bridges, all of which originate from single myonemes running towards the apex of the epithelium. The myonemes are connected to each other by spot-like adherens junctions (Figure 4G) and are anchored to the mesoglea by focal adhesion sites (Figure 4E inlet).

Likewise, the parietal muscle consists of a serial arrangement of myonemes (Figure 4B). The folding of the muscle sheets is less complex in this region, making it easier to follow individual cells from the base to the apex on single sections. Similar to the retractor muscle, the apical part of the cell in the parietal muscle is connected to neighboring cells to form an epithelium; the myonemes are connected to the apical part by cytoplasmic bridges (Figure 4D). A prominent neural plexus is situated on both sides of the parietal muscle; neurites run mainly along the oral-aboral axis (Figure 4C). The myonemes are again anchored by adherens junctions to neighboring cells and by focal adhesion sites to the mesoglea (Figure 4C, inlet 1).

Ring muscle cells are oriented perpendicular to the longitudinal muscles. They span the body wall gastrodermis (adult endoderm) between the parietal muscle portions of all eight mesenteries and show a clear integration into the epithelium, having their myofilaments located at their bases (Figure 4H). In brief, all three muscle types of the body column show an epithelial organisation, but the retractor and parietal muscles are highly modified.

The tentacle muscle system consists of basiepithelial myocytes and epitheliomuscular cells

Tentacles are basically tube-like expansions of the body column at the oral pole with continuation of the gastric cavity (Figure 1). Therefore they are structured in the

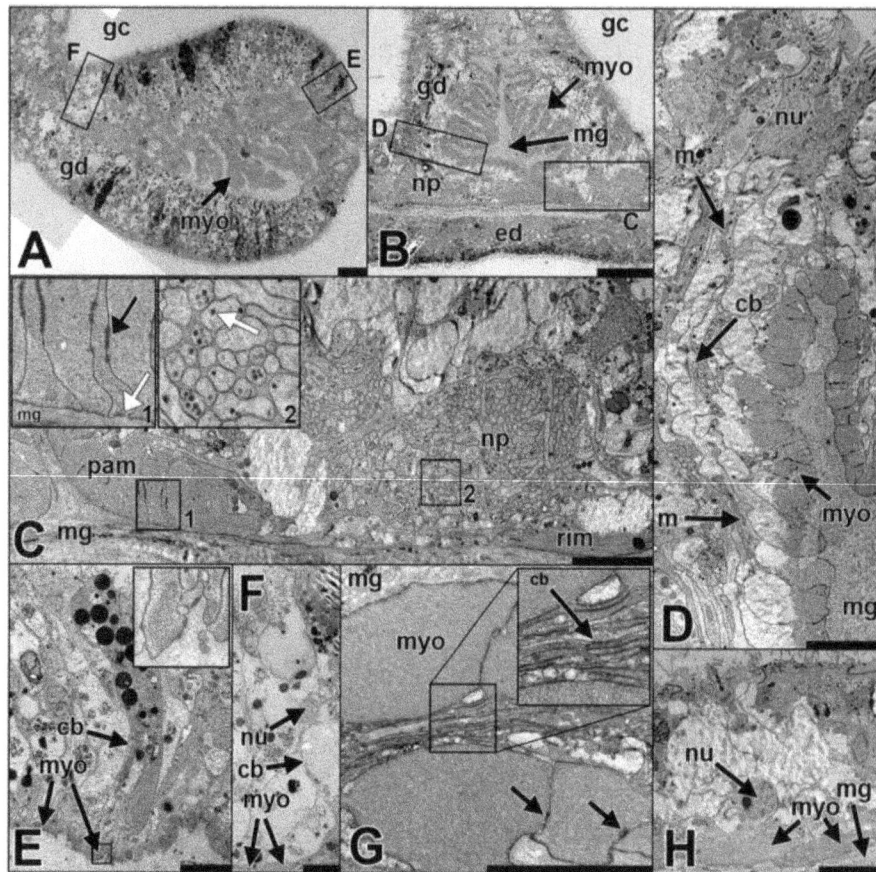

Figure 4 Epithelial organization of the columnar muscles. All sections are cross-sections **A** Overview of the retractor muscle region. **B** Overview of the parietal muscle region. **C** Detail of **B**. A distinct neural plexus consisting of numerous neurites is located next to the parietal muscle and close to the transition zone to the ring muscle. Inlet 1: spot-like adherens junctions between the basal part (black arrow) connect muscle cells to each other. Focal adhesions (white arrow) connect the cells to the mesoglea. Inlet 2: Detail image of neurites containing neural vesicles (white arrow). **D** Detail of **B**. Parietal muscle cells consist of a cell body involved in building up the epithelium. Via cytoplasmic bridges they are connected to the myofilament-containing basal part of the cell, adjacent to the mesoglea (cell highlighted in yellow); this yields a consistent sheet of myonemes. **E** and **F** Details of **A**. Muscle cells (highlighted in yellow) remain epithelial at least in the proximal and distal boundaries of the retractor muscle. **E** (inlet): Focal adhesion between muscle cell and mesoglea. **G** Detailed image of well-formed myonemes in the central region of the retractor muscle. Thin cytoplasmic bridges (inlet, black arrow) are projected by every cell. Basal parts of the cells are connected by spot-like adherens junctions (black arrows). **H** Detailed image of a ring muscle cell showing its participation in building up the epithelium. The myofilaments are located at the basal part of the cell. cb, cytoplasmic bridges; gc, gastric cavity; gd, gastrodermis; m, mitochondrium; mg, mesoglea; myo, myoneme; np, neural plexus; nu, nucleus; pam, parietal muscle; rim, ring muscle. Scale: **A, B**: 20 µm, **C-F, H**: 5 µm, **G**: 2 µm.

same way as the double-layered body column, lacking any endodermal invaginations such as mesenteries in the body column (Figure 5A). Most of the cell types present in the tentacles are situated in the epidermis (adult ectoderm), making it most important for the functions of the tentacle. At least two different types of gland cells are present, one with electron-dense and one with electron-light vesicles. One of which has recently been shown to secrete a toxin [12]. Since the main function of the tentacle is to capture prey, it is unsurprising that it is very rich in spirocysts. The longitudinal muscle consists of a discontinuous sheet of myonemes located at the base of the epidermis, forming a ring around the mesoglea. A prominent neural plexus lies on top of that muscle layer; it also shows a ring-like

structure (Figure 5B). To clarify whether the basally located myonemes exhibit a connection to the apex of the epithelium, as is the case in the columnar muscles, we produced serial longitudinal- and cross-sections of tentacles. We found that the nuclei of the muscle cells are located near the myofilaments and that no cytoplasmic bridges between the basal myonemes and the apical part of the epithelium are formed. The elongated muscle cells are arranged side by side, connected laterally by spot-like adherens junctions (Figure 5C + D). Through the discontinuous sheet of muscle cells, overlying epithelial cells send their processes and attach to the mesoglea (Figure 5C + F).

The gastrodermis of the tentacle is mainly composed of circular muscle cells. They resemble the circular muscles

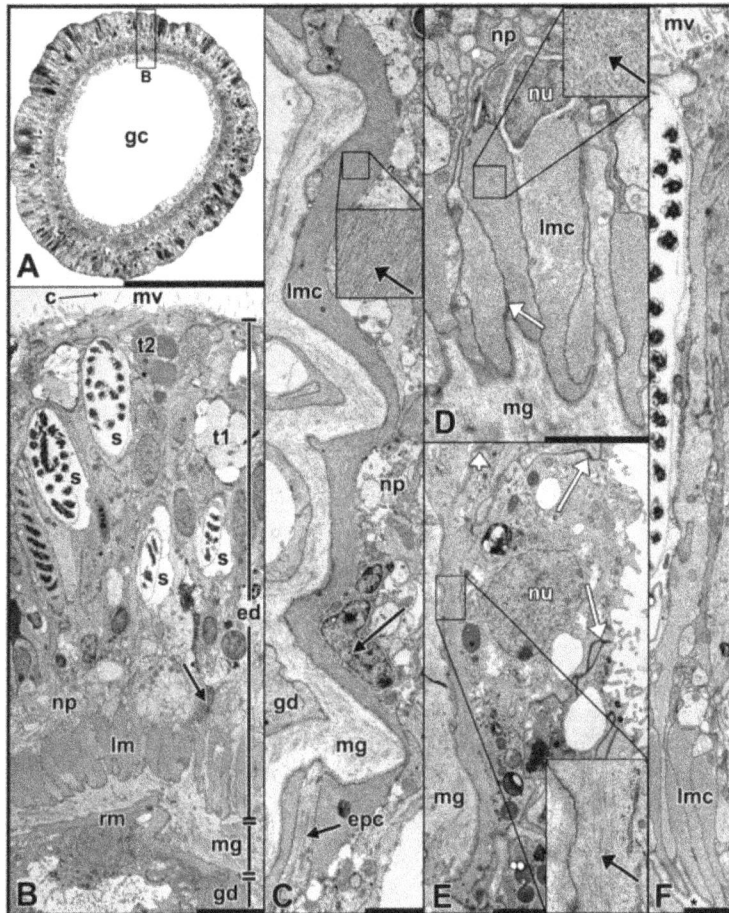

Figure 5 Organization of the tentacle muscle system in an adult polyp. A Tentacle cross-section. **B** Detail of **A**. Typical appearance of a tentacle epithelium. Longitudinally orientated muscle cells are situated side by side at the base of the epidermis and are connected to the mesoglea (highlighted in red). A neural plexus (highlighted in green) is located on top of the muscle layer. Epithelial cells are still connected to the mesoglea by thin processes, which find their way through the discontinuous sheet of muscle cells. In the gastrodermis the epithelial muscle cells are oriented circularly. **C** Longitudinal section of a longitudinal muscle cell pointing out the basiepithelial organization. Note the close location of the nucleus to the myofilaments. Inlet: parallel arrangement of thick myofilaments (arrow). **D** Detail of a cross-sectioned longitudinal muscle cell with adjacent nucleus. White arrow: spot-like adherens junction connecting muscle cells to each other. Inlet: Thick myofilaments (black arrow) are distributed irregularly between thin ones. **E** Ring muscle cell of the gastrodermis (cross-section) with myofilaments located at the base of the cell adjacent to the mesoglea. White arrow: Belt-like apical junctional complex. Inlet: In comparison to the longitudinal muscle the myofilaments of the ring muscles are weakly developed. Black arrow: Thick myofilament. **F** Epithelial cell (highlighted in yellow) spanning from the external surface to the mesoglea through the sheet of longitudinal muscle cells. Asterisk denotes mesoglea. c, cilium; ed, epidermis; epc, epithelial cell; gc gastric cavity; gd; gastrodermis; lm, longitudinal muscle layer, lmc, longitudinal muscle cell; mv, microvilli; np, neural plexus; nu, nucleus; muscle; rm, ring muscle layer; s, spirocyst; t1, type 1 gland cell with electron-light vesicles; t2, type 2 gland cell with electron-dense vesicles. Scale: **A**: 50 μm, **B**: 5 μm, **C-F**: 2 μm.

of the body column in that their myofilaments are located basally and in that the connection between the apex and the base of their cells is not drawn into long and thin cytoplasmic bridges. Due to the low myofilament density, the tentacle ring muscle is comparatively weakly developed (Figure 5E).

Development of columnar longitudinal muscles

To investigate the development of columnar longitudinal muscles, subpharyngeal cross-sections of different stages ranging from mid-planula to juvenile polyps were produced. Columnar longitudinal muscles are located in the

mesentery, which corresponds to a folding of the endodermal epithelium. The first pair of mesenteries becomes visible on a macroscopic level at the primary polyp stage. They are called primary mesenteries and can be easily distinguished from the remaining three pairs of secondary mesenteries because of their early-formed mesenterial filaments at the distal tip. By comparison the secondary mesenteries form their mesenterial filaments only later during polyp growth. However, when reaching adulthood, all eight mesenteries have developed to a similar degree and they cannot be distinguished anymore on a macroscopic level.

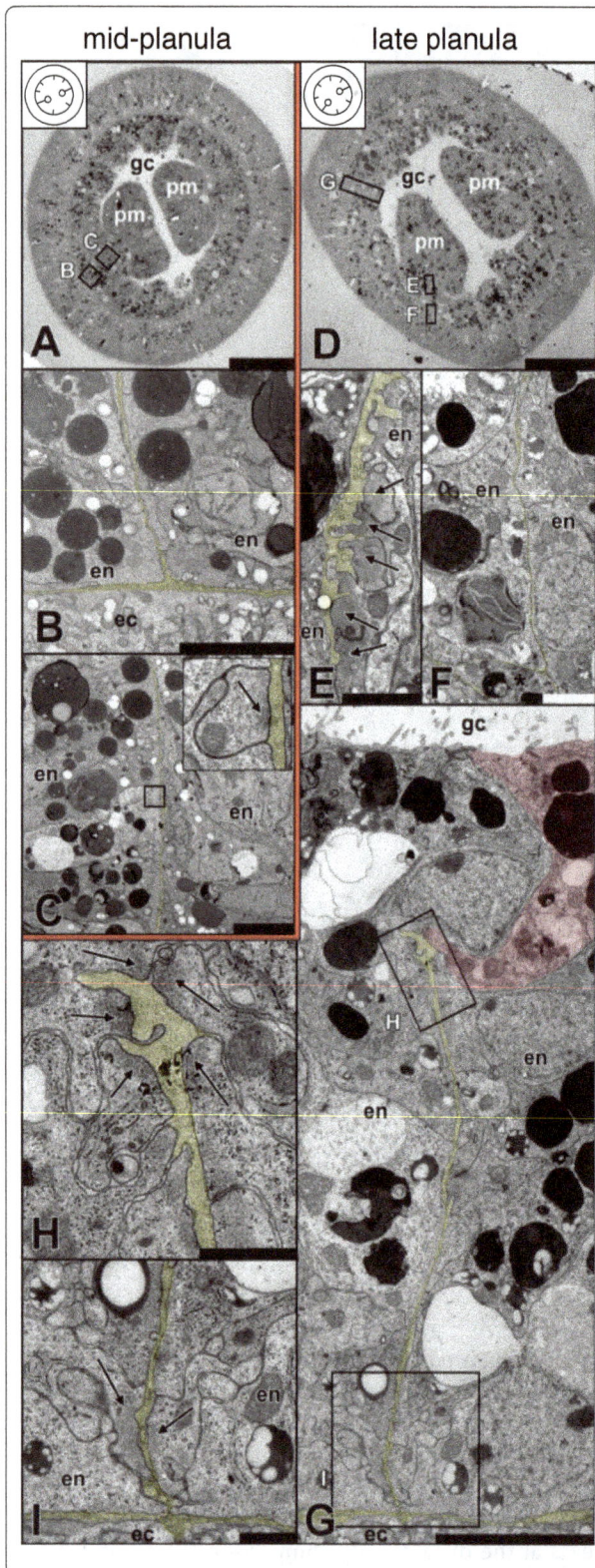

Figure 6 Development of columnar longitudinal muscles in mid and late planula. **A** Mid-planula (~4 d), overview. The first two emerging mesenteries (primary mesenteries) are formed opposite each other. Note that the location of the secondary mesenteries is already set at this stage (inlet). **B** Detail of **A**. The base of the mesentery (corresponding to the future site of parietal muscle formation) shows no distinct accumulation of myofilaments. **C** Detail of **A**. Single cell (inlet, black arrow: myofilaments) showing first sign of muscle formation. **D** Late planula (~5 d), overview. Primary mesenteries gradually shift to one side. **E** Detail of **D**. The developing retractor muscle in the primary mesentery can already be identified. Myonemes are formed exclusively on one side of the mesoglea. Black arrows: accumulating myofilaments. **F** Detail of **D**. Basal part of the primary mesentery. No myofilaments have accumulated yet. Note an amoeboid cell at the branching of the mesoglea, which could be detected occasionally (asterisk). **G** Detail of **D**. Secondary mesentery. Future retractor muscle cell highlighted in red. **H** Detail of **G**. Retractor muscle cells forming at the tip of the mesentery (in contrast to the primary mesentery), having no bias to one side at this stage. Black arrows: accumulating myofilaments. **I** Detail of **G**. Cells at both sides of the base of the mesentery start to accumulate myofilaments (black arrows) adjacent to the mesoglea. All sections are cross-sections of the subpharyngeal region. For easier understanding the mesoglea of all close-ups is highlighted in yellow. Inlets in A and D indicate the location of primary (lines with circles) and secondary mesenteries (lines without circles). ec, ectoderm; en, endoderm; gc, gastric cavity; pm, primary mesentery. Scale: **A, D**: 50 μm; **B, C, G**: 5 μm; **E, F**: 2 μm; **H, I**: 1 μm.

Needless to say, a mesenterial anlage is required before muscle cells can develop. A first morphological indication of muscle differentiation was observed in the primary mesenteries of the mid-planula stage (Figure 6A-C). The base of the mesentery (future parietal muscle site) lacks any clear signs of muscle cells (Figure 6B). At the same time, a few cells containing a small number of myofilaments were already visible along the mesentery, lying adjacently to the mesoglea (Figure 6C, inlet). Secondary mesenteries at this stage have not developed any morphologically visible muscular structures yet (data not shown). In late planula stages (Figure 6D-I) the future muscle regions become apparent. The retractor muscle cells of the primary mesenteries have formed additional myofilaments and are positioned side by side (Figure 6E). They emerge slightly distal of the mesentery base, already formed exclusively on one side of the mesentery. At this stage, no clear myofilament formation is visible at the base of the mesentery (Figure 6F). In secondary mesenteries, both muscle groups – parietal and retractor muscles – are identifiable. Unlike the situation in primary mesenteries, here the retractor muscle cells form at the tip of the mesentery, without any bias to one side yet (Figure 6G-I). Generally, muscle cells do not differentiate synchronously in all eight mesenteries. This process is usually quite variable, and neither primary, nor secondary mesenteries are distinctly ahead in the differentiation of muscle cells.

In early primary polyps the morphology has changed noticeably. The body wall endoderm has become thinner and the secondary mesenteries are easier visible within

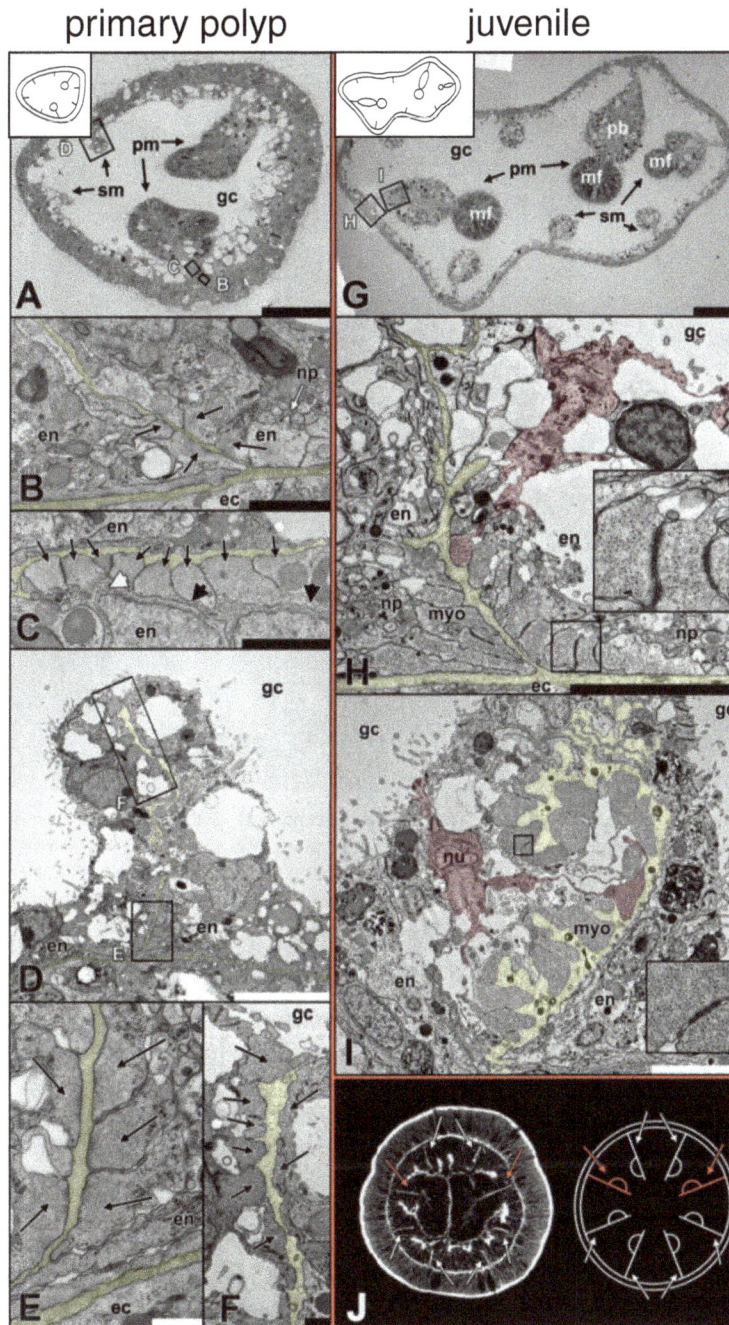

Figure 7 Development of columnar longitudinal muscles in primary polyps and juveniles. A Early primary polyp (~5 d), overview. **B** Detail of **A**. Cells located next to each other at the base of the mesentery increasingly accumulate myofilaments (black arrows) in their basal part. **C** Detail of **A**. Some retractor muscle cells (long black arrows) start to constrict the basal from the apical part of the cell, leaving behind thin cytoplasmic bridges (short black arrows), while in some the nucleus still lies near the myoneme (short white arrow). **D** Detail of **A**. Secondary mesentery. **E** Detail of **D**. As in primary mesenteries more cells accumulate myofilaments (black arrows) at their bases. **F** Detail of **D**. In the secondary mesenteries, myonemes (black arrows) are still located at the tip of the mesentery, but are more pronounced on one side. **G** Juvenile (~3 months) polyp, overview. **H** Detail of **G**. Despite the constriction of the myoneme, parietal muscle cells retain an epithelial organization (single cell highlighted in red). Inlet: Arrangement of myofilaments and adherens junctions between myonemes. **I** Detail of **G**. In the retractor muscle the sheet of myonemes starts to fold. Muscle cells are still epithelial (single cell highlighted in red). Inlet: Arrangement of myofilaments and adherens junctions between myonemes. **J** Optical cross-section of a mid-planula. The orientation of the retractor muscle is species-specific and predetermined for every mesentery. This allows predicting the future side of the retractor muscle in all mesenteries, as soon as all mesenteries have emerged. Red arrows: primary mesenteries; white arrows: secondary mesenteries. ec, ectoderm; en, endoderm; gc, gastric cavity; pb, proximal bulge; pm, primary mesentery; sm, secondary mesentery; mf, mesenterial filament, myo, myoneme; np, neural plexus; nu, nucleus. Scale: **A, G**: 50 μm; **B, C**: 2 μm; **D, H,** I: 5 μm, **E**: 500 nm; **F**: 1 μm.

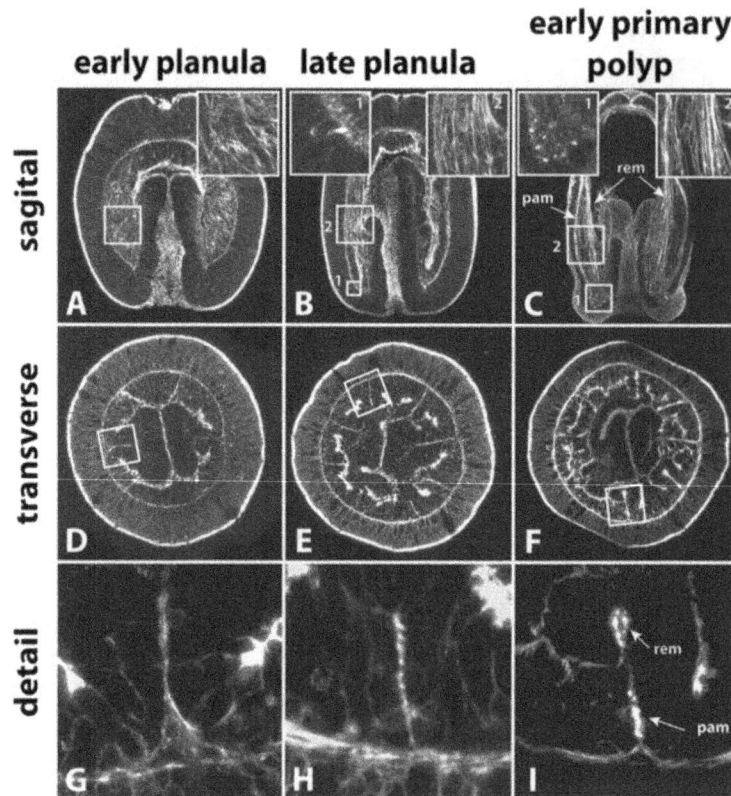

Figure 8 Condensation of actin filaments during the differentiation of muscle cells. A, D, G: Early planula. F-actin is weakly pronounced in the mesenteries but not orientated uniformly. **B, E, H**: In the late planula, F-actin becomes increasingly accumulated and oriented along the oral-aboral axis adjacent to the mesoglea. Moreover, future tentacle cells become enriched in F-actin (**B,** inlet 1). **C, F, I**: Early primary polyps show distinct longitudinal muscle strands and increased F-actin accumulation in tentacle muscle cells (**C,** inlet 1). The parietal muscle and retractor muscle can clearly be identified. Images in **A** (6 sections), **B** (8 sections) and **C** (20 sections) constitute maximum projections. pam, parietal muscle; rem, retractor muscle.

the endoderm. The primary mesenteries no longer lie opposite each other and become clearly shifted to one side (Figure 7A). In the retractor- and parietal muscle region of primary and secondary mesenteries, the number of cells with myofilaments increases. At this stage the formation of a neural plexus next to the parietal muscle begins (Figure 7B). In contrast to the parietal muscle, retractor muscle cells of the primary mesenteries already start to constrict the myoneme from the rest of the cell, and cytoplasmic bridges become visible (Figure 7C). In secondary mesenteries (Figure 7D-F), myonemes of the retractor muscle are still located at the tip, but they are already more pronounced on one side of the mesentery (Figure 7F).

In the juvenile polyp (Figure 7G-I) the relative position of the primary to the secondary mesenteries becomes clear (although already set in the early planula). The primary mesentery at this stage is composed of a distal mesenterial filament and a proximally located bulged tissue (extending from the stalk to the mesenterial filament) harbouring the retractor muscle at its base. In one of the six secondary mesenteries the mesenterial filament has started to form at the tip; in the remaining ones

the retractor muscle is still the most distal structure (Figure 7G). In fully differentiated polyps the secondary mesenteries will have grown to the same size as the primary ones; then, the origin of every single mesentery can only be reconstructed by determining the position of the retractor muscles. The morphology of the parietal muscle resembles that of early stages with the exception that more cells are involved in building up the muscle sheet and that the myonemes become increasingly constricted from the apical epithelial part. The basal part maintains a connection to the apical one and neighboring muscle cells are anchored to each other by adherens junctions at their basal sides. In comparison to the situation in adults (Figure 2B, Figure 3B) the muscle sheets have not folded yet (Figure 7H). The myonemes of the retractor muscle are more densely packed with myofilaments than in the parietal muscle and the cytoplasmic bridge between the base and apex of the cell is already constricted. Like in the parietal muscle, the myonemes are connected to each other by adherens junctions. The muscle sheet is no longer arranged in a straight way, but has started to produce the first folding (Figure 7I). This process will

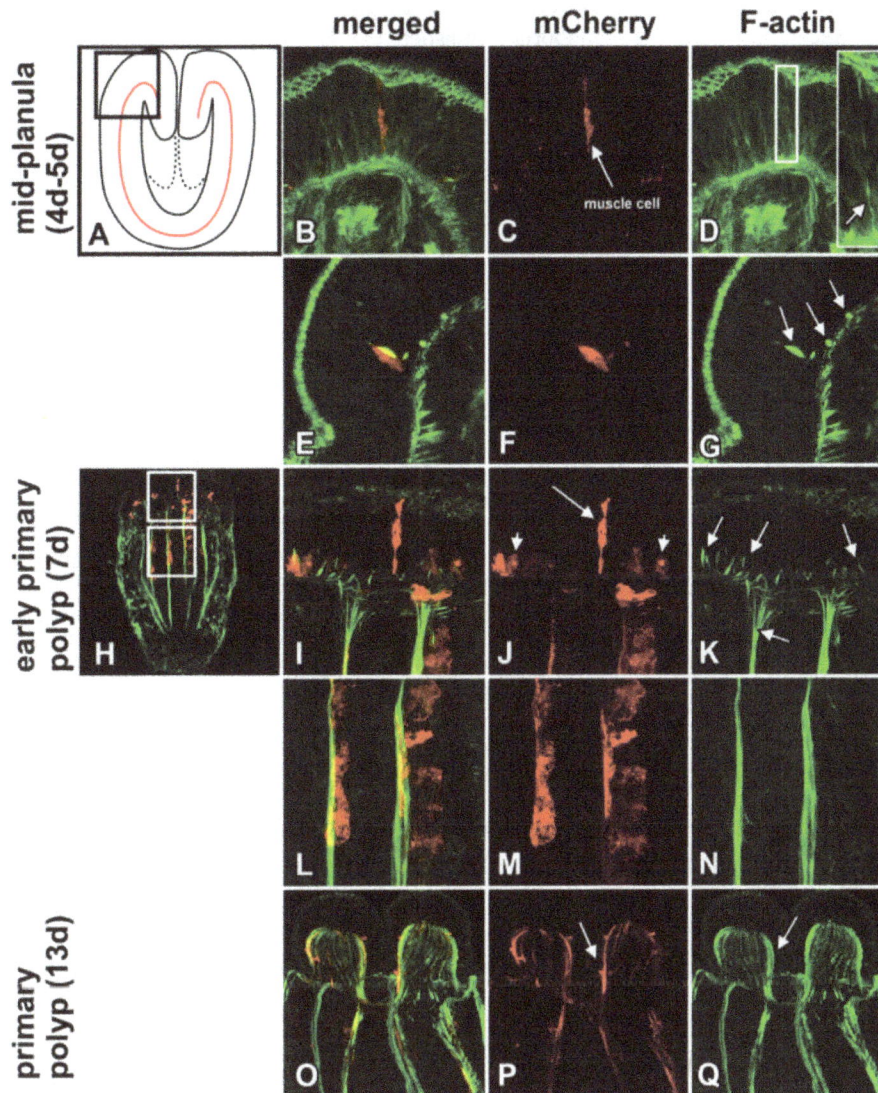

Figure 9 Differentiation of tentacle longitudinal muscle cells in a muscle specific transgenic (MHC::mCherry) reporter line. F-actin was stained with phalloidin (green), mCherry with an α-RFP antibody (red). **A-G** Optical longitudinal sections of mid-planula stages (4–5 d) in the future tentacle bulb region. A Overview. **B-D** Maximum projection of six layers. An early undifferentiated tentacle muscle cell (labelled by the α-RFP antibody) spans throughout the ectodermal epithelium and already shows a weak accumulation of actin filaments (**D**, inlet). E-G Maximum projection of seven layers. Tentacle muscle cells lose connection to the apex of the epithelium. Actin filament accumulation at the base of the epithelium increases in several cells (**G**, arrows). **H-N** Optical longitudinal sections of early primary polyp stages (7 d). Maximum projection of 21 layers. Different stages of detachment of tentacle longitudinal muscles from the epithelium co-occur at the same stage (**J**, long arrow: epithelial muscle cell, short arrow: submerged muscle cell). Actin filaments accumulate in several cells before mCherry becomes visible (**K**, arrows). **L-N** Columnar retractor muscle cells expressing mCherry. **O-Q** Optical longitudinal sections of primary polyps (13 d). Maximum projection of 53 layers. Numerous actin filaments are enriched in the outgrowing tentacle bulbs (**Q**, arrow). Tentacle longitudinal muscles have lost their connection to the apex of the epithelium and become situated at the base of the ectoderm, increasing their length with tentacle elongation (**P**, arrow).

proceed during the maturation of the juvenile, ultimately yielding the typically multiple-folded, one-sided muscle "flag".

The orientation of the retractor muscles leads to an internal bilateral symmetry and is consistent within different individuals. Primary mesenteries always have a predetermined place in this system. This makes it possible to predict the site of retractor muscle formation as soon as the

mesenteries have set and before there are any morphological signs of muscle cell differentiation (Figure 7J).

Staining with phalloidin additionally demonstrated the accumulation of F-actin in presumptive muscle cells during the development from the planula to the primary polyp. Endodermally located F-actin is already present in the early planula but does not yet show a clear concentration and uniform orientation (Figure 8A, D, G). In the late planula those actin threads become increasingly concentrated and exhibit a clear orientation along the oral-aboral axis. In the future tentacle bulb region, single cells have accumulated F-actin in their basal part (Figure 8B, E, H). In the early primary polyp an increasing number of future tentacle cells show concentrated F-actin; the columnar muscles already are differentiated into a parietal- and a retractor muscle region (Figure 8C, F, I).

Development of tentacle longitudinal muscles

As mentioned above, fully differentiated longitudinal muscle cells of the tentacle are situated at the base of the epidermis and show no connection to the apex of the epithelium. In order to determine whether they emerge from originally epithelial cells, we used a transgenic line (MHC:: mCherry) that expresses a fluorescent reporter gene (mCherry) under the control of a muscle-specific promotor (striated-*myhc*, formerly *myhc1*) specifically in retractor and tentacle muscles [13]. Expression of mCherry was first detected in 4 day mid-planula stages (Figure 9A-G). It starts with several single columnar cells at the oral pole close to the pharynx, spanning the whole thickness of the ectodermal epithelium. Small accumulations of actin filaments are already co-localized with the expression of mCherry in those cells (Figure 9B-D). At this stage most tentacle muscle cells have already abandoned their apical cell contacts and have sunken down to a basiepithelial position. At this stage of cellular differentiation their concentrated F-actin has increased (Figure 9E-G). Although most of the mCherry positive cells are already situated at the base, a few epithelial muscle cells may be present in early primary polyps. Many cells already formed spindle shaped, apical-basal orientated actin-filament accumulations at their bases before mCherry can be detected (Figure 9I-K). The expression of mCherry in the retractor muscle starts in the late planula (data not shown) and becomes more apparent in early primary polyps. In the latter, the nuclei and remaining cell bodies are visibly located on top of the actin-enriched contractile elements (Figure 9L-N). In older primary polyps (13d) the tentacle bulbs have started to grow out and thin actin-filaments have become aligned along the longitudinal axis of the tentacle (Figure 9O-Q). In comparison to earlier stages (7d) the orientation of actin-filaments has shifted by 90° from a basal-apical to a lateral character (Figure 9Q). Correspondingly, the muscle cells become spindle-shaped and oriented along the longitudinal axis of the tentacle, having lost their connection to the apical surface of the epithelium (Figure 9P).

Discussion

In the present study we show that the muscular system of *N. vectensis* consists of two systems, both with a circular and at least one longitudinal portion. The muscle cells building up these two systems exhibit different levels of epithelial organization. Three main types can be distinguished: (1) The type 1 epitheliomuscular cell, in which the nucleus-containing part is connected to neighboring cells by apical cell-cell junctions, ensuring a complete integration into the epithelial cell complex. Their relatively weak myofilaments are housed in the elongated basal part, which lies adjacent to the extracellular matrix. This cell type is represented in the ring muscle cells of the column and tentacles (Figure 4H, Figure 5E, Figure 10A.1). This type is also found in Hydrozoa, which lack more elaborated muscle cell types. (2) The type 2 epitheliomuscular cell. It basically resembles type 1, with the difference that the apical-basal axis of the cell is extremely elongated, and that the apical and the basal part are connected only by an elongated and very slim cytoplasmic bridge. The longitudinal muscles of the column (parietal and retractor muscle) belong to this type (Figure 4D-F, Figure 10A.2). (3) The type 3 basiepithelial myocytes of the tentacle epidermis. In this type, the apical cell-cell junctions have been lost and the muscle cells are therefore positioned at the base of the epithelium. They constitute a spindle-shaped, elongated cell with a central thickening containing the nucleus. Like in other muscle types the myofilaments are still situated adjacent to the mesoglea (Figure 5C + D, Figure 10A.3).

Type 2 and 3 represent specialized muscle cells and they underlie different courses of development. In both cases, myofilaments initially are formed at the bases of the cells. The type 2 muscle cell starts differentiating by constricting the basal from the apical part, maintaining a connection to neighboring cells by apical cell junctions. This process proceeds as the number of muscle cells increases, allowing the mesoglea and the basal part to undergo increasing folding of the continuous basal muscular sheet and the formation of elongated and slim cytoplasmic bridges, which connect the base to the apex (Figure 10B). Hence, the apical epithelial part does not have to follow the muscle folding or contraction. This arrangement allows maximal flexibility and mobility in the context of a one-cell layered epithelial organization. Type 3 muscle cells, in contrast, lose their apical cell junctions. This results in a localization at the base of the epithelium. Here, these cells form a discontinuous layer of myocytes (Figure 10C). Both types of muscle cells are anchored to the mesoglea basally. Type 1 and 2 muscle cells are classical epitheliomuscular cells. Type 3 muscle cells have lost their apical integration and therefore constitute basiepithelial myocytes.

Biomechanical considerations

The three types of muscle cell formation presented here are interpreted to be a direct consequence of biomechanical constraints. Ring muscle cells are exclusively situated in the gastrodermis of either the column or tentacles. Beyond their function as contractile units they mainly serve

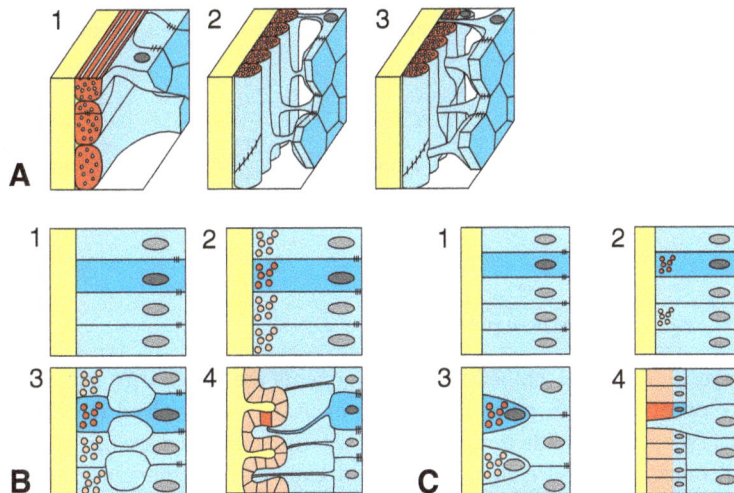

Figure 10 Schematic representation of organization and development of muscle cells. A Modes of muscle cell organization. The type 1 muscle cell corresponds to the classical epitheliomuscular cell. The cell is integrated into the epithelium by apical cell-cell junctions, connecting them to neighboring cells, while the basal part is elongated and houses myofilaments. Type 2 muscle cells are still epitheliomuscular; their connections between the apical and the basal part, however, are drawn out to form thin and elongated cytoplasmatic bridges. In type 3 muscle cells the apical cell junctions have been lost, resulting in a sinking to the base of the epithelium, with the nucleus situated near the myofilaments. **B + C** Development of type 2 & 3 muscle cells. In both types myofilaments start to become accumulated at the base of epithelial cells. **B** In type 2 muscle cells the apical cell-cell junctions remain and the connection between the apex and the base becomes constricted to form thin and elongated cytoplasmic bridges. The basally situated myonemes build up a continuous and folded muscle sheet. **C** Type 3 muscle cells lose their apical cell-cell junctions and sink down to the base of the epithelium, where they are arranged in a discontinuous muscle layer, enabling overlying epithelial cells to attach to the mesoglea. Color code: yellow: mesoglea, blue: epithelium, red: myofilaments, grey, nucleus.

to build up the epithelium. Ring muscle cells are the predominant cell type of that region. The epithelium is relatively flat and a long and thin cytoplasmic bridge is mechanically not necessary to maintain a connection between the base and the apex. In the longitudinal muscles of the column (parietal and retractor muscle) the situation is different. In both muscles the basally situated muscle sheet is pleated accordion-fashioned, enabling more myonemes to attach to the mesoglea; this increases the contractile strength per area [14]. As a result the surface area of the basal side of the epithelium considerably exceeds that of the apex. This requires modifying the interconnecting part if integration into the epithelium is to be maintained. This is exactly what we observed when examining the elongated and slim cytoplasmic bridges: they become progressively apparent when the muscle sheets start to fold.

Unlike in the column, longitudinal muscle cells of the tentacles are located in the epidermis. Tentacles serve to catch prey and are therefore rich in epidermal cell types such as gland cells and nematocytes, which facilitate this function. All cell types are more or less equally distributed in a ring-like manner around the mesoglea. This would leave less space for epitheliomuscular cells. For this reason we interpret the muscle cells to have lost their apical cell-cell junctions to acquire a basiepithelial position. As they have lost their apical cell junctions, the overlying cells must assume the role of building an

epithelial barrier. In addition overlying cells require an anchoring to the mesoglea: this is now provided because the muscle layer is no longer continuous.

Comparison of muscle cell types within Cnidaria
Different groups within the Cnidaria show a very high variability in terms of how and where contractile cells are formed. In hydropolyps the muscular system is exclusively composed of smooth epitheliomuscular cells. While longitudinally orientated myonemes are located in the epidermis, circularly arranged myonemes are located in the gastrodermis. Both epithelia mainly consist of epitheliomuscular cells [15]. In the hydromedusa of *Obelia* the ectodermal subumbrella is composed of epitheliomuscular cells with both radial helical myofibrils and submerged myocytes, which exhibit true striation [16]. In scyphopolyps the muscular system is exclusively formed by the ectoderm. Here the different muscle fields are physically continuous at the peristomial pits and are either epitheliomuscular (tentacle) or formed as myocytes (column) embedded within the mesoglea. Myofibrills are striated at least in some parts of the polyp before they merge into smooth fibers [17]. The muscle cells of Scyphomedusae are located in the ectodermal subumbrella and formed as circular striated and radial smooth epitheliomuscular cells [18,19]. Werner and colleagues have shown that the muscular system in cubopolyps mainly consists of ectodermal muscle cells [20]. They

Table 1 Origin, striation and cellular organization of cnidarian muscles

	Ectodermal	Endodermal	Striated	Smooth	Epitheliomuscular	Myocytes
Cubopolyp	+	-	(+)	+	+	+
Cubomedusa	+	-	+	+	+	-
Scyphopolyp	+	-	(+)	+	+	+
Scyphomedusa	+	-	+	+	+	-
Hydropolyp	+	+	-	+	+	-
Hydromedusa	+	-	+	+	+	+
Anthozoa	+	+	(+)	+	+	+

"+": trait present, "-": trait absent, "(+)": trait present, but only reported in single cases.

are either composed of myocytes, which are embedded into the mesoglea (*Carybdea*), or of myocytes and additional epitheliomuscular cells (*Tripedalia*). With the exception of the tentacle tips, all muscle cells have smooth fibers. In Cubomedusae, muscles are subumbrellar (ectodermal) and epitheliomuscular with circular, striated fibers that turn radially in the frenulae of the velarium [21,3]. From anthozoan polyps, predominantly smooth and epitheliomuscular cells have been reported. Striated muscle fibers, however, have been found in a small subset of actiniarian anthozoans [22]. Hyman [23] and Muscatine & Lenhoff [14] report the existence of additional epidermal muscle cells that are independent from the supporting cells and have adopted a subepidermal position. The latter descriptions are not consistent with our findings because we demonstrate the myocyte-nature of longitudinal tentacle cells: they take on a basiepithelial position, connected on the basal side to the mesoglea. In fact they represent basiepithelial myocytes.

In summary, all groups within the Cnidaria have been described to possess ectodermally derived muscle cells, while endodermal ones are restricted to Anthozoa and hydropolyps. Hydropolyps seem to be the only group lacking striated muscle fibers. Note, however, that striated muscles have been documented only in single cases in Anthozoa, cubopolyps and scyphopolyps, and their absence is rather the rule. In general, the muscle fibers of most polypoid cnidarians are smooth, while medusae possess additional striated fibers [23]. Epitheliomuscular cells are present in all groups and myocytes at least in either the polyp or the medusa stage of every group (Table 1).

In general, cnidarian tissues show a low grade of division of labor [14]. It has been hypothesized that multifunctionality is a general feature of ancient cell types and that during the course of evolution these multiple functions were segregated among sister cell types. A case in point is epitheliomuscular cells [24]. Our observations support this hypothesis, as we were able to observe the process of de-epithelialization of initially epithelial muscle cells to a basiepithelial location in a single animal. In contrast, a comprehensive comparative analysis of the evolutionary origin of all known muscle proteins revealed that many

components are either very ancient, pre-dating the origin of animals, or they evolved rather late in specific lineages [5]. No molecular synapomorphy has been found that would explain the emergence of striated muscles. This suggests that striated muscles have evolved independently in cnidarians, ctenophores and bilaterians, using a core set of ancient muscle proteins [5].

Cnidaria are still thought to exhibit the most primitive state of an interconnected muscular system [23]. The fact that both epitheliomuscular cells and myocytes are present in all cnidarian classes indicates that they must have already been present in the last common ancestor of Cnidaria and Bilateria. While in Cnidaria epitheliomuscular cells still play a major role in building up a contractile apparatus, this function is progressively adopted by mesodermal myocytes in Bilateria. Clearly, the evolution of mesoderm constitutes a crucial step in facilitating the emergence of complex muscular systems. Myocytes themselves, however, might have played a vital role in the early evolution of mesoderm. Nemertodermatid and acoel myocytes, like other mesodermal cells, apparently originate from the gastrodermis before they eventually emigrate into a subepithelial position [25]. A continuous sequence from epitheliomuscular cells to a subperitoneal musculature has been discovered in annelids and echinoderms, suggesting the epithelial organization of mesodermal muscle cells is an ancestral state [25-30].

In this context *N. vectensis* might represent an additional model for demonstrating the early stage of emigration of epitheliomuscular cells. *N. vectensis* myocytes have not been observed to entirely detach from the epithelium and to result in a mesenchymal, subepithelial position. Nonetheless, one can easily envisage such a process, which would ultimately lead to the situation in cubo- and scyphopolyps, where certain myocytes are completely embedded in the mesoglea [20,17].

Competing interests
The authors declare that they have no competing interests.

Authors' contributions
SMJ carried out the experimental work, analyzed the data. MW contributed to the histological preparations, and critically evaluated the data and the ms.

UT conceived of the study, analyzed the data and together with SMJ wrote the manuscript. All authors read and approved the final manuscript.

Acknowledgment
We thank the Core Facility Cell Imaging and Ultrastructure Research of the Faculty of Life Sciences for support with the confocal and electron microscopy. We are grateful to Michael Stachowitsch for critically reading the manuscript. This work was supported by the Austrian Science Fund (FWF) grant (P21108-B17) to U.T.

Author details
[1]Department of Molecular Evolution and Development, Centre for Organismal Biology, Faculty of Life Sciences, University of Vienna, Althanstrasse 14, 1090 Wien, Austria. [2]Department of Integrative Zoology, Centre for Organismal Biology, Faculty of Life Sciences, University of Vienna, Althanstrasse 14, 1090 Wien, Austria.

References

1. Medina M, Collins AG, Silberman JD, Sogin ML: Evaluating hypotheses of basal animal phylogeny using complete sequences of large and small subunit rRNA. *Proc Natl Acad Sci U S A* 2001, **98**:9707–9712.
2. Collins AG: Phylogeny of Medusozoa and the evolution of cnidarian life cycles. *J Evol Biol* 2002, **15**:418–432.
3. Gruner H, Hartwich G: *Lehrbuch der Speziellen Zoologie. Band 1: Wirbellose Tiere. 2.Teil: Cnidaria, Ctenophora, Mesozoa, Plathelminthes, Nemertini, Entoprocta, Nemathelminthes, Priapulida.* Jena: Hrsg. von Hans-Eckhard Gruner. Bearb. von G. Hartwich. Gustav Fischer; 1984.
4. Seipel K, Schmid V: Evolution of striated muscle: jellyfish and the origin of triploblasty. *Dev Biol* 2005, **282**:14–26.
5. Steinmetz PRH, Kraus JEM, Larroux C, Hammel JU, Amon-Hassenzahl A, Houliston E, Woerheide G, Nickel M, Degnan BM, Technau U: Independent evolution of striated muscles in cnidarians and bilaterians. *Nature* 2012, **487**:231–234.
6. Darling JA, Reitzel AR, Burton PM, Mazza ME, Ryan JF, Sullivan JC, Finnerty JR: Rising starlet: the starlet sea anemone, *Nematostella vectensis.* *Bioessays* 2005, **27**:211–221.
7. Genikhovich G, Technau U: The starlet sea anemone Nematostella vectensis: an anthozoan model organism for studies in comparative genomics and functional evolutionary developmental biology. *Cold Spring Harb Protoc* 2009, doi:10.1101/pdb.emo129.
8. Technau U, Steele RE: Evolutionary crossroads in developmental biology: Cnidaria. *Development* 2011, **138**:1447–1458.
9. Frank P, Bleakney JS: Histology and sexual reproduction of the anemone *Nematostella vectensis* Stephenson 1935. *J Nat Hist* 1976, **10**:441–449.
10. Fritzenwanker JH, Technau U: Induction of gametogenesis in the basal cnidarian Nematostella vectensis (Anthozoa). *Dev Genes Evol* 2002, **212**:99–103.
11. Genikhovich G, Technau U: Induction of spawning in the starlet sea anemone Nematostella vectensis, in vitro fertilization of gametes, and dejellying of zygotes. *Cold Spring Harb Protoc* 2009, doi:10.1101/pdb.prot5281.
12. Moran Y, Genikhovich G, Gordon D, Wienkoop S, Zenkert C, Ozbek S, Technau U, Gurevitz M: Neurotoxin localization to ectodermal gland cells uncovers an alternative mechanism of venom delivery in sea anemones. *Proc Biol Sci* 2012, **279**:1351–1358.
13. Renfer E, Amon-Hassenzahl A, Steinmetz PRH, Technau U: A muscle-specific transgenic reporter line of the sea anemone, Nematostella vectensis. *Proc Natl Acad Sci U S A* 2010, **107**(1):104–108.
14. Muscatine L, Lenhoff HM: *Coelenterate biology. Reviews and new perspectives.* New York: Academic Press; 1974.
15. Hess A, Cohen AI, Robson EA: Observations on the Structure of Hydra as seen with the electron and light microscopes. *Q J Microsc Sci* 1957, **98**:315–326.
16. Chapman DM: A new type of muscle cell from the Subumbrella of obelia. *J Mar Biol Ass* 1968, **48**:667–688.
17. Chia F, Amerongen HM, Peteya DJ: Ultrastructure of the neuromuscular system of the polyp of *Aurelia aurita* L., 1758 (Cnidaria, Scyphozoa). *J Morphol* 1984, **180**:69–79.
18. Gladfelter WB: Structure and function of the locomotory system of the Scyphomedusa *Cyanea capillata. Mar Biol* 1972, **14**:150–160.
19. Anderson PAV, Schwab WE: The organization and structure of nerve and muscle in the jellyfish *Cyanea capillata* (Coelenterata; Scyphozoa). *J Morphol* 1981, **170**:383–399.
20. Werner B, Chapman DM, Cutress CE: Muscular and nervous systems of the cubopolyp (Cnidaria). *Experientia* 1976, **32**:1047–1049.
21. Satterlie RA, Thomas KS, Gray GC: Muscle organization of the cubozoan jellyfish *Tripedalia cystophora* Conant 1897. *Biol Bull* 2005, **209**:154–163.
22. Amerongen HM, Peteya DJ: Ultrastructural study of two kinds of muscle in sea anemones: the existence of fast and slow muscles. *J Morphol* 1980, **166**:145–154.
23. Hyman L: *The invertebrates: Protozoa through Ctenophora.* New York: McGraw-Hill; 1940.
24. Arendt D: The evolution of cell types in animals: emerging principles from molecular studies. *Nat Rev Genet* 2008, **9**:868–882.
25. Rieger RM, Ladurner P: The significance of muscle cells for the origin of mesoderm in bilateria. *Integr Comp Biol* 2003, **43**:47–54.
26. Rieger RM: Über den Ursprung der Bilateria: Die Bedeutung der Ultrastrukturforschung für ein neues Verstehen der Metazoenevolution. *Verh Dtsch Zool Ges* 1986, **79**:31–50.
27. Rieger RM, Lombardi J: Ultrastructure of coelomic lining in echinoderm podia: significance for concepts in the evolution of muscle and peritoneal cells. *Zoomorphology* 1987, **107**:191–208.
28. Westheide W, Hermans CO: *The Ultrastructure of Polychaeta.* New York: Fischer; 1988.
29. Stauber M: The lantern of Aristotle: organization of its coelom and origin of its muscles (Echinodermata, Echinoida). *Zoomorphology* 1993, **113**:137–151.
30. Bartolomaeus T: On the ultrastructure of the coelomic lining in the Annelida, Sipunculida and Echiura. *Microfauna Marina* 1994, **9**:171–220.

2

Semi-automatic landmark point annotation for geometric morphometrics

Paul A Bromiley[1*], Anja C Schunke[2], Hossein Ragheb[1], Neil A Thacker[1] and Diethard Tautz[2]

Abstract

Background: In previous work, the authors described a software package for the digitisation of 3D landmarks for use in geometric morphometrics. In this paper, we describe extensions to this software that allow semi-automatic localisation of 3D landmarks, given a database of manually annotated training images. Multi-stage registration was applied to align image patches from the database to a query image, and the results from multiple database images were combined using an array-based voting scheme. The software automatically highlights points that have been located with low confidence, allowing manual correction.

Results: Evaluation was performed on micro-CT images of rodent skulls for which two independent sets of manual landmark annotations had been performed. This allowed assessment of landmark accuracy in terms of both the distance between manual and automatic annotations, and the repeatability of manual and automatic annotation. Automatic annotation attained accuracies equivalent to those achievable through manual annotation by an expert for 87.5% of the points, with significantly higher repeatability.

Conclusions: Whilst user input was required to produce the training data and in a final error correction stage, the software was capable of reducing the number of manual annotations required in a typical landmark identification process using 3D data by a factor of ten, potentially allowing much larger data sets to be annotated and thus increasing the statistical power of the results from subsequent processing e.g. Procrustes/principal component analysis. The software is freely available, under the GNU General Public Licence, from our web-site (www.tina-vision.net).

Background

Anatomical point landmarks are useful features for a wide range of tasks in medical image analysis and machine vision, and are of particular relevance to morphometrics. Traditional approaches to morphometrics focused on the measurement and analysis of specific lengths, angles, areas etc., and were limited to a relatively small number of such features. Since the pioneering work of Bookstein [1], methods based on the application of statistical shape analysis to large numbers of point landmarks have become increasingly popular. One such approach to landmark-based shape analysis is to perform Procrustes superimposition of a set of annotated specimens, in order to remove non-shape variation (translation, rotation and scale) according to Kendall's definition [2]. A principal component analysis (PCA) can then be performed on the superimposed landmarks in order to identify the main modes of shape variation. Such methods are supported by modern data acquisition methodologies, mainly high resolution CT scans, which provide a multitude of characters, and so potential landmark locations, on outer and inner surfaces. Landmark-based geometric morphometrics can provide a quantitative measurement of the shape of an entire structure or organism. The results can provide a more thorough understanding of forms, e.g. through functional morphology or shape spaces, than could be achieved through traditional morphometrics, and provide a route to phylogeny reconstruction (e.g. [3,4]). In combination with other data, they can also be used to establish correlations between ecological factors and shape (e.g. [5,6]), or to quantify genetic parameters of shape. In all cases, quantitative measurement of multiple shape parameters allows powerful, statistical tests of morphometric hypotheses. However, landmark-based morphometric methods have a significant drawback, in that they require

*Correspondence: paul.bromiley@manchester.ac.uk
[1] Centre for Imaging Sciences, University of Manchester, Stopford Building, Oxford Road, Manchester M13 9PT, UK
Full list of author information is available at the end of the article

the annotation of large numbers of landmarks across multiple specimens, a task that is both difficult and time consuming when performed manually.

Bookstein [1] divided landmarks into three classes according to their relationship to local features. Type one are anatomical points that are defined locally through the juxtaposition of distinct tissues, for example the intersection of cranial sutures, or of veins in insect wings. Type two are intermediate, for example points of locally maximal curvature. Type three are defined by distant, rather than local, features, for example the centre of a circle that is tangent to a structure at more than one point. In a limited number of cases, for example insect wings (e.g. [7]), annotation can be performed on a 2D image of a 2D structure, such that the entire image can be viewed simultaneously. Manual annotation of type one landmarks is then relatively straightforward, although still time consuming if large numbers of landmarks are involved. However, the majority of anatomical structures are three dimensional, and modern tomographic imaging methodologies can provide 3D data. Landmark annotation then becomes more challenging, for two reasons. First, common display technologies are limited to two dimensions, such that only a sub-set of the 3D image e.g. a 2D slice or a projection such as a surface rendering, can be viewed at any time. The display must therefore be repeatedly manipulated during annotation in order to view the location of each landmark. Second, 2D images of specimens allow for intersections of a structure, e.g. a suture, with the background of the image, while 3D objects have more degrees of freedom in rotation, so the same points would need to be specified e.g. as the anterior-most point of a suture. The result is that the process of manual landmark annotation is more difficult, and so more time consuming, in 3D than in 2D data. This has significant implications for subsequent analysis of the data, since the statistical power of any analysis technique, and so the confidence limits on the conclusions, will be dictated in part by the number of data sets that are used.

In [8] the authors described a software package designed to support the process of manual landmark annotation on 3D medical image volumes, with particular reference to micro-CT images. The software presents both 2D and 3D renderings of the image volume, the latter using a fast volume rendering algorithm [9,10] in order to provide the most informative view of the data possible whilst not imposing requirements for specific graphics hardware, and provides numerous functions specifically designed to accelerate the process of landmark annotation for geometric morphometrics. However, the manual input required to annotate a significant number of landmark points is still considerable. In this work, the authors describe an extension to the software package that supports semi-automatic localisation of morphological landmarks in a query image. As described below, the algorithm described here was specifically designed for use in geometric morphometrics, avoiding techniques that could introduce shape-dependent biases into the results.

The problem addressed here was to find the locations of landmarks in a 3D query image volume given a database of example image volumes containing similar structures in which the required landmarks had been manually annotated i.e. to find the mappings from the landmarks in the database images to the corresponding positions in the query image. The literature includes landmark detection techniques, e.g. [11,12], in which a query image is analysed to locate points of maximal surface curvature, maximal intensity gradient etc., that would constitute potential landmarks. Correspondences between landmarks in different images can then be established either manually or automatically. Such methods typically require a surface segmentation, and multiple rules regarding which points constitute potential landmarks. A potentially simpler alternative when manually annotated training data is available, and the approach adopted here, is to consider the mappings between landmark points as sparse transformations from the coordinate systems of the database images to that of the query image, such that the problem falls into the general domain of registration. Registration, the estimation of a transformation that maps one image (the source) into the coordinate system of another image (the target) is a core problem in machine vision and medical image analysis with a correspondingly extensive literature; general reviews of medical image registration are provided by [13-15] and [16], whilst [17,18] and [19] provide recent reviews of surface registration algorithms, with particular reference to surfaces represented by point clouds or meshes.

Image registration is performed by optimising the parameters of a transformation model using a cost function that quantifies the similarity of the transformed source and target images. This can be viewed as a model fitting process, in which the transformed source image constitutes a model, and the target image the data to which the model is fitted. Transformation models can be divided into two classes. The first are global, such as rigid, similarity or affine transformations, where a single set of parameters specifies the transformation. The second are deformable, where the transformation can be characterised as a vector field that varies across the source image. By Kendall's definition [2], shape variation between the structures in the source and target images cannot be modelled using a global transformation model. By definition, the specimens included in a landmark-based morphometric analysis will have differences in shape, requiring a deformable transformation. Significant effort has been applied to the problem of deformable registration of medical images; [20] provides a recent review.

However, deformable registration is an ill-posed problem [20]. Therefore, such methods frequently use a cost function based on two terms; a data term based on the comparison of image intensities or derived features, and a regularisation term that constrains the deformation using an assumed physical model such as viscous fluid flow, elasticity, diffusion etc. in order to make the problem well-posed.

Methods based on free-form deformations of image patches or sub-regions, which attempt to minimise or eliminate the influence of assumed models by using a piecewise rigid or affine transformation, have also been investigated. Lau et al. [21] described a hierarchical approach, in which overlapping sub-regions on a regular grid were independently registered using cost functions based on mutual information (MI; [22-24]), normalised mutual information (NMI; [25]) and the correlation coefficient (CC; [13]), without a regularisation term. A dense deformation field was then estimated from the sparse field of displacement vectors at the region centres by median filtering and Gaussian interpolation, introducing an assumption of smoothness. Malsch et al. [26] described a method in which only sub-regions with high information content were used. Irregularly spaced sub-regions with high local variance were selected and registered using the CC; again, a dense deformation field was estimated from the sparse field of displacement vectors at the region centres by interpolation with thin-plate splines (TPS; [27]), introducing a smoothness assumption based on minimising the bending energy. Söhn et al. [28] extended this approach by analysing the quality of the optimum alignment found for each sub-region using the second derivative of the cost function. Sub-regions on a regular grid were independently aligned using a NMI cost function, and an elastic relaxation was applied depending on the alignment quality: when the cost function exhibited a clear optimum, no relaxation was applied; a combination of data and elastic terms were applied when the optimum was degenerate; and the relaxation was performed with no data term when the optimum was indistinct or absent. B-spline interpolation was then applied to estimate a dense deformation field. Erdt et al. [29] adopted a similar approach, and provided a mathematical framework for the use of eigenvectors of the Hessian matrix of the cost function for estimation of alignment quality, in terms of a Taylor series expansion of the shape of the cost function at the optimum. This method can also be derived within a statistical framework in terms of the minimum variance bound [30,31], and such error information has also been utilised within regularised registration techniques [32]. Finally, [33] described a patch-based registration method inspired by patch-based, multi-atlas segmentation algorithms. The method assumed that the deformation fields between a set of training images and

a template were known; a dictionary of patches from the training images, and their deformations, was then constructed. A query image could then be registered to the template by selecting patches around points of high information (high Canny edge detector [34] responses were used), finding the most similar patches in the dictionary, and constructing a weighted combination of the deformations of those dictionary patches. A dense deformation field was then constructed using TPS interpolation, again introducing a smoothness assumption. The assumption of smoothness was the only model-based constraint on the allowable deformation in these free-form, non-rigid registration methods, and was required only to interpolate a dense deformation field. Therefore, for applications requiring only the identification of landmarks, such methods require no assumed model of the deformation.

Significant effort has also been applied to the problem of automatic landmark point localisation in 2D images within the computer vision field. One popular approach has been the use of statistical shape models, in algorithms such as the Active Shape Model (ASM) [35,36] and related work. In the original work, a set of training images were aligned into a common coordinate system using Procrustes analysis, and the coordinates of the landmarks from each training image, in this reference frame, were concatenated to form a single, high-dimensional vector, such that the complete set of training images defined a point cloud in a high-dimensional space. A principal component analysis (PCA) was applied to extract the major modes of variation of this point cloud. The shape model then consisted of two components; a linear combination of these modes, weighted by a set of shape parameters, and a global rigid transformation that located the model in an image. The model was fitted to a query image by optimising the weights and transformation model parameters in order to maximise the image intensity gradient at the landmark point positions i.e. assuming that landmarks would be located on edges in the image. The Active Appearance Model (AAM) [37] extended the same approach to include image intensities, thus producing a model of both shape and appearance. Later developments, for example the Constrained Local Model (CLM) [38], replaced the global appearance model with a set of texture patches around the landmark points. Fitting consisted of a registration of image patches learned from the training data, or reconstructed from the appearance model, to the query image, with the shape model used as a constraint during optimisation in order to ensure that the relative locations of the patches represented a high-probability shape given the training data. AAMs and related algorithms learn a model directly from training data with no other input. However alternative approaches that incorporate a model of an object as a collection of interconnected parts, e.g. the Pictorial Structure Model (PSM) [39], have also been developed. Whilst most effort has been focused on the application of these techniques to 2D images of faces or objects in natural scenes, they have also been applied to 3D medical image data for purposes such as segmentation; see [40] for a recent review.

The methods described above all perform a registration by optimising a cost function that measures the similarity between the intensities of two images, or image patches, regularised using a model that describes the probability of a given deformation. They exist on a spectrum of model complexity, from very limited models assuming only that the deformation field is smooth and continuous, through full physical (e.g. elastic) models of the allowable deformations, to AAMs that are bootstrapped from training data. However, a model-based regularisation could not be used in the work described here, for the following reasons. Most importantly, the aim was to produce landmarks for geometric morphometrics, which would be analysed with the standard techniques used in that area of research, including Procrustes analysis followed by PCA and interpolation between landmarks using thin-plate splines. The same techniques are used to build the shape models used in AAMs and related algorithms. Therefore, the subsequent analysis would be capable of regenerating the shape model used in automatic annotation i.e. any mode of shape variation present in the query image, but not included in the annotation model, would not be found in PCA analysis of the results. Since the aim of many experiments in landmark-based geometric morphometrics is to quantify the modes of shape variation, this form of bias would be unacceptable. The training data for an AAM would need to exhibit all possible shape variation within the relevant shapes in order to guarantee that such biases were not present; this would make the training set prohibitively large. Similarly, a smoothness assumption would not allow points of infinite curvature in the deformation field. However, such points could be present at the interface between sliding surfaces e.g. the points of contact between the upper and lower molar rows, which would be relevant landmark locations for many studies comparing morphology with ecology. Furthermore, the bootstrapped models used in AAMs and related algorithms require extensive offline training; more manually annotated images would be used to train the model than would typically be included in landmark-based geometric morphometrics experiments. Since the aim here was to maximally accelerate the landmark annotation process, a method without this requirement was needed.

The work described here adopted the hierarchical, patch-based registration used in methods based on free-form deformations, as described above. Patches of image data around each landmark in each database image were registered, using an affine transformation, to the query image. This avoided the use of any model-based shape constraint. Furthermore, since there was no need to interpolate a dense deformation field, no assumption of smoothness was required. A similar approach proved successful in earlier work on automatic landmark annotation for morphometric analysis of microscope images of fly wings, i.e. 2D images of planar objects with no out-of-plane rotations [41,42]. The software described here required a small database of images similar to the query image, in which the required landmarks had been manually annotated. In the absence of regularisation, a multi-stage registration approach was developed in order to compensate for shape variations between the database and query images, which would necessarily be present. The initial stages operated on the whole image, and so were affected by shape variations. However, the results from the initial stages were used only to initialise later stages operating on texture patches around each landmark. Multiple patch-based stages with reducing patch sizes were used, with the intention that the effect of global shape variation would be reduced as the patches became smaller. Automatic image registration algorithms are typically incapable of dealing with gross misalignments e.g. where the images are rotated by 180° with respect to one another, and so manual intervention was required to provide an initialisation. In the work described here, four non-coplanar landmark points in each volume were used for this purpose (see the Conclusions for further comments), and this stage of the algorithm can be omitted completely if care is taken during sample preparation, such that the specimens are in approximately the same orientation in all image volumes used. The point-based stage of the registration minimised the RMS distances between the registration points. All image-based stages minimised the χ^2 of the scaled intensity gradients in the database and query images; the scaling provided some independence to variations in scanner parameters and average bone density, and was estimated using maximum likelihood after the point-based stage of registration.

Each database image produced one estimate of the location of each landmark in the query image; these were combined to generate the final estimated landmark location. A robust alternative to simple approaches such as averaging was implemented, allowing sub-selection of only the most reliable estimates i.e. those for which the database image provided the best model of the query image. Furthermore, this did not require a shape model and could operate with a small number of examples. Since the introduction of the generalised Hough transform [43], array-based voting schemes have been shown to be effective for locating structures in images. In particular, several recent papers (e.g. [44,45]) have used Random Forests [46] to locate structures in images, combining the estimates from each tree in a voting array. A similar approach was adopted here; the estimated positions of a given landmark from each database entry cast votes into a 3D array, which was then convolved with a Gaussian kernel to approximate the random error on the estimates. Outliers resulting from failed registrations formed a broad background distribution, whilst points from the signal distribution (i.e. successful registrations) were randomly scattered around the true location of the landmark point in the query image according to the random error on the estimation process, and so contributed to a single, dominant peak in the voting array. The most significant mode in the smoothed array was taken as the estimated location for the landmark. This provided a degree of robustness to outliers; however, situations might still occur in which no element of the database provided a good estimate of the landmark location. Therefore, a robust outlier detection method was applied to the final estimated locations, based on testing for consistency between the result from the array-based voting and the estimated χ^2 per degree of freedom on the points that contributed to the array.

Methods

The automatic landmark annotation process was based on a hierarchical, free-form registration of image patches from the database to the query image. The sequence of processes involved is shown in Figure 1. An initial alignment was derived from four landmark points, by minimising the root-mean-squared (RMS) distances between corresponding points over a nine-parameter affine transformation (i.e. 3D translation, rotation and scale) using the simplex algorithm [47]. Figure 2 shows a typical image volume for experiments in landmark-base geometric morphometrics, a 3D micro-CT image of a *Mus*

musculus specimen, together with manual annotations of a typical set of landmarks, described in detail in Table 1. Four typical global registration points are shown in Figure 2b. An automated check was implemented to ensure that the points were not coplanar. This point-based registration could have been decomposed into individual transformations, deriving the translation parameters by aligning the centroids of the four points and the scaling parameters from the standard deviations of the points, leaving only the rotation to be obtained through optimisation. However, in practice all nine parameters of the transformation model were obtained via the optimisation, so that the distance between the centroids of the points in each registered image volume could be used as a semi-independent, automated check.

Once the initial manual alignment had been obtained, it was used to initialise a multi-stage automatic registration process. The first stage was performed on the entire image volume, and optimised a nine-parameter affine transformation using the simplex algorithm [47]. The latter stages operated on patches of image data around each landmark point. As described above, the intention was to terminate the process with patches small enough that the effects of shape variation between the database and query images were minimised. However, registration of small patches had a correspondingly small capture range: in preliminary work, a single stage of patch-based registration proved to be insufficient to attain accuracies equivalent to manual annotation. Therefore, additional stages of patch-based registration were included, with the patch size reduced between each stage; an empirical evaluation of accuracy and time versus the number of registration stages was performed, and the optimal number of patch-based stages was shown to be three, for a total of five stages of registration including the point-based initialisation and the global, image-based stage (see Additional file 1). Some experiments in geometric morphometrics

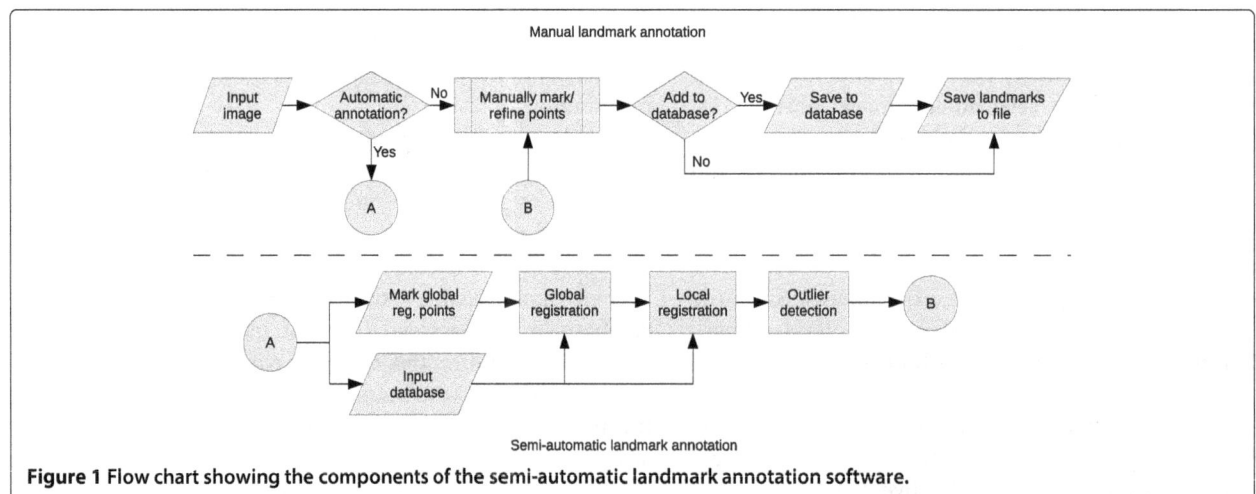

Figure 1 Flow chart showing the components of the semi-automatic landmark annotation software.

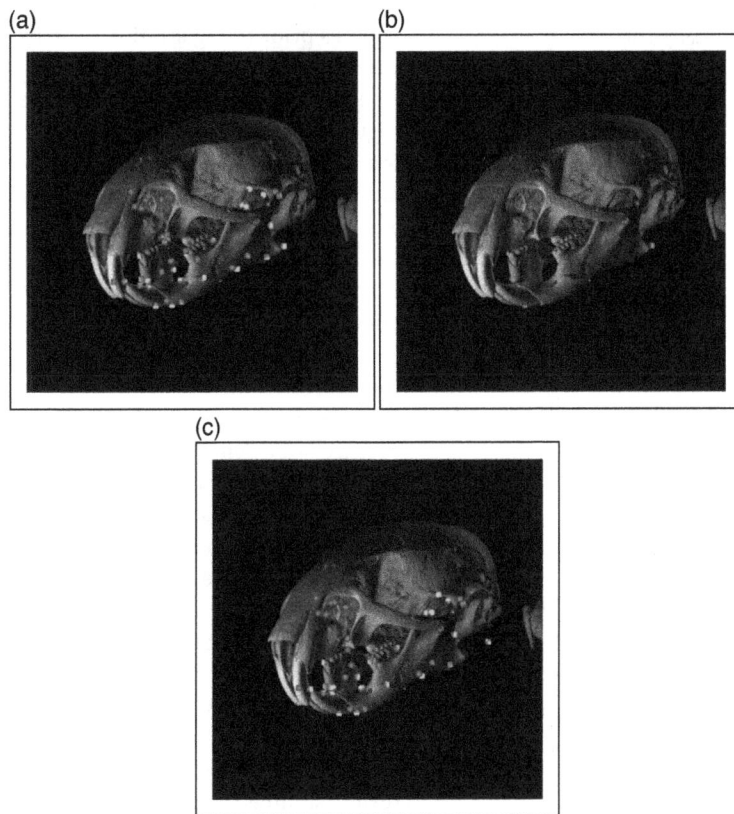

Figure 2 3D renderings of the automatic landmark point annotation process on a *Mus musculus* specimen. **(a)** Manual annotation of the 40 mandible landmarks. **(b)** Four of the manual landmarks used in the global registration. **(c)** Automatically annotated landmarks, derived using a database of seven *Mus* specimens. Landmarks passing the outlier test are shown in green: those failing are shown in chequered green and red.

may include landmarks specified as the extremal points on curved surfaces from certain view directions. The inclusion of rotation in the patch-based registration would destabilise the registration in such cases (e.g. registration of an arc of a circle to that circle is degenerate over rotation, but not translation or scaling). Therefore, the transformation model used in the patch-based stages of registration consisted only of 3D translation and scaling. Each stage of registration was initialised using the concatenation of the transformation matrices from all previous stages. However, a simple check was included to prevent problems with fit failures; if the result after any stage of registration generated a projected location that lay outside the boundaries of the query image volume, then the parameters of that stage were set to an identity transformation.

Since a nine-parameter affine transformation model was used, the problem was over-determined with realistic patch sizes. Therefore, rather than using cuboids from the data as image patches, 2D slices aligned to the major axes of the image volume and centred on the landmark points were used; in the first stage, operating on the entire image volumes, the slices were taken through the centre points

of the volumes. This considerably reduced the amount of data included in the cost function calculation, and so reduced the processor time required whilst still achieving sufficient levels of accuracy.

The cost function for all automatic registration stages was the χ^2 per degree of freedom of the scaled images or image patches (a derivation is provided in Appendix A)

$$\chi^2 = \frac{1}{(N-D)} \sum_v \frac{(J_v - \gamma I_v)^2}{\sigma_J^2 + \sigma_I^2 \gamma^2} \quad where \quad \gamma = \frac{|J|}{|I|} \quad (1)$$

where, assuming without loss of generality that I is the source (database) image and J the target (query) image, J_v is the value of voxel v in the target image, I_v is the value of the corresponding voxel in the transformed source image after re-sampling onto the voxel grid of the target image using an interpolation algorithm, N is the number of voxels in the images or image patches, σ_I and σ_J are the standard deviations of the noise on the scaled images or patches, D is the number of transformation model parameters being optimised, and γ is a scaling factor, providing some degree of independence to differences in scanner

Table 1 The 50 skull landmark set

Landmark no.	Description
1	Anterior end of nasal suture.
2	Posterior end of nasal suture.
3	Posterior end of frontal suture.
4	Posterior end of parietal suture.
5	Posterior-most point of occipital.
6	Dorsal-most point of foramen magnum.
7	Anterior-most point of premaxilla behind incisivi.
8	Posterior-most point of palatal suture.
9	Anterior-most medial point of occipital.
10	Anterior-most point of foramen magnum.
11	Anterior end of molar row.
12	Posterior end of molar row.
13	Tip of incisor.
14	Anterior tip of premaxilla.
15	Anterior end of incisive foramen.
16	Posterior end of incisive foramen.
17	Anterior-most dorsal point of infraorbital foramen.
18	Anterior-most lateral point of infraorbital foramen.
19	Anterior-most ventral point of infraorbital foramen.
20	Anterior-most dorsal point of orbita.
21	Anterior-most ventral point of orbita.
22	Dorsal-most point of lateral-most point of zygomatic arch.
23	Ventral-most point of lateral-most point of zygomatic arch.
24	Posterior end of orbita.
25	Anterior-most point of bulla.
26	Anterior-most point of acoustic meatus.
27	Dorsal-most point of bulla.
28	Ventral-most point of bulla.
29	Posterior-most point of bulla.
30	Dorsal end of condyle.

These landmarks were identified on the 12-element data set described in Table 4. Points 11 to 30 were identified on the left-hand-side of the specimen; points 31 to 50 were the equivalent points on the right-hand side of the specimen.

parameters and average bone density. Trilinear interpolation was used in the resampling. The scaling factor was obtained through a maximum-likelihood based approach (see Appendix A). Inevitably, the scaling could be computed only from aligned images. Therefore, it was computed after the point-based stage of registration, and then remained fixed through all of the image-based registration stages. Preliminary work on this topic was described in [48].

In order to reduce the noise on the images and thus provide a smoother cost function, reducing the probability that the optimisation would become trapped in a local minimum, Gaussian smoothing was applied to all image patches prior to registration. The kernel was truncated at three standard deviations from the mean and, in order to ensure that no edge effects were present, a boundary region equal to three standard deviations of the smoothing kernel was added around all stored image patches and included in the smoothing, but excluded from the χ^2 calculation, thus ensuring that all smoothed voxels contributing to the χ^2 were calculated from equal numbers of un-smoothed voxels and avoiding truncation effects.

Rather than operating directly on the image intensities, the cost function was applied to image intensity gradients in order to provide further robustness to differences in average bone density or scanner parameters. This strategy has been found useful in other applications that require matching of similar but non-identical images e.g. stereo pairs of natural scenes [49]. Images showing the gradient components in the x and y directions of each patch were calculated, for each of the three orthogonal patches passing through each landmark point, using finite differencing i.e. taking the intensity difference between neighbouring voxels in the relevant direction. This gave a total of six gradient patches for each landmark. The cost function was applied to each gradient patch separately and the results summed to produce a single χ^2 per degree of freedom. Note that N in Eq. 1 refers to the total number of voxels included in the three orthogonal patches, not the total number in the six gradient patches, since the x- and y-gradients of each patch were obtained from the same original data.

Explicit inclusion of the noise term in the cost function allowed the χ^2 per degree of freedom at the end of the registration to be used as a check on registration error, through comparison to the χ^2 distribution. The noise on the original images was estimated from the width of zero crossings in horizontal and vertical gradient histograms [50]. Smoothing reduced the noise by a factor of $4\pi\sigma_K^2$ where σ_K was the standard deviation of the smoothing kernel [51] (see Appendix B). The calculation of image gradients by finite differencing introduced a further factor of $\sqrt{2}$ into the noise calculation.

Particular care was required in cases where part of an image patch lay outside the volume. Simply ignoring such regions would bias the registration towards moving the patches almost completely outside the query image volume in cases where there was any shape difference. Therefore, any regions lying outside the image volumes were zero-padded and included in the χ^2 calculation. Furthermore, image masks were generated for all patches, recording the zero-padded regions. The cost function was then calculated as three separate terms i.e. one for voxels lying inside both volumes, one for voxels where the I image voxel was zero padded, and one for voxels where the

J image voxel was zero padded; the fourth combination, where both voxels were zero-padded, was identically zero. The correct noise term was used in each case ($\sigma_J^2 + \gamma^2 \sigma_I^2$, $\gamma^2 \sigma_I^2$, and σ_J^2 respectively, divided by the correction factors for smoothing and differentiation). These three χ^2 terms were then summed prior to division by the number of degrees of freedom, the total number of voxels included in all three terms minus the number of parameters in the transformation model.

Array-based voting

The result from the various stages of registration was a projected location, in the coordinate system of the query image, for each landmark point in each database entry. These constituted multiple estimates of the position of each landmark in the query image, with the number of estimates equal to the number of database entries. The multiple estimates for each landmark were then combined in order to generate a single, final estimate of the location of that landmark in the query image. However, it could not be guaranteed that all estimates were accurate; if one of the database images exhibited significant shape difference, compared to the query image, in the region around one of the landmark points, then it would provide a poor model for that landmark. The corresponding registration would then be likely to fail, resulting in an outlier amongst the multiple estimates for the landmark. Any simple method of combination, such as taking the mean of all estimates, would be affected by the presence of such outliers.

Instead, a method that was robust to outliers was implemented, based on the array-based voting used in methods such as the Hough transform; the same approach has proven reliable in shape-model based approaches to the annotation of landmark points on both clinical and non-clinical images (e.g. [45,52,53]). The voting method was applied to each landmark independently. The multiple estimates for the position of a landmark were analysed to find the maximum and minimum values of their x, y and z coordinates; this specified a 3D volume large enough to contain the estimates. An array of this size was created, and a value of unity was entered into the array at the position of each estimate. The array was then smoothed using a Gaussian kernel, approximating the random error on the registration results. The kernel size was a free parameter of the algorithm (see Additional file 1 and see the Conclusions for further comments). Assuming that the majority of the database images provided good models of the query image, the entries in the voting array would include a compact distribution located close to the correct position for the landmark, representing successful registrations. The width of this distribution would be dictated by the random error on the registration, which would in turn be dictated by the noise on the original images. In contrast, outliers would by definition be scattered far from the correct position. Therefore, smoothing the array with a kernel corresponding to the random error on successful registrations would result in a single, main peak at the position of the compact group of accurate estimates, and a number of smaller, secondary peaks at the positions of outliers. The position of the highest peak in the smoothed array was taken as the final estimate of the landmark location, producing a method that combined the multiple estimates of each landmark location whilst being robust to the presence of outliers.

Prior to creation of the voting array for each landmark point, the projected locations of the corresponding database points in the coordinate system of the query image were compared to the boundaries of that image; any point lying outside the boundary plus a border of three times the standard deviation of the smoothing kernel was omitted from the voting process, in order to prevent problems with severe outliers, on the basis that these points could not contribute to a valid final estimate in any case. The size of the array was then determined by finding the range of the projected locations for the remaining points.

Outlier detection

In order for the final algorithm to have any utility, it was essential that it should provide a reliable indication of the accuracy of automatic annotations; otherwise, much of the manual interaction that it was intended to replace would be re-introduced through a requirement for manual inspection of the results. The array-based voting described in the previous section required that the majority of the multiple estimates of the position of each landmark were located close to the correct position; if this were not the case, due to significant differences in shape between the query image and all of the database images, then the voting would produce an incorrect result. An outlier detection method with an extremely low false positive rate, i.e. an extremely low number of outliers flagged as accurate annotations, was therefore required.

If a given database image patch formed a perfect model of the corresponding query image patch, then the only source of errors on the optimised transformation model parameters would be the random noise on the images. Since a maximum likelihood estimator was used as the registration cost function, the minimum variance bound (MVB; [30,31]) could be applied to estimate this distribution; error propagation could then be applied to find the distribution of the estimated landmark locations. However, in practice there will always be small shape differences between the database and query images, introducing a systematic error onto each patch registration. Assuming these systematic errors to be uncorrelated across the database images, they introduce a secondary distribution, i.e. the multiple estimated landmark locations generated from the database images will be randomly scattered around the true landmark location in the query image according to a distribution dependent on random shape differences (which could be termed "shape noise"). This will add to the distribution due to random image noise, and so the MVB will provide an underestimate of the true distribution of the estimated landmark locations.

As stated above, [28] and [29] developed a method that measured the shape of the cost function around the optimum using the eigenvectors of the Hessian matrix, allowing the rejection of results where shape differences destabilised the registration to the point where there was no clear optimum. However, this method was limited to analysis of the cost functions for individual registrations. In the work described here, the multiple registration results for each landmark supported an analysis of the distribution of registration parameters induced by shape variation between the database images. The number of samples available was too small to allow a full characterisation of the distribution; therefore, unreliable results were identified through an analysis of the available point estimates. The individual registration results were first sorted into order based on the χ^2 per degree of freedom at the end of the registration process. The distances between the results and the final location generated by the array-based voting were then compared to a threshold, treating each distance as an estimate of the standard deviation of the distribution. The threshold and the number of comparisons performed were free parameters of the algorithm (see Additional file 1); in practice, comparison to three points proved optimal. Any final location for which any one of the three registration results with the lowest χ^2 per degree of freedom was more distant than the threshold was flagged as a potential outlier. The technique therefore imposed a requirement that the patches which were estimated to provide the best models of the query image patch, based on their χ^2 per degree of freedom, formed a distribution within the voting array no broader than the threshold.

Software

The semi-automatic landmark point localisation algorithm was implemented within the TINA Geometric Morphometrics toolkit, which also includes the TINA Manual Landmarking tool [8], and algorithms that perform quantitative shape analysis with weighted covariance estimates for increased statistical efficiency [54]. This package has been made available as free and open source software (FOSS) under the GNU General Public Licence (www.gnu.org), and can be obtained via the TINA web-site (www.tina-vision.net). The User Manuals for the TINA Geometric Morphometrics toolkit and the TINA

software are included Additional files 2 and 3, respectively. Figure 1 shows the sequence of operations involved in semi-automatic landmarking, and how the algorithm interfaces with the manual annotation software. Figure 3 shows a screen-shot of the software in operation. The 3D renderings and associated landmark annotations are shown in more detail in Figure 2.

Results and discussion

Evaluation of the algorithm was performed using micro-CT images of rodent skulls with an in-plane resolution of 635×635 voxels and between 1000 and 1500 slices, with voxel dimensions of 0.035 mm along all axes. A detailed description of all image volumes and landmark points used is provided in Tables 1, 2, 3 and 4. The algorithm included a number of free parameters, such as image patch and smoothing kernel sizes for each stage of registration. As described in Additional file 1, these were optimised using a data set of 8 *Mus* specimens of varying species (see Table 2) with expert manual annotation of 40 mandible landmarks (see Table 3). All manual annotations were performed using the TINA software [8].

In order to avoid any bias, the evaluation of automatic point localisation accuracy was performed on a data set of 12 *Mus musculus* specimens from consomic strains, independent of those used in the parameter optimisation experiments (see Table 4). Two sets of expert manual annotations of 50 skull landmarks were performed on each (see Table 1), allowing estimation of manual annotation repeatability; the repeat annotation was performed after an interval of one week, in order to minimise bias due to training effects. Separate experiments were performed for each set of manual annotations in sets of leave-one-out experiments, using 11 specimens to construct the training database and the 12th as the query image, repeating for all 12 image volumes. Four manually annotated points in each volume were used to provide the initial, global alignment; however, the locations of these points were re-estimated by the automatic annotation algorithm, such that results were not contaminated with manual annotations.

A number of extreme outliers (Euclidean distances of over 100 voxels between automatic and manual annotations) were observed in the initial results; detailed investigation revealed that, in all cases, these were due to errors in the manual landmark annotations, consisting of transpositions of equivalent points on either side of the plane of bilateral symmetry. Since these were systematic rather than random errors, they were reported to, and corrected by, the expert; the correction process was limited to a single pass in order to prevent the possibility of experimenter effect. After this, one point transposition error remained in the first set of manual landmarks, and none remained in the second. All results reported here were generated

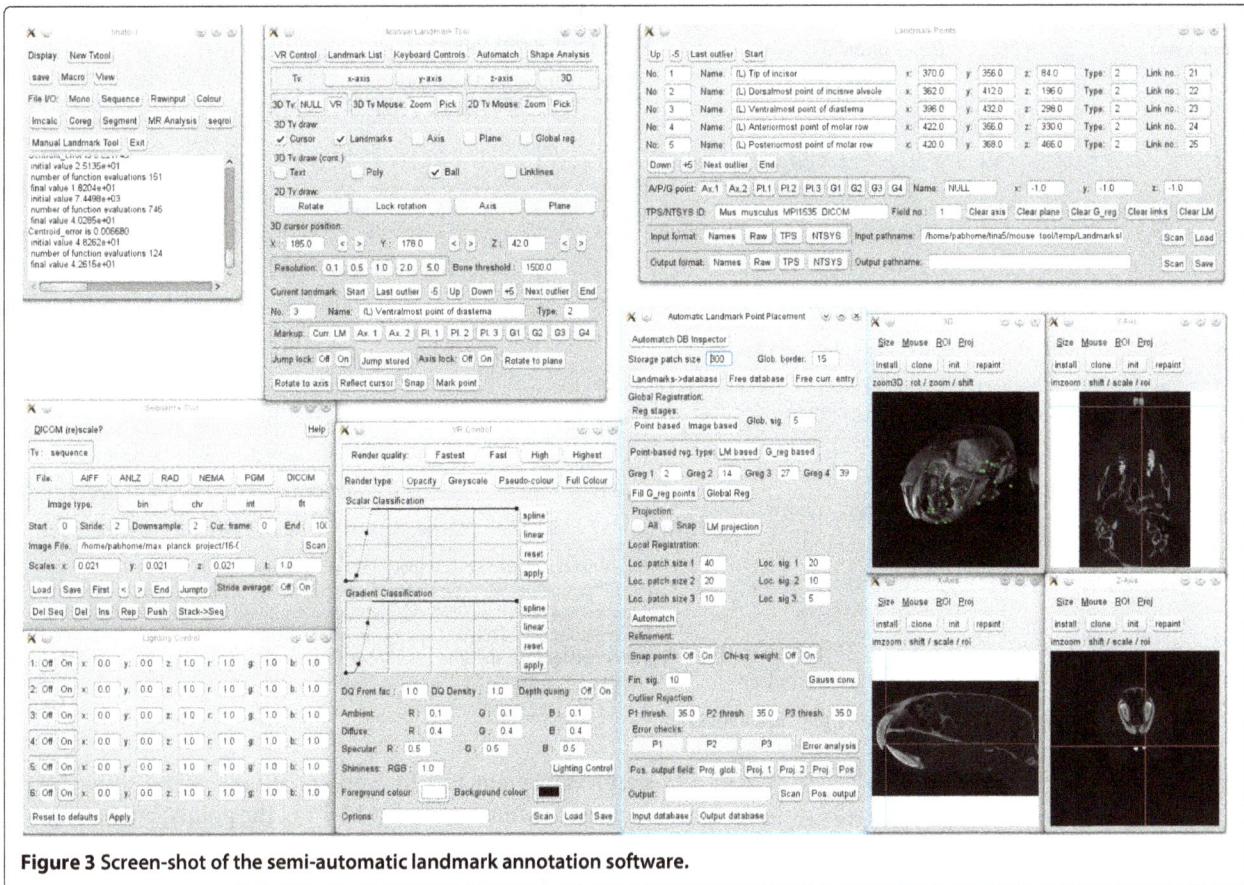

Figure 3 Screen-shot of the semi-automatic landmark annotation software.

Table 2 The 8- and 14-element data sets

Specimen	Description
1	*Mus macedonicus* (Macedonian mouse)
2	*Mus musculus domesticus* (House mouse)
3	*Mus musculus domesticus* (House mouse)
4	*Mus musculus domesticus* (House mouse)
5	*Mus musculus musculus* (House mouse)
6	*Mus musculus musculus* (House mouse)
7	*Mus musculus musculus* (House mouse)
8	*Mus musculus musculus* (House mouse)
9	*Apodemus flavicollis* (Yellow-necked mouse)
10	*Apodemus sylvaticus* (Wood mouse)
11	*Apodemus sylvaticus* (Wood mouse)
12	*Meriones unguiculatus* (Mongolian gerbil)
13	*Microtus fortis* (Reed vole)
14	*Phodopus sungorus* (Djungarian hamster)

The 8-element data set consisted of the first 8 specimens, all of which were various species from the genus *Mus*. The 14-element data set also included specimens from other genera. Expert manual annotation of 40 mandible landmarks was performed for each specimen; see Table 3 for details.

from the corrected landmarks. However, the number of such errors, 4% of the points in the first set of manual landmarks and 0.5% of the points in the second set, was recorded, and the lower value used as a target for the false positive rate of the error detection stage of the automatic point localisation algorithm.

In order to calculate true and false positive and negative rates for the outlier test, a threshold value on point localisation error in voxels was required. This is referred to in the following sections as the error threshold, and was determined using the repeatability of the manual annotations on the 12 consomic *Mus musculus* specimens. As described in Additional file 1, the worst outlier amongst the 50 points in each of the 12 specimens was found by comparing the two sets of manual annotations (after the correction process described in the previous paragraph), and the mean of those values used as the threshold. This gave a numerical value of 30 voxels, equivalent to 1.05 mm at the resolution of the image volumes used here. An automatic annotation was therefore defined as erroneous if its displacement from the corresponding manual annotation was greater than the largest displacement, on average, that would be seen in repeated manual annotations.

Table 3 The 40 mandible landmark set

Landmark no.	Description
1	Tip of incisor.
2	Dorsal-most point of incisive alveole.
3	Ventral-most point of diastema.
4	Anterior-most point of molar row.
5	Posterior-most point of molar row.
6	Lateral point of mandibular foramen.
7	Tip of coronoid process.
8	Anterior-most point of curve between coronoid and articular process.
9	Ventral-most point of curve between coronoid and articular process.
10	Anterior-most point of condyle.
11	Lateral-most point of condyle.
12	Posterior-most point of condyle.
13	Anterior-most point of curve between articular and angular process.
14	Tip of angular process.
15	Medial-most point of angular process.
16	Lateral point of masseteric crista at dorsal-most ventral point.
17	Lateral-most inner point of ventral border.
18	Anterior end of attachment area of transverse mandibular muscle.
19	Posterior-most point of incisive alveole.
20	Anterior end of masseteric crista.

These landmarks were identified on the 8- and 14-element data sets described in Table 2. Landmarks 1–20 consisted of these points on the left-hand side of the specimen; landmarks 21–40 consisted of the same points on the right-hand side of the specimen.

Table 4 The consomic *Mus musculus* specimens included in the 12-element data set

Specimen	Description
1	PWD/Ph (wild-derived *Mus musculus musculus* strain)
2	PWD/Ph (wild-derived *Mus musculus musculus* strain)
3	PWD/Ph (wild-derived *Mus musculus musculus* strain)
4	PWD/Ph (wild-derived *Mus musculus musculus* strain)
5	C57BL/6J-10d$^{PWD/Ph}$
6	C57BL/6J-10d$^{PWD/Ph}$
7	C57BL/6J-7$^{PWD/Ph}$
8	C57BL/6J-7$^{PWD/Ph}$
9	C57BL/6J-7$^{PWD/Ph}$
10	C57BL/6J (*Mus musculus domesticus* background)
11	C57BL/6J (*Mus musculus domesticus* background)
12	C57BL/6J (*Mus musculus domesticus* background)

This data set included several specimens of each pure strain (C57BL/6J and PWD/Ph), together with specimens in which chromosomes 7 or 10 from the *Mus musculus musculus* strain PWD/Ph had been substituted into the *Mus musculus domesticus* strain C57BL/6J. Expert manual annotation of 50 skull landmarks was performed for each specimen; see Table 1 for details.

Figure 4 shows the results of the leave-one-out experiments performed on the data set of 12 image volumes of consomic *Mus musculus* specimens using the second set of manual landmarks to produce the database and using four manual annotations to perform the initial, manual stage of alignment. The results are presented as box-and-whisker plots of the Euclidean distance in voxels between the automatic and manual annotations for each point, showing the median, minimum, maximum, 25th and 75th percentiles. The ROC curve, generated by varying the outlier test threshold, shows that the threshold and number of points included, derived from an independent data set, were approximately optimal for these data; there were no false positives at the operating point used, i.e. all 525 points passing the test were within the error threshold of 30 voxels. Across all twelve volumes, 87.5% of the automatic annotations passed the outlier test, and 98.3% were within the error threshold. In a hypothetical annotation process with a pre-built database, this level of performance equates to a potential user of the software having to manually inspect 12.5% of the points, but having to correct the positions of fewer than 2% of the points, i.e. fewer than 12 points. Therefore, combined with the four points per specimen required for the initial, point-based stage of registration, the user would be required to manually annotate an average of 5 points per specimen out of a total of 50 landmark points i.e. one tenth of the total number of points.

The results show that the automatic landmark localisation algorithm placed points to within a median of 3.60 voxels of the manual annotations. However, this value contains contributions from the errors on both the manual and automatic annotations, i.e. it does not represent the random error on the automatic landmark annotations. In order to quantify this random error, the repeatability of the automatic annotation process was evaluated by comparing the automatic landmarks generated from databases constructed from the two sets of manual annotations. Figure 5 shows the result of this comparison, together with the repeatability of manual annotation. The median Euclidean distance between the two sets of manual annotations across all points in all image volumes was 3.34 voxels: assuming that the error distributions on both sets of manual annotations were the same, this implies that the median manual annotation error on a single point was approximately 2.4 voxels. The mean of the worst outliers in all 12 volumes was 29.3 voxels. The median distance between the two sets of automatically located landmarks across all points in all volumes was 1.4 voxels;

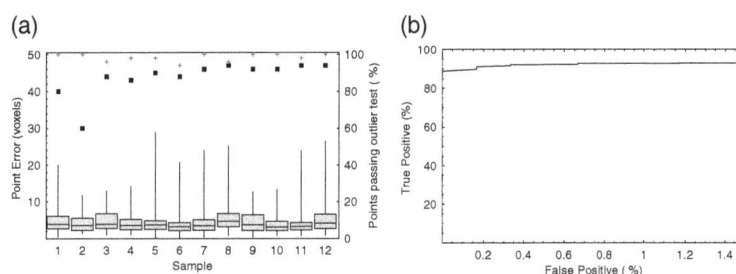

Figure 4 Automatic landmark annotation accuracy using an initial, point-based registration. (a) Box-and-whisker plots of the point localisation errors for automatic annotation of 50 skull landmarks on the 12 consomic *Mus musculus* specimens (read against the left-hand scale), using the second set of manual landmarks and an initial, point-based registration stage. The black squares and red crosses show, respectively, the percentage of points passing the outlier test and the percentage within the error threshold (read against the right-hand scale); only points passing the outlier test have been included in the box-and-whisker plots. **(b)** ROC curve of the true and false positive rates of points passing the outlier test; the operating point of the test is coincident with the y-axis.

again assuming equal errors on the two sets of results, the median of the automatic annotation error on a single point was approximately 1.0 voxels. The mean of the worst outliers in all twelve volumes was 7.4 voxels. Only automatically located points passing the outlier test were included in the results; this was 81.8% of the points across all 12 volumes. Due to the non-Gaussian distribution of the annotation errors, the manual and automatic repeatabilities were compared using a Mann-Whitney U test, which conclusively demonstrated ($U = 72116$, $U_\mu = 147300$, $U_\sigma = 5187$, $Z = -14.52$, $p \approx 0$) that the automatic repeatability was significantly better than the manual repeatability.

Evaluating the repeatability of the automatic annotation did not evaluate its accuracy; such an evaluation was not possible without a set of gold-standard (i.e. error-free) landmark locations. However, comparison of the median Euclidean distance in voxels between the two sets of manual annotations (3.34 voxels) and the median Euclidean distance between the second set of manual annotations and the automatic annotations derived from them (3.6 voxels) indicated no statistically significant difference

(Mann-Whitney $U = 149405$, $U_\mu = 157200$, $U_\sigma = 5429$, $Z = -1.44$, $p = 0.15$). This indicated that automatic annotation was not significantly less accurate than manual annotation.

Double iteration without point-based registration

The reliability of the outlier test, demonstrated in the experiments described above, allows an alternative mode of operation for the algorithm that can potentially eliminate the need for manual annotation of the four points used in the initial stage of global registration. The gross misalignments between image volumes, for which this stage of registration was designed, can be avoided if care is taken during the preparation of specimens, such that they are all scanned in approximately the same orientation. The algorithm can then be applied in two stages; a first pass, with no point-based registration, generates an intermediate set of automatic annotations. Fewer points will pass the outlier test; however, the points that do can be used to perform the point-based stage of registration in a second pass of the algorithm. A second set of experiments, identical to those described above but using this

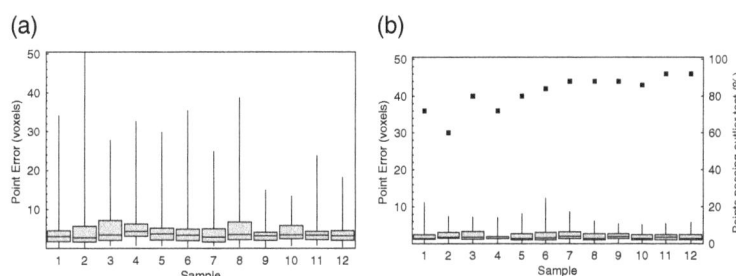

Figure 5 Repeatability of manual and automatic landmark annotation. Box-and-whisker plots of the repeatability of manual **(a)** and automatic **(b)** (read against the left-hand scale) localisation of 50 skull landmarks on the 12 consomic *Mus musculus* specimens. The automatic landmarks were generated using an initial, point-based stage of global registration. Only points passing the outlier test were included in the automatic annotation results; the black squares show the percentage of points passing (read against the right-hand scale).

"double-pass" mode of operation, was performed in order to test this approach. The results for automatic annotation using a database built from the second set of manual annotations are shown in Figure 6. The median Euclidean distance between the manual and automatic annotations across all twelve volumes was 3.49 voxels, compared to 3.60 voxels for the equivalent experiment in single-pass mode, but the difference was not statistically significant (Mann-Whitney $U = 134236$, $U_\mu = 137550$, $U_\sigma = 4906$, $Z = -0.68$, $p = 0.50$). Similarly, there was no significant difference between the number of points passing the outlier test (87.5% in single-pass and 87.3% in double-pass mode) or the percentage of points with errors lower than the error threshold (98.3% in single-pass and 97.8% in double-pass mode). The ROC curve shows that the outlier test parameters, derived from an independent data set, were approximately optimal for these data; the false positive rate of the outlier test was 0.17%, i.e. across the 600 points in the 12 volumes, i.e. only one point with an error larger than the threshold was not flagged as an outlier. Figure 7 shows a comparison of the repeatability of manual and automatic landmark annotation when the software was used in double-pass mode; as with the single-pass mode, the automatic repeatability was significantly better (Mann-Whitney $U = 77551$, $U_\mu = 146700$, $U_\sigma = 5162$, $Z = -13.39$, $p \approx 0$). However, it should be noted that the double-pass mode of the algorithm is dependent on the alignment of the specimens within the scanner, and is likely to fail if significant misalignments are present.

Multiple-genera database

In order to evaluate the robustness of the algorithm to databases containing specimens with significant shape differences, a further data set consisting of 14 micro-CT image volumes of rodent skulls from multiple genera was used (see Table 2), with manual annotations of 40 mandible landmarks on each (see Table 3). This was a superset of the 8 *Mus* skull data set used in the parameter optimisation experiments, and was therefore not completely independent (i.e. the free parameters of the algorithm were partially derived from this data set), although the evaluation of the free parameters showed a high degree of independence between performance and parameter values (see Additional file 1 for details). In addition to the 8 *Mus* specimens, the data set included one *Apodemus flavicollis*, two *Apodemus sylvaticus*, one *Meriones unguiculatus*, one *Microtus fortis* and one *Phodopus sungorus* specimen. This combination of specimens was chosen to exhibit a range of shape variation, i.e. the *Apodemus* specimens were more similar in shape to the *Mus* specimens than were the *Microtus*, *Phodopus* or *Meriones* specimens. The skull constituted the majority of the bone surfaces in the images and so dominated the registration result; the mandible is not rigidly fixed to the skull, and so the use of mandible landmarks provided a more significant challenge to the algorithm and was more suitable to illustrate failure modes. As above, the automatic point localisation algorithm was applied to the data in a set of leave-one-out experiments, using 13 image volumes to construct the database and predict landmark locations in the 14th volume, repeating for all 14 volumes. A second set of experiments was conducted using only the eight *Mus* specimens from the data set. The results are shown in Figures 8 and 9.

The effects of building the database from multiple genera can clearly be seen in the breakdown of the results by genus. During annotation of the *Microtus*, *Phodopus* and *Meriones* specimens, which varied significantly in shape from the *Mus* and *Apodemus* specimens, there were no similar specimens in the database and consequently the outlier test rejected all points. The median landmark error across the *Mus* specimens was slightly lower using a mixed-genera database than using a *Mus*-only database

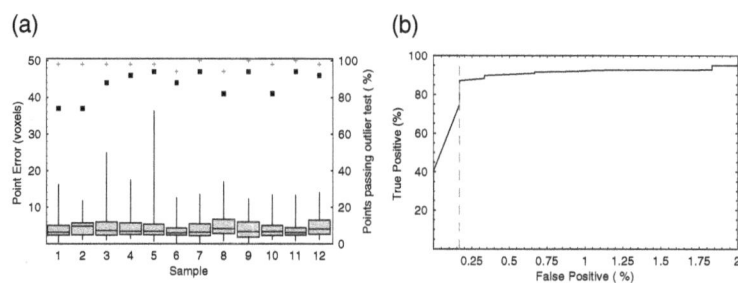

Figure 6 Automatic landmark annotation accuracy in "double-pass" mode. (a) Box-and-whisker plots of the point localisation errors for automatic annotation of 50 skull landmarks on the 12 consomic *Mus musculus* specimens (read against the left-hand scale), using the second set of manual landmarks. The algorithm was applied in "double-pass" mode with no initial, point-based stage of registration. The black squares and red crosses show, respectively, the percentage of points passing the outlier test and the percentage within the error threshold (read against the right-hand scale); only points passing the outlier test have been included in the box-and-whisker plots. **(b)** ROC curve of the true and false positive rates of points passing the outlier test; the dashed line shows the operating point.

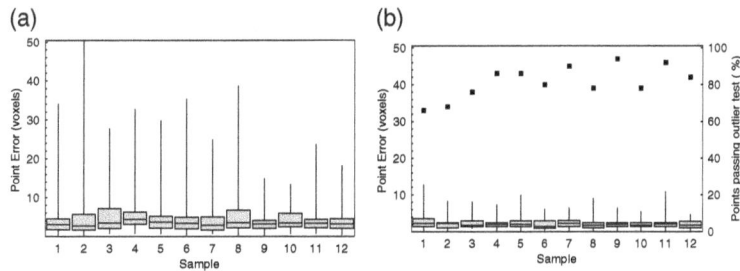

Figure 7 Repeatability of manual and "double-pass" automatic landmark annotation. Box-and-whisker plots of the repeatability of manual **(a)** and automatic **(b)** (read against the left-hand scale) localisation of 50 skull landmarks on the 12 consomic *Mus musculus* specimens. The automatic landmarks were generated using the "double-pass" mode with no initial, point-based stage of global registration. Only points passing the outlier test were included in the automatic annotation results; the black squares show the percentage of points passing (read against the right-hand scale).

(5.11 voxels compared to 5.24 voxels); however, the difference was not statistically significant (Mann-Whitney $U = 21464$, $U_\mu = 21830$, $U_\sigma = 1239$, $Z = -0.30$, $p = 0.77$), indicating that the additional specimens added little information. However, their presence did lead to a marked reduction in the number of points passing the outlier test (57.8% compared to 73.8%), although the percentage of points with errors lower than the error threshold was not significantly different (95.0% vs. 95.3%). The median annotation errors on the *Apodemus* specimens were larger than those on the *Mus* specimens (6.83 voxels vs. 5.11 voxels), but the difference was on the borderline of statistical significance (Mann-Whitney $U = 4545$, $U_\mu = 5428$, $U_\sigma = 506$, $Z = -1.75$, $p = 0.08$).

These results serve to indicate the robustness of the algorithm to variations in the database. When specimens from multiple genera with significant variation in shape were entered into the database, only those with similar shape to the query image contributed information to the final landmark location estimate. Conversely, for those specimens where the database provided no usable information, the algorithm successfully indicated that the automatic annotation was not reliable and rejected all

points. Contamination of the database with multiple genera did not result in a significant decrease in landmark annotation accuracy, but did result in a large reduction in the number of points passing the outlier test, reflecting the bias of the outlier test towards low false positive rates.

Database size and processor time requirements

The dependence of algorithmic performance on the number of image volumes in the database was evaluated by repeating the leave-one-out experiments on the 12 consomic *Mus musculus* specimens with fewer database entries. In order to avoid confounding the results by varying several features of the experimental procedure simultaneously, databases with fewer than three entries (the smallest number required to perform the outlier test) were not considered, and the image volumes included were selected randomly. The results are shown in Figure 10; each box-and-whisker shows the point localisation errors in voxels across all 12 volumes. Only points passing the outlier test were included, and the percentage of such points is also shown. The results demonstrate that, as would be expected, the point localisation errors decreased

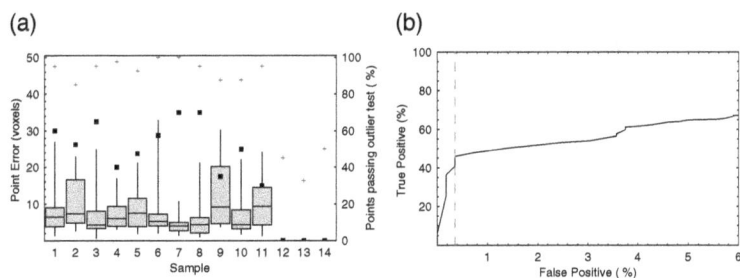

Figure 8 Automatic landmark annotation accuracy using a mixed-genera database. (a) Box-and-whisker plots of the point localisation errors for automatic annotation of 40 mandible landmarks on the data set of 14 rodent specimens (read against the left-hand scale). The black squares and red crosses show, respectively, the percentage of points passing the outlier test and the percentage within the error threshold (read against the right-hand scale); only points passing the outlier test have been included in the box-and-whisker plots. **(b)** ROC curve of the true and false positive rates of points passing the outlier test; the dashed line shows the operating point.

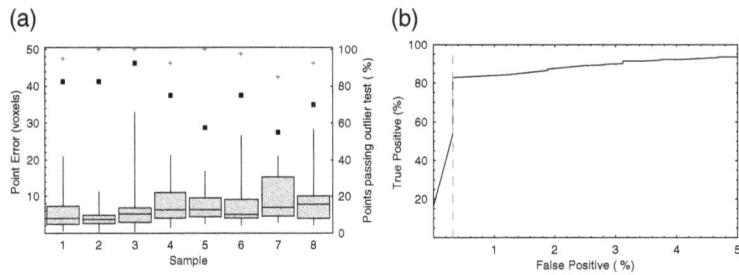

Figure 9 Automatic landmark annotation accuracy using 8 *Mus* specimens from the mixed-genera database. (a) Box-and-whisker plots of the point localisation errors for automatic annotation of 40 mandible landmarks on the data set of 8 *Mus* specimens (read against the left-hand scale). The black squares and red crosses show, respectively, the percentage of points passing the outlier test and the percentage within the error threshold (read against the right-hand scale); only points passing the outlier test have been included in the box-and-whisker plots. **(b)** ROC curve of the true and false positive rates of points passing the outlier test; the dashed line shows the operating point.

with increasing database size and the number of points passing the outlier test increased. However, both dependencies were relatively weak with this data set; there was little improvement in performance with database sizes larger than eight entries. This is significantly fewer images than would be required for the construction of an appearance model.

The processor time requirements of the algorithm were also evaluated during the tests on database depth. Experiments were performed on a Dell Precision workstation with 2 Intel Xeon 5670 processors and 24 Gb of main memory, running OpenSuse 11.3×64 (Linux kernel 2.6.34). It was anticipated that, in practical use, a single database would be constructed and then used to annotate multiple image volumes. Therefore, the database and image loading times were ignored and the wall-clock time required to perform the registration, array-based voting and outlier test was measured. Figure 11 shows the results, averaged over all experiments performed at each database size, as the number of seconds required to annotate a single point. The vast majority of the time taken to run the algorithm was accounted for by the registrations and so, since the registrations were performed independently for each database entry, there was

a linear dependence on the number of entries. The gradient of the linear fit, i.e. the time taken per point per database entry, was 1.64 seconds. In [8], the time required to manually annotate 10 landmarks on the mandible in micro-CT images of three rodent specimens (*Microtus, Mus* and *Pachyuromys*) was evaluated in both TINA and AMIRA (www.vsg3d.com/amira), both by a non-expert and an expert AMIRA user. Average timings were 13s per point using AMIRA and 30s per point using TINA for the expert, and 54s per point using AMIRA and 65s per point using TINA for the non-expert, although strong training effects were observed with TINA, as would be expected of users handling unfamiliar software. Therefore, assuming a reasonable database size of around 8 entries and thus approximately 13s per point for automatic annotation, the time required to perform automatic annotation was comparable to the time required to perform manual annotation with AMIRA. However, unlike manual annotation with AMIRA, automatic annotation with TINA does not require continuous user input. Comparison between TINA and AMIRA in terms of manual annotation accuracy is provided by [8]; note that AMIRA does not provide automatic annotation.

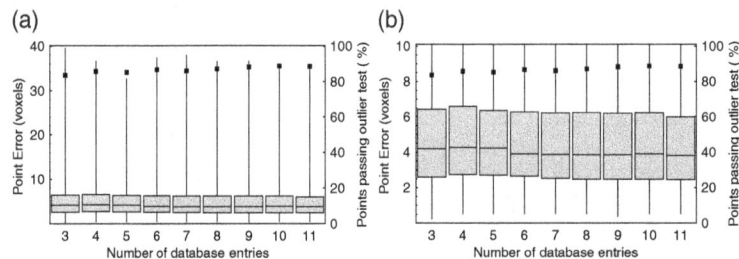

Figure 10 Dependence of automatic annotation accuracy on database size. Box-and-whisker plots of the point localisation errors for automatic annotation of 50 skull landmarks in the 12 consomic *Mus musculus* specimens, (read against the left-hand scale), against the number of image volumes in the database. The black squares show the percentage of points that passed the outlier test (read against the right-hand scale); only points passing the outlier test have been included in the box-and-whisker plots. Both graphs show the same data, plotted on different ranges.

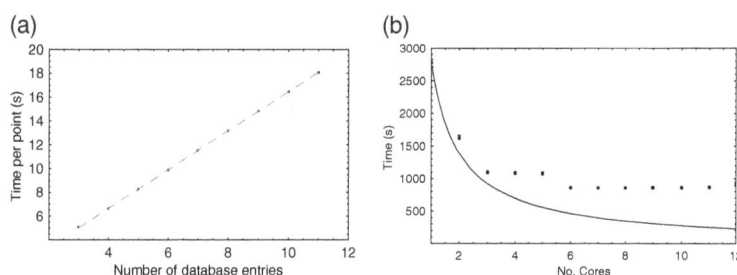

Figure 11 Dependence of run time on database size and the number of cores in use. (a) The average wall-clock time required to perform all registration stages, array-based voting and the outlier test on the 50 skull landmarks in each of the 12 consomic *Mus musculus* specimens, plotted against the number of image volumes in the database. The dashed line shows a linear fit to the data. **(b)** The average wall-clock time required to perform all registration stages, array-based voting and outlier test on 50 skull landmarks in each of the the 12 consomic *Mus musculus* specimens with 11 database entries, plotted against the number of processor cores used. The 1/cores curve that would be achieved with 100% parallelisation efficiency is also shown.

The software made extensive use of parallelisation, and so the dependence of the run time on the number of processor cores in use was also evaluated. Figure 11 shows the time taken to perform the registration, array-based voting and outlier test stages of the algorithm in leave-one-out tests on the 12 consomic *Mus musculus* specimens (i.e. with a database size of 11), averaged over the 12 experiments, against the number of processor cores, together with the 1/cores dependency that would be expected if the parallelisation efficiency was 100%. The results showed good parallelisation efficiency up to three cores, little further reduction in the time taken until six cores were in use (allowing full parallelisation over the six image patches for each landmark point; see the Methods section for details), and then no further reduction. The loss of parallelisation efficiency above three cores indicates that memory bandwidth was the main limiting factor, due to the algorithm performing large numbers of relatively simple operations on small blocks of data. However, these results indicate that the timings described above should be achievable on most relatively modern hardware.

Conclusions

Geometric morphometric analyses, consisting of manual annotations of landmark points followed by Procrustes analysis, are a popular way to quantify biological shapes for comparison to genetic, phylogenetic or ecological factors. The requirement for extensive and time-consuming manual annotation places practical limitations on the number of specimens that can be included in such analyses, in turn limiting their statistical power both in terms of the magnitude of shape variation that can be measured and the confidence intervals on any conclusions. In this paper, a semi-automatic landmark annotation algorithm designed to accelerate the landmark annotation process has been presented.

The aim was to produce landmarks for use in shape analysis, and so no constraint based on a shape model, or assumptions about shape, could be used in the algorithm. These models or assumptions would be recovered by the subsequent analysis and therefore potentially bias the results if they were not a perfect fit to the data. Instead, a free-form, intensity-based registration approach was adopted. Multiple example images, each with manually annotated landmarks, were registered to the query image using a multi-scale approach. The final stages of registration operated on small patches of image data around each landmark in order to minimise the effects of global shape variation. This resulted in one set of estimated landmark locations for each database image; these were combined using a Hough-like voting array in order to introduce robustness to biases and outliers whilst minimising assumptions about their distribution.

Since the algorithm was designed to replace manual annotation, it would have little utility if it required extensive manual checking of the results. Therefore, it had to provide some indication of the reliability of the automatic annotation in order to guide manual checking and, where necessary, correction of the results. Furthermore, the operating point on the ROC curve of the outlier test had to be biased such that the false positive rate was minimal, i.e. it was essential to flag all incorrect annotations as errors, even at the expense of flagging a significant proportion of correct annotations, in order that the user could trust the outlier test and thus avoid having to check all automatic annotations. The presence of outliers indicated either convergence to a local minimum or a database image that provided a poor model of the query image; the latter possibility rendered all statistical measures based on the assumption that the model fits the data, unreliable. Therefore, a method based on consistency between the multiple estimates for each landmark was developed. This treated each database image as an independent model of the query image, and used a minimal assumption that good models should produce a compact peak in the Hough-like voting array, whilst poorly fitting models should produce a broad outlier distribution.

In order to have utility within the target user community, the algorithm had to perform landmark annotation as quickly and as accurately as manual annotation; therefore, evaluation focused on these two properties. Comparisons were performed between two independent sets of manual annotations of 50 skull landmarks in 12 *Mus musculus* specimens, and two sets of automatic annotations generated from them. Automatically annotated points failing the outlier test were not included in the comparisons. For the 87.5% of the points that were included, the results showed that automatic annotation was at least as accurate as, and more repeatable than, manual annotation, i.e. had a lower random error and no significant systematic error. The repeatability of the automatic annotations indicated that voxel-level accuracy was achieved. The outlier test proved extremely reliable, rejecting 12.5% of the automatic annotations to achieve a false positive rate of less than 0.5%, due to the care taken in the estimation of noise distributions.

The evaluation on a wider variety of rodent specimens, including multiple genera, demonstrated the reliability of the outlier test. In cases where the database contained no specimen that provided a good model of the query image, the algorithm correctly flagged all points as outliers. Conversely, the presence of database specimens that provided poor models of the query image did not significantly affect the accuracy of the automatic annotation process, even where they formed the majority of the database. This suggests an iterative mode of operation; any query image that generates large numbers of outliers is not well modelled by the existing database. Therefore, after manual correction of the outliers, it can be added to the database in order to expand the number of specimens that can be annotated accurately.

Timing tests demonstrated that the automatic landmark localisation process required approximately the same wall-clock time as manual annotation on reasonably modern hardware. However, no user input was required for this stage of the process. Therefore, actual manual annotation required for each image volume in a hypothetical landmark annotation process would be limited to the four registration points, visual check of the detected outliers, the majority of which would be in the correct location due to the low false-positive bias of the outlier detector, and correction of the true outliers, which averaged around 1 to 2 points per volume with a single-genera image database. For realistic landmark list sizes of 40 to 50 points, the automatic landmark localisation software could therefore reduce the required number of manual landmark annotations by a factor of ten. Further

improvements were achieved using the double-pass mode of operation, although this would require care during sample preparation to ensure a reasonably good alignment of the samples within the image volumes. Annotation of the training images constitutes the majority of the manual annotation required when using the proposed technique. However, the evaluation of database size in the experiments on consomic *Mus musculus* specimens indicated that only eight images were required, far fewer than would be required by alternative techniques based on statistical shape models.

Several potential routes exist for future improvement of the software. For instance, it may be possible to automatically estimate patch sizes using techniques such as those described in [12], reducing the need to optimise the free parameters of the algorithm prior to application to new image types. Furthermore, annotation of the four points used to initialise the global registration requires significant user interaction. This could be simplified using the 3D rendering of the data provided by the software, by prompting the user to rotate the rendering into a standard orientation, which would then provide the initialisation. A similar orientation would have to be stored for each of the images in the database.

The software described in this paper has been made available as free and open source (FOSS) software under the GNU General Public Licence (www.gnu.org), and can be obtained via our web-site (www.tina-vision.net).

Appendix A: derivation of the cost function and scaling factor

Let I_v and J_v be corresponding voxels in two identical, noise-free images or image patches I and J. Scale the intensities of one of the images by a factor γ; without loss of generality, assume that this is image J. Add Gaussian random noise with standard deviations of σ_I and σ_J to the two images. Find an estimator for γ.

The derivation given here follows that given in [55]. An estimator of γ can be obtained as the gradient of a linear fit to J_v vs. I_v over all v with an intercept of zero, as shown in Figure 12. Since there is noise on both I and J, the noise-free intensities of a point $A = I_v, J_v$ could correspond to any point C along the linear fit. However, if $\sigma_I = \sigma_J = \sigma$, the probability that a datum at A was generated from C is given by

$$
\begin{aligned}
P(C \to A) &= \frac{1}{2\pi\sigma^2} exp\left[-\frac{(I_C - I_A)^2}{2\sigma^2}\right] exp\left[-\frac{(J_C - J_A)^2}{2\sigma^2}\right] \\
&= \frac{1}{2\pi\sigma^2} exp\left[-\frac{r^2}{2\sigma^2}\right] \\
&= \frac{1}{2\pi\sigma^2} exp\left[-\frac{h^2}{2\sigma^2}\right] exp\left[-\frac{u^2}{2\sigma^2}\right] \quad (2)
\end{aligned}
$$

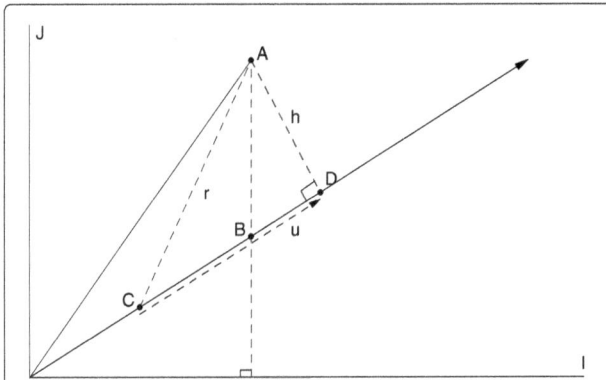

Figure 12 The construction of the χ^2 between two scaled image patches. A linear fit to the joint intensity histogram of a pair of images I and J. Since both images contain noise, a datum at point A could be generated from any point C along the linear fit. However, if the standard deviations of the noise on I and J are equal, then the probability of generating data at A depends only on the perpendicular distance h to the linear fit, not on u.

Integrating over u gives a constant $\sigma\sqrt{2\pi}$, so the probability depends only on h i.e. only on the perpendicular distance to the line. In the event that $\sigma_I \neq \sigma_J$, the images can be scaled to $I' = I/\sigma_I$ and $J' = J/\sigma_J$.

The distance h can be obtained from the vectors $A = (I_v, J_v)$ and $B = (I_v, \gamma I_v)$ using

$$A.B = |A||B|\cos\theta \ , \ \ \frac{|h|}{|A|} = \sin\theta \ \ and \ \ \cos^2\theta + \sin^2\theta = 1$$

where θ is the angle between the vectors A and B. After some manipulation, these give

$$h = \frac{J_v - \gamma I_v}{\sqrt{1 + \gamma^2}}$$

Substituting this into Eq. 2 gives

$$P(J_v|I_v, \sigma, \gamma) = \frac{1}{\sqrt{2\pi}\sigma} exp\left[-\frac{(J_v - \gamma I_v)^2}{2\sigma^2(1 + \gamma^2)}\right] \quad (3)$$

or equivalently

$$\chi^2 = \sum_v \frac{(J_v - \gamma I_v)^2}{\sigma^2(1 + \gamma^2)} \propto \sum_v \frac{(J_v - \gamma I_v)^2}{1 + \gamma^2}$$

In order to obtain an estimator for γ, differentiate the χ^2 w.r.t. γ and set the result equal to zero. After some manipulation, this gives

$$\sum_v \frac{\gamma^2 I_v J_v + \gamma(I_v^2 - J_v^2) - I_v J_v}{(1 + \gamma^2)^2} = 0$$

The numerator in this fraction must be equal to zero, since the denominator cannot be regardless of the value of γ. Therefore, substituting

$$|I|^2 = \sum_v I_v^2 \ , \ \ |J|^2 = \sum_v J_v^2 \ \ and \ \ I.J = \sum_v I_v J_v$$

gives

$$\gamma^2 I.J + \gamma\left(|I|^2 - |J|^2\right) - I.J = 0$$

This quadratic can be solved in the usual way to give $\hat{\gamma}$, an estimator for γ

$$\hat{\gamma} = \frac{-b \pm \sqrt{b^2 - 4ac}}{2a} \quad where \quad a = I.J \ , \ \ b = |I|^2 - |J|^2$$
$$and \ \ c = -I.J$$

This gives

$$\hat{\gamma} = \frac{|J|^2 - |I|^2 \pm \sqrt{\left(|I|^2 - |J|^2\right)^2 + 4(I.J)^2}}{2I.J}$$

Now, $I.J = |I||J|cos\phi$ where, if the intensities from I and J were concatenated to form two vectors in a v-dimensional space, ϕ would be the angle between those two vectors. This angle will be small if the signal-to-noise ratio is high, and so assuming that $\phi \approx 0$,

$$\hat{\gamma} = \frac{|J|^2 - |I|^2 \pm \sqrt{\left(|I|^2 + |J|^2\right)^2}}{2|I||J|}$$

which gives

$$\hat{\gamma} = \frac{|J|}{|I|} \ \ or \ \ \hat{\gamma} = -\frac{|I|}{|J|}$$

The two solutions are perpendicular and give the best and worst linear fit to the data. The positive estimator is used when the correlation between I and J is positive, and the negative estimator when the correlation is negative. A negative correlation may be seen with some image modalities, such as MR images acquired with different pulse sequences; however, in the work presented here micro-CT images were used, and so only the positive solution is relevant. The result is both a maximum likelihood and minimum χ^2 estimator.

The above derivation also provides the cost function for registration of I and J used in the work described here as Equation 3

$$\chi^2 = \sum_v \frac{(J_v - \gamma I_v)^2}{\sigma^2(1 + \gamma^2)}$$

This is valid only if the noise on I and J are equal. In the more general case where they are not equal, the images can be scaled to to $I' = I/\sigma_I$ and $J' = J/\sigma_J$; let γ' represent γ estimated in the I', J' space, so that the cost function becomes

$$\chi^2 = \sum_v \frac{(J_v/\sigma_J - \gamma' I_v/\sigma_I)^2}{(1 + \gamma'^2)}$$

Rearranging gives

$$\chi^2 = \sum_v \frac{\left(J_v - \gamma'\frac{\sigma_J}{\sigma_I}I_v\right)^2}{\sigma_J^2(1 + \gamma'^2)}$$

and, since

$$\gamma' = \frac{|J'|}{|I'|} = \frac{\sigma_I}{\sigma_J} \frac{|J|}{|I|}$$

letting $\gamma = |J|/|I|$, the cost function becomes

$$\chi^2 = \sum_v \frac{(J_v - \gamma I_v)^2}{\sigma_J^2 + \sigma_I^2 \gamma^2}$$

The requirement for the images to be similar before the scaling factor can be estimated is not an issue when the manual, point-based stage of registration is included, since this will achieve the required approximate alignment independently of the voxel intensities. Less obviously, it is also not a problem when operating the algorithm in double-pass mode, due to a feature of registration using images consisting primarily of step edges between uniform regions. Since the cost function being optimised is

$$\chi^2 = \sum_v \frac{(J_v - \gamma I_v)^2}{\sigma^2 (1 + \gamma^2)}$$

and assuming without loss of generality that J is the source image, the solution will be obtained where

$$\frac{\partial \chi^2}{\partial \mathbf{T}} = 0 = \frac{\partial \chi^2}{\partial J_v} \frac{\partial J_v}{\partial \mathbf{T}} \propto \sum_{i=1}^{N} (J_v - \gamma I_v) \frac{\partial J_v}{\partial \mathbf{T}}$$

where $\mathbf{T}$ represents the parameter vector of the transformation model. The $\partial J_v/\partial \mathbf{T}$ term ensures that only regions with a significant image gradient contribute to the alignment process. In images with large regions of smooth gradients, errors in the estimate of γ therefore have a significant effect on the accuracy of the registration result. However, micro-CT images consist primarily of significant step edges between regions that are comparatively uniform except for the effects of noise, and so the optimisation will attain a (possibly local) minimum that aligns the edges regardless of the estimate of γ used. Therefore, the first, image-based stage of global registration can be performed with $\gamma = 1$, in order to attain an approximate alignment. The equations given above can then be used to estimate γ, prior to the later, patch-based stages of registration.

Appendix B: the effects of smoothing on noise

If an image $I(x,y)$ with a uniform Gaussian noise field of standard deviation σ_I is smoothed through convolution with a Gaussian kernel of standard deviation σ_k, what is the standard deviation of the noise field on the smoothed image?

The full form for the 2D Gaussian distribution is

$$f(x,y) = \frac{1}{2\pi \sigma_x \sigma_y \sqrt{1-\rho^2}} exp \left[\frac{-1}{2(1-\rho^2)} \left(\frac{(x-\mu_x)^2}{\sigma_x^2} \right. \right.$$
$$\left. \left. + \frac{(y-\mu_y)^2}{\sigma_y^2} - \frac{2\rho(x-\mu_x)(y-\mu_y)}{\sigma_x \sigma_y} \right) \right]$$

where σ_x and σ_y are the standard deviations in the x and y directions, ρ is the correlation coefficient, and μ_x and μ_y give the position of the mean. Assuming that the smoothing kernel will be an isotropic Gaussian, $\sigma_x = \sigma_y = \sigma_k$ and $\rho = 0$, so

$$G(x,y) = \frac{1}{2\pi \sigma_k^2} exp \left[\frac{-(x-\mu_x)^2 - (y-\mu_y)^2}{2\sigma_k^2} \right]$$

The smoothed image $I_G(x,y)$ is given by

$$I_G(x,y) = \sum_{x,y} I(x,y).G(x,y)$$

Error propagation [55] gives the standard deviation σ_f of the noise on a function of several random variables n_i,

$$\sigma_f^2 = \sum_i \frac{\partial f}{\partial n_i} \sigma_{n_i}^2$$

and so the standard deviation σ_G of the noise on $I_G(x,y)$ is given by

$$\sigma_G^2 = \sum_{x,y} \left[\frac{\partial}{\partial I} I(x,y) G(x,y) \right]^2 \sigma_I^2 = \sigma_I^2 \sum_{x,y} G^2(x,y)$$

$$\approx \sigma_I^2 \int_{-\infty}^{\infty} \int_{-\infty}^{\infty} G^2(x,y) dx dy$$

The definite integral of the Gaussian can be performed by converting to polar coordinates; let

$$A = \int_{-\infty}^{\infty} e^{-ax^2} dx \Rightarrow A^2 = \int_{-\infty}^{\infty} e^{-ax^2} dx \int_{-\infty}^{\infty} e^{-ay^2} dy$$

$$= \int_{-\infty}^{\infty} \int_{-\infty}^{\infty} e^{-a(x^2+y^2)} dx dy$$

Transform to polar coordinates $r^2 = x^2 + y^2$, $dx dy = r dr d\theta$

$$A^2 = \int_0^{\infty} \int_0^{2\pi} e^{-ar^2} r dr d\theta = 2\pi \int_0^{\infty} e^{-ar^2} r dr$$

$$= \pi \int_0^{\infty} e^{-ar^2} d(r^2) = \pi \left[\frac{-1}{a} e^{-ar^2} \right]_0^{\infty} = \frac{\pi}{a}$$

so

$$A = \sqrt{\frac{\pi}{a}} \tag{4}$$

Therefore, the more general form

$$B = \int_{-\infty}^{\infty} k e^{\frac{-(x+b)^2}{c^2}} dx$$

can be written, putting $y = x + b$, as

$$B = k \int_{-\infty}^{\infty} e^{\frac{-y^2}{c^2}} dx$$

or, putting $z = y/|c|$, as

$$B = k|c| \int_{-\infty}^{\infty} e^{-z^2} dz$$

Using Eq. 4 gives

$$B = \int_{-\infty}^{\infty} ke^{\frac{-(x+b)^2}{c^2}} dx = k|c|\sqrt{\pi}$$

Returning to the original problem,

$$G^2(x, y) = \left(\frac{1}{2\pi\sigma_k^2}\right)^2 \left(exp\left[\frac{-(x - mu_x)^2 - (y - \mu_y)^2}{2\sigma_k^2}\right]\right)^2$$

and, since $(e^a)^2 = e^{2a}$

$$G^2(x, y) = \frac{1}{4\pi^2\sigma_k^4} exp\left[\frac{-(x - mu_x)^2 - (y - \mu_y)^2}{\sigma_k^2}\right]$$

so

$$\sigma_G^2 = \int_{-\infty}^{\infty}\int_{-\infty}^{\infty} G^2(x, y) dx dy = \frac{1}{4\pi\sigma_k^2}\sigma_I^2$$

Note that smoothing will introduce correlations between neighbouring voxels over a range dependent on the standard deviation of the smoothing kernel. Therefore, the standard deviation of the noise after smoothing will no longer be a reliable indication of noise-induced intensity differences between neighbouring voxels. However, in the present case the result was used as an overall scaling for the cost function, and so remained valid.

Additional files

Additional file 1: Semi-automatic landmark point annotation for geometric morphometrics: parameter optimisation. This file provides a full description of, and results from, the experiments performed to optimise the free parameters of the automatic landmark point annotation algorithm.

Additional file 2: The TINA geometric morphometrics tool. This file is the manual for the TINA Geometric Morphometrics Tool, which includes the TINA Manual Landmarking Tool and the automatic landmark point annotation software described in this paper. It contains detailed information on the download, installation and use of the software. Updates are available at http://www.tina-vision.net/docs/memos.php, file 2010-007.

Additional file 3: Tina 5.0 user's guide. The file gives additional information on the TINA software in general, including features not used for the landmarking procedure. Updates are available at http://www.tina-vision.net/docs/memos.php, file 2005-002.

Competing interests
The authors declare that they have no competing interests.

Authors' contributions
PAB developed and tested the manual and automatic landmark localisation software and drafted the manuscript. ACS provided all data sets and corresponding manual landmark annotations, and provided specifications and feedback on software requirements and performance both before and during development. HR developed a preliminary version of the image registration functions used in automatic landmark annotation. DT and NAT initiated the project and provided technical supervision, respectively. All authors read and approved the final manuscript.

Acknowledgements
The project was funded by the Max Planck Society, Project "Automatic Identification of 3D Landmarks in Micro-CT Mouse Skull Data".

Author details
[1]Centre for Imaging Sciences, University of Manchester, Stopford Building, Oxford Road, Manchester M13 9PT, UK. [2]Department for Evolutionary Genetics, Max Planck Institute for Evolutionary Biology, August-Thienemann-Str. 2, 24306 Plön, Germany.

References
1. Bookstein F: *Morphometric Tools For Landmark Data.* Cambridge: Cambridge University Press; 1991.
2. Kendall DG: **Shape manifolds, procrustean metrics, and complex projective spaces.** *Bull Lond Math Soc* 1984, **16**(2):81–121.
3. Boell L, Tautz D: **Micro-evolutionary divergence patterns of mandible shapes in wild house mouse (Mus musculus) populations.** *BMC Evol Biol* 2011, **11**:306.
4. Cardini A, Elton S: **Does the skull carry a phylogenetic signal? Evolution and modularity in the guenons.** *Biol J Linn Soc* 2008, **93**:813–834.
5. von Cramon-Taubadel N: **Global human mandibular variation reflects differences in agricultural and hunter-gatherer subsistence strategies.** *PNAS* 2011, **108**(49):19546–19551.
6. Meloro C, O'Higgins P: **Ecological adaptations of mandibular form in fissiped carnivora.** *J Mamm Evol* 2011, **18**:185–200.
7. Breuker CJ, Patterson JS, Klingenberg CP: **A single basis for developmental buffering of Drosophila wing shape.** *PLoS ONE* 2006, **1**:e7.
8. Schunke AC, Bromiley PA, Tautz D, Thacker NA: **TINA manual landmarking tool: software for the precise digitization of 3D landmarks.** *Front Zool* 2012, **9**(6):1-5.
9. Lacroute P, Levoy M: **Fast volume rendering using a shear-warp factorization of the viewing transform.** In *Proc. SIGGRAPH '94, July 24–29 Orlando, Florida.* New York USA: ACM; 1994:451–458.
10. Lacroute P, Levoy M: **The Volpack Volume Rendering Library.** [http://graphics.stanford.edu/software/volpack/]
11. Frantz S, Rohr K, Stiehl HS: **Development and validation of a multi-step approach to improved detection of 3D point landmarks in tomographic images.** *Image Vis Comput* 2005, **233**(11):956–971.
12. Liu J, Gao W, Huang S, Nowinski WL: **A model-based, semi-global segmentation approach for automatic 3-D point landmark localization in neuroimages.** *IEEE Trans Med Imaging* 2008, **27**(8):1034–1044.
13. Hill DLG, Batchelor PG, Holden M, Hawkes DJ: **Medical image registration.** *Phys Med Biol* 2001, **46**:R1–R45.
14. Wyawahare MV, Patil PM, Abhyankar HK: **Image registration techniques: an overview.** *Int J Signal Process Image Process Pattern Recogn* 2009, **2**(3):11–28.
15. Oliveira FPM, Tavares JMRS: **Medical image registration: a review.** *Comput Methods Biomech Biomed Engin* 2012, **17**(2):1–21.
16. Mani VRS, Arivazhagan S: **Survey of medical image registration.** *J Biomed Eng Tech* 2013, **1**(2):8–25.
17. Audette MA, Ferrie FP, Peters TM: **An algorithmic overview of surface registration techniques for medical imaging.** *Med Image Anal* 2000, **4**:201–217.
18. van Kaick O, Zhang H, Hamarneh G, Cohen-Or D: **A survey on shape correspondence.** *Comput Graph Forum* 2011, **30**(6):1681–1707.

19. Tam GKL, Cheng ZQ, Lai YK, Langbein FC, Liu Y, Marshall D, Martin RR, Sun XF, Rosin PL: **Registration of 3D point clouds and meshes: a survey from rigid to nonrigid.** *IEEE Trans Vis Comput Graph* 2013, **19**(7):1199–1217.
20. Sotiras A, Davatzikos C, Paragios N: **Deformable medical image registration: a survey.** *IEEE Trans Med Imaging* 2013, **32**(7):1153–1190.
21. Lau YH, Braun M, Hutton BF: **Non-rigid image registration using a median-filtered coarse-to-fine displacement field and a symmetric correlation ratio.** *Phys Med Biol* 2001, **46**:1297–1319.
22. Collignon A, Maes F, Delaere D, Vandermeulan D, Suetens P, Marchal G: **Automated multi-modality image registration based on information theory.** In *Information Processing in Medical Imaging.* Edited by Bizais Y, Barillot C, Paola RD. Dordrecht: Kulwer, Academic; 1995:263–274.
23. Viola P, Wells WM: **Alignment by maximisation of mutual information.** In *Proceedings ICCV'95.* Cambridge, MA, USA: IEEE Computer Society Press; 1995:16.
24. Viola P, Wells WM: **Alignment by maximisation of mutual information.** *Int J Comput Vis* 1997, **24**(2):137–154.
25. Studholme C, Hill DLG, Hawkes DJ: **An overlap invariant entropy measure of 3D medical image alignment.** *Pattern Recogn* 1999, **32**:71–86.
26. Malsch U, Thieke C, Huber PE, Bendl R: **An enhanced block matching algorithm for fast elastic registration in adaptive radiotherapy.** *Phys Med Biol* 2006, **51**(19):4789–4806.
27. Bookstein F: **Principal warps: thin-plate splines and the decomposition of deformations.** *IEEE Trans Pattern Anal Mach Intell* 1989, **11**:567–585.
28. Söhn M, Birkner M, Chi Y, Wang J, Yan D, Berger B, Alber M: **Model-independent, multimodality deformable image registration by local matching of anatomical features and minimization of elastic energy.** *Med Phys* 2008, **35**(3):866–878.
29. Erdt M, Steger S, Wesarg S: **Deformable registration of MR images using a hierarchical patch based approach with a normalized metric quality measure.** In *Biomedical Imaging (ISBI), 2012 9th IEEE International Symposium on.* Barcelona, Spain: IEEE Computer Society Press; 2012:1347–1350.
30. Bromiley PA, Pokrić M, Thacker NA: **Computing covariances for mutual information coregistration.** In *Proc. MIUA'04.* UK: BMVA Press; 2004:77–80.
31. Bromiley PA, Pokrić M, Thacker NA: **Emprical evaluation of covariance estimates for mutual information coregistration.** In *Proc. MICCAI'04.* Saint-Malo, France: Springer-Verlag; 2004:607–614.
32. Rohr K, Stiehl HS, Sprengel R, Buzug TM, Weese J, Kuhn MH: **Landmark-based elastic registration using approximating thin-plate splines.** *IEEE Trans Med Imaging* 2001, **20**(6):526–534.
33. Kim M, Wu G, Shen D: **Sparse patch-guided deformation estimation for improved image registration.** In *Machine Learning in Medical Imaging, Volume 7588 of Lecture Notes in Computer Science.* Edited by Wang F, Shen D, Yan P, Suzuki K. Berlin, Heidelberg, Germany: Springer-Verlag; 2012:54–62.
34. Canny J: **A computational approach to edge detection.** *IEEE Trans Pattern Anal Mach Intell* 1986, **8**(6):679–698.
35. Cootes TF, Taylor CJ: **Active shape models - 'smart snakes'.** In *Proc. British Machine Vision Conference (BMVC'92).* London: Springer-Verlag; 1992:266–275.
36. Cootes TF, Taylor CJ, Cooper D, Graham J: **Active shape models - their training and application.** *Comput Vis Image Understand* 1995, **61**:38–59.
37. Cootes TF, Edwards GJ, Taylor CJ: **Active appearance models.** *IEEE Trans Pattern Anal Mach Intell* 2001, **23**:681–685.
38. Cristinacce D, Cootes TF: **Automatic feature localisation with constrained local models.** *J Comput Vis* 2005, **61**:55–79.
39. Felzenswalb P, Hutenocher D: **Pictorial structures for object recognition.** *Int J Comput Vis* 2005, **61**:55–79.
40. Heimann T, Meinzer H: **Statistical shape models for 3D medical image segmentation: a review.** *Med Image Anal* 2009, **13**(4):543–563.
41. Palaniswamy S, Thacker NA, Klingenberg CP: **Automatic identification of morphometric landmarks in digital images.** In *Proc. BMVC'07, 10–13 September, Warwick, U.K.* UK: BMVA Press; 2007:112.
42. Palaniswamy S, Thacker NA, Klingenberg CP: **Automated landmark extraction in digital images - performance evaluation.** In *ProcVIE'08, July 19 - Aug 1, Xi'an, China.* UK: IET; 2008.
43. Ballard DH: **Generalizing the hough transform to detect arbitrary shapes.** *Pattern Recogn* 1981, **13**:111–122.
44. Gall J, Lempitsky V: **Class-specific hough forests for object detection.** In *Proc. IEEE Conference on Computer Vision and Pattern Recognition (CVPR'09).* New York USA: IEEE Computer Society Press; 2009:1022–1029.
45. Cootes TF, Ionita MC, Lindner C, Sauer P: **Robust and accurate shape model fitting using random forest regression voting.** In *Proc. ECCV'12*; 2012:278–291.
46. Breiman L: **Random forests.** *Mach Learn* 2001, **45**:5–32.
47. Nelder JA, Meade R: **A simplex method for function minimisation.** *Comput J* 1965, **7**:308–313.
48. Ragheb H, Thacker NA: **Quantitative localisation of manually defined landmarks.** In *Proc. MIUA'11, 14–15 July, London, U.K.* UK: BMVA Press; 2011:221–225.
49. Lane RA, Thacker NA, Seed NL: **Stretch-correlation as a real-time alternative to feature-based stereo matching algorithms.** *Image Vis Comput* 1994, **12**(4):203–212.
50. Olsen SI: **Estimation of noise in images: an evaluation.** *CVGIP: Graph Models Image Process* 1993, **55**:319–323.
51. Condon JJ: **Errors in elliptical Gaussian fits.** *Publ Astron Soc Pacs* 1997, **109**(732):166–172.
52. Valstar MF, Martinez B, Binefa X, Pantic M: **Facial point detection using boosted regression and graph models.** In *Proc. CVPR.* New York USA: IEEE Computer Society Press; 2010:2729–2736.
53. Lindner C, Thiagarajah S, Wilkinson JM, arcOGEN Consortium T, Wallis GA, Cootes TF: **Fully automatic segmentation of the proximal femur using random forest regression voting.** *IEEE Trans Med Imag* 2013, **32**(8):1462–1472.
54. Ragheb H, Thacker NA, Bromiley PA, Tautz D, Schunke AC: **Quantitative shape analysis with weighted covariance estimates for increased statistical efficiency.** *Front Zool* 2013, **10**(16):1–23.
55. Barlow R: *Statistics: A Guide to the use of Statistical Methods in the Physical Sciences, 1st edition.* Chichester: John Wiley and Sons; 1989.

3

Detailed reconstruction of the musculature in *Limnognathia maerski* (Micrognathozoa) and comparison with other Gnathifera

Nicolas Bekkouche[1], Reinhardt M Kristensen[2], Andreas Hejnol[3], Martin V Sørensen[4] and Katrine Worsaae[1*]

Abstract

Introduction: *Limnognathia maerski* is the single species of the recently described taxon, Micrognathozoa. The most conspicuous character of this animal is the complex set of jaws, which resembles an even more intricate version of the trophi of Rotifera and the jaws of Gnathostomulida. Whereas the jaws of *Limnognathia maerski* previously have been subject to close examinations, the related musculature and other organ systems are far less studied. Here we provide a detailed study of the body and jaw musculature of *Limnognathia maerski*, employing confocal laser scanning microscopy of phalloidin stained musculature as well as transmission electron microscopy (TEM).

Results: This study reveals a complex body wall musculature, comprising six pairs of main longitudinal muscles and 13 pairs of trunk dorso-ventral muscles. Most longitudinal muscles span the length of the body and some fibers even branch off and continue anteriorly into the head and posteriorly into the abdomen, forming a complex musculature. The musculature of the jaw apparatus shows several pairs of striated muscles largely related to the fibularium and the main jaws. The jaw articulation and function of major and minor muscle pairs are discussed. No circular muscles or intestinal musculature have been found, but some newly discovered muscles may supply the anal opening.

Conclusions: The organization in *Limnognathia maerski* of the longitudinal and dorso-ventral muscle bundles in a loose grid is more similar to the organization found in rotifers rather than gnathostomulids. Although the dorso-ventral musculature is probably not homologous to the circular muscles of rotifers, a similar function in body extension is suggested. Additionally, a functional comparison between the jaw musculature of *Limnognathia maerski*, Rotifera and Gnathostomulida, emphasizes the important role of the fibularium in *Limnognathia maerski*, and suggests a closer functional resemblance to the jaw organization in Rotifera.

Keywords: CLSM, 3D reconstructions, Jaw apparatus, F-actin, Trophi, Mastax

Introduction

Limnognathia maerski Kristensen & Funch, 2000, is a minute animal living in fresh water ponds and lakes [1-3]. The animal was discovered in 1994 at Disko Island, Greenland, but not described before 2000, and it has subsequently been reported from the Sub Antarctic Crozet Island [1], in a stream from southern Wales, United Kingdom, and in the river Lambourn (Berkshire), United Kingdom (P. E. Schmid and J.M. Schmid-Araya, personal communication). With a unique combination of characters, it is considered the only member of the recently described Micrognathozoa [2-5], belonging to Gnathifera. However, the phylogenetic relationships within Gnathifera are still debated, and the molecular studies are based on very limited information [5]. So far, the complex jaw apparatus of *L. maerski* has received the main attention in studies, leading to several disputed homology hypotheses for each sclerite of the trophi [1,3,6,7]. However, no detailed studies have addressed the overall morphology of organs systems and further anatomical knowledge on *L. maerski* is warranted in order to compare this unique evolutionary lineage with the other gnathiferan groups, as well as other animals.

Limnognathia maerski measures 80-150 μm, possesses a complex set of jaws, a conspicuously arranged ventral ciliation and, so far, only females are known. The

* Correspondence: kworsaae@bio.ku.dk
[1]Marine Biological Section, Department of Biology, University of Copenhagen, Universitetsparken 4, 2100 Copenhagen Ø, Denmark
Full list of author information is available at the end of the article

ventrally ciliated head consists of a forehead with ciliary sensory organs and a more posterior part containing the pharyngeal apparatus. The trunk is composed of an accordion-like thorax and a large abdomen with ventral ciliophores and a posterior adhesive pad [3]. In the original description, the overall musculature of *L. maerski* is briefly described. It is composed of several longitudinal and dorso-ventral muscles, minute muscles articulating the dorsal plates and a dense pharyngeal musculature. No circular musculature has been found. However, precise information on the number, configuration and relative size of each set of muscles was not provided. Ultrastructural data provided information on the structure of muscles attachment sites, the absence of myosyncytia and myoepithelia, the cross-striated nature of the pharyngeal musculature, and the mainly obliquely striated longitudinal musculature [3].

Following Sørensen [6], the jaws of *L. maerski* are composed of six main elements: i) The median, ventralmost basal plate with posterior stems and anterior flattened and toothed manus, ii) the large and conspicuous ventral fibularium made of different chambers containing cells, extending dorso-laterally, iii) the latero-ventral ventral jaws (pseudophalangia) that articulate posteriorly with the associated accessory sclerites, iv) the mediodorsal main jaws, each with a posteriorly projecting cauda, surrounded by the fibularium, v) the dorso-lateral dorsal jaws also confined to the fibularium area, vi) and the pharyngeal lamellae, a pair of lamellate structures positioned antero-laterally to the rest of the jaw apparatus. Additionally, Kristensen and Funch [3], describe the lamella orales as a paired structure similar to the lamellae pharyngea, situated dorso-laterally, inside the fibularium. However, the presence of these structures has not been confirmed in any subsequent studies [1,6].

The animal lives in limnic mosses or in the sediment of relatively calm springs and lakes, and was first recognized for its unusual 'ciliate-like' swimming in the water column. It also uses ciliary motion to glide over surfaces. Occasionally, it performs muscular contractions during lateral bending and longitudinal accordion like contractions for directional change, ventral bending while egg laying and dorsal contraction during vomit behavior [3]. Foraging of *L. maerski* involves fine movements of the jaw apparatus as well as larger movements of the head. While feeding, the ventral jaws are protruded and involved in substrate grasping. During the vomit behavior, the forehead is moved upward and backward, and most of the jaw apparatus is protruded through the mouth opening, while it performs fast snapping movements of the jaw elements and forward and backward movements of the main jaws (see reference [3] and Figure 1B). Accessory sclerites and pseudophalangia may move independent of the rest of the jaw apparatus, allowing the

ventral jaws to move from a rostro-caudal orientation to a dorso-ventral orientation without moving the other jaws elements [3,6,7].

The body wall musculature differs between the putatively closest micrognathozoan relatives: Gnathostomulida and Rotifera. In Gnathostomulida, the overall musculature consists of numerous circular and diagonal muscles and several bundles of longitudinal muscles (six to nine pairs [8-10]) extending the entire body length, where the superimposition of longitudinal, diagonal and circular muscles forms a dense grid like body wall musculature [9,10]. In the majority of rotifers, most of the longitudinal muscles do not extend through the entire body, but are limited to certain body regions, e.g., coronal retractors in the head or muscles in the posterior part of the trunk, being involved in the contraction of the head and foot, respectively [11-13]. Circular muscles are few and usually incomplete transverse, rather than circular (*e.g.*, [11-15]), although some Gnesiotrocha have complete rings [12,16,17]. Most of the diagonal and transverse muscles are usually absent (*e.g.*, [12,18]), and if present they are only few and/or inconspicuous [14,19]. Splanchnic muscles surrounding the gut are not found in Gnathostomulida [9], whereas they present a very thin musculature in Rotifera. This muscular grid is documented for Seisonidae [11] and Monogononta (*e.g.*, [13,19]) but visceral muscles are not found in Bdelloidea [12,14]. Dorso-ventral musculature has not been described for Gnathostomulida [9], and most of the functionally dorso-ventral muscles in Rotifera are supposedly modified incomplete circular muscles [12,19] meaning "true" dorso-ventral muscles, as reported by Kristensen and Funch [3], seem to be unique for Micrognathozoa.

The jaw musculature also differs between Gnathostomulida and Rotifera, due to the organization of their jaws. In gnathostomulids, the jaw apparatus consists of i) a set of main jaws, and in some taxa ii) an unpaired basal plate [20-22]. In rotifers, the jaw apparatus (trophi) includes 7 main elements: the i) unpaired posteriorly directed fulcrum, ii) paired rami, iii) paired unci, and iv) paired manubria. The fulcrum and rami together form the central element, the incus, whilst the unci and manubria form the mallei (*e.g.*, [23-25]). The rotifer incus has been considered homologous with the gnathostomulid main jaws [21,26]. However, it also has been suggested that some parts of the gnathostomulid articularium (antero-lateral parts of the main jaws) are homologous with the rotifer manubria [27]. The gnathostomulid basal plate is considered autapomorphic for the group, and no homologous counterpart has been identified in the rotiferan trophi. The structural differences in the musculature of gnathostomulid and rotiferan jaw apparatuses clearly relate to the differences in the hard parts and the additional number of rotifer jaw elements. Indeed, most of the musculature supplying the

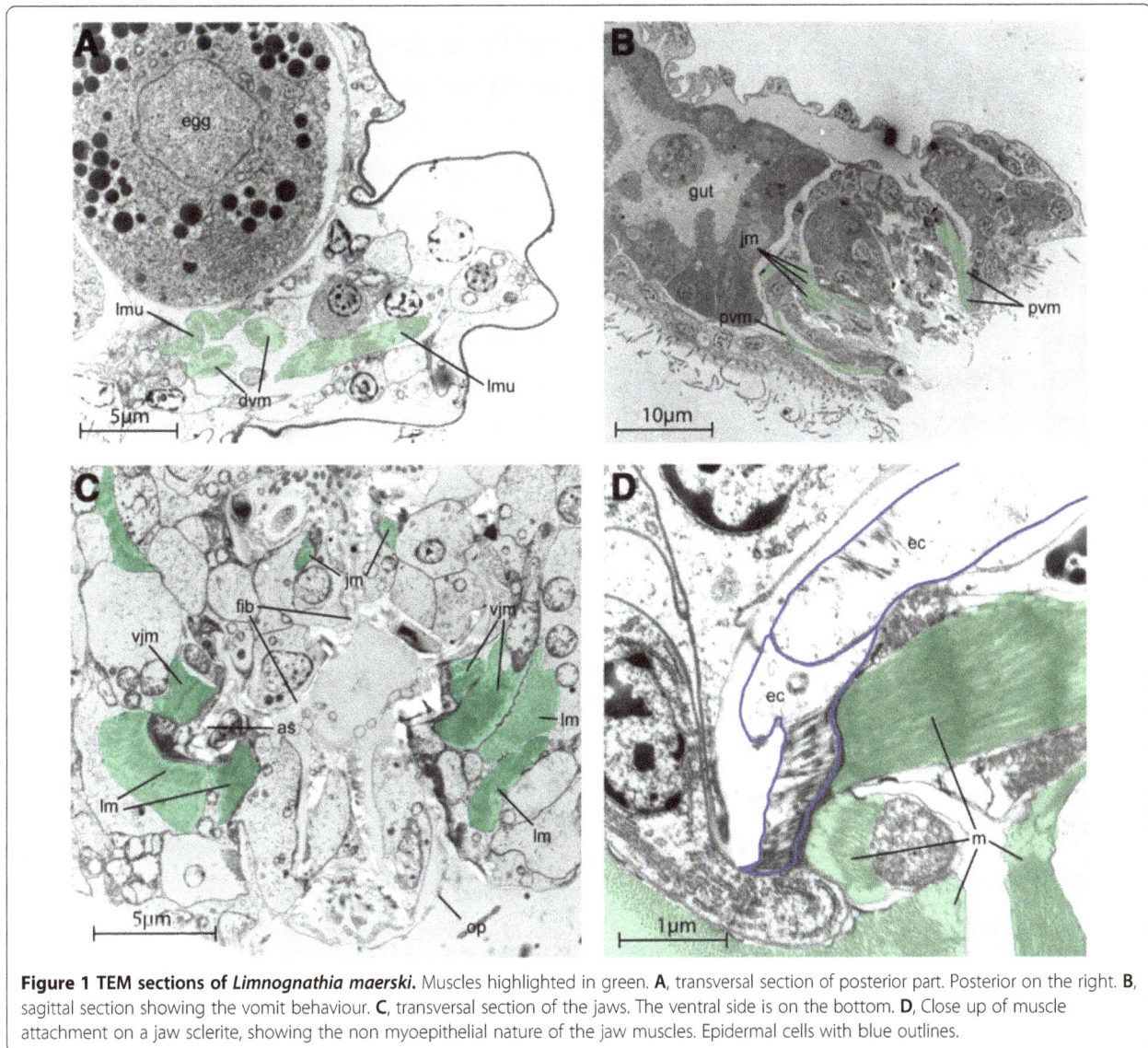

Figure 1 TEM sections of *Limnognathia maerski*. Muscles highlighted in green. **A**, transversal section of posterior part. Posterior on the right. **B**, sagittal section showing the vomit behaviour. **C**, transversal section of the jaws. The ventral side is on the bottom. **D**, Close up of muscle attachment on a jaw sclerite, showing the non myoepithelial nature of the jaw muscles. Epidermal cells with blue outlines.

rotifer trophi consists of relatively small paired muscles connecting the different jaw elements (sclerites), while, in Gnathostomulida, the main jaws are mainly moved together by large muscles attached to the pharynx wall. The movement between jaw elements in Gnathostomulida is consequently achieved by U-shaped muscles (bent transversal muscles) and laterally attached transversal muscles.

Recently, several CLSM studies of phalloidin-stained musculature have been carried out on a great number of microscopic animals, revealing comprehensive information on their overall musculature [9,11,28-30] and also, in the case of gnathiferans, on the musculature of the rotifer mastax [31,32] or gnathostomulid pharynx [26]. Combined with TEM, many details can be inferred on the relative position of muscles and their ultrastructure, but also connections to the other part of the body. In order to compare the general muscular organization as well as jaw musculature of *L. maerski* with other animals, we here describe its musculature employing F-actin staining and confocal laser microscopy (CLSM) as well as transmission electron microscopy (TEM).

Results

The overall musculature is organised into seven main pairs of longitudinal muscles extending from head to abdomen and 13 oblique dorso-ventral muscles localised in the thoracic and the abdominal part (Figures 2, 3, and 4). No circular muscles are present. The musculature furthermore comprises the dense pharyngeal muscle and the fine anterior forehead muscle. Cross striated muscles are found in the body wall musculature (Figure 1A) as well as in the jaw musculature (Figures 1B,C,D and 5C,D).

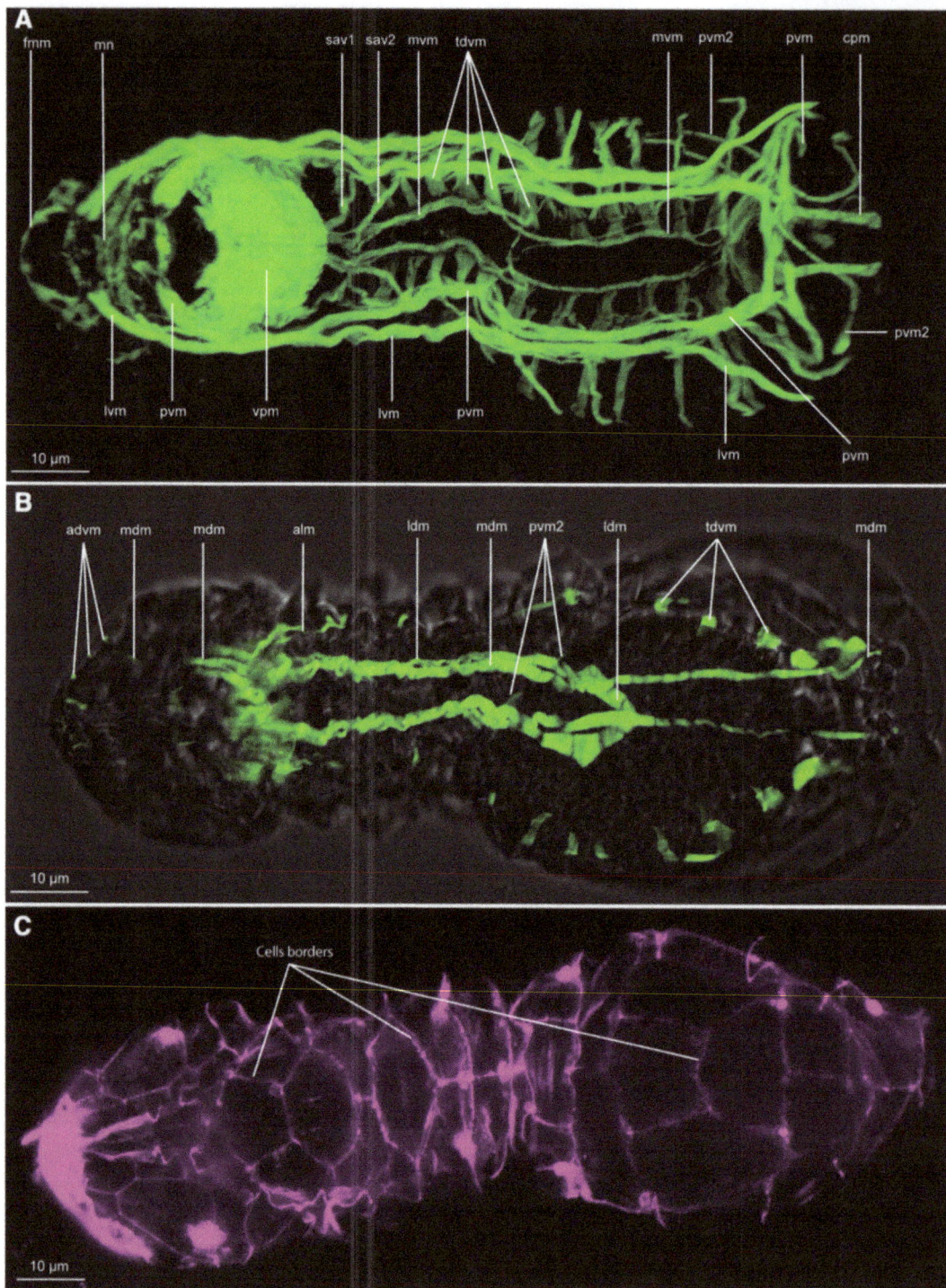

Figure 2 CLSM of phalloidin stained muscle system and light microscopy of *Limnognathia maerski*. Anterior end is positioned left on all pictures. **A**: Ventral view, Z-stack of the ventral portion, showing only the muscle system. **B**: Single section showing CLSM of the dorsal muscle system and the contour of the specimen, visualized with transmitted light. **C**: Synapsin2 staining of *L. maerski*, maximum intensity projection of a dorsal substack. Lines show the border of the dorsal cells to which the dorso-ventral muscles attach (illustrated in Figure 4B). advm, anterior dorso-ventral muscles; alm, anterior lateral muscle; cpm, ciliated adhesive pad muscle; fmm, front margin muscle; ldm, lateral dorsal muscle; lvm, lateral ventral muscle; mdm, median-dorsal muscle; mvm, medio-ventral muscle; mn, muscle network; pvm, paramedian ventral muscle; pvm2, posterior lateral muscle; sav1,2, small anterior ventral longitudinal muscles; tdvm, trunk dorso-ventral muscles; vpm, ventral pharyngeal muscles.

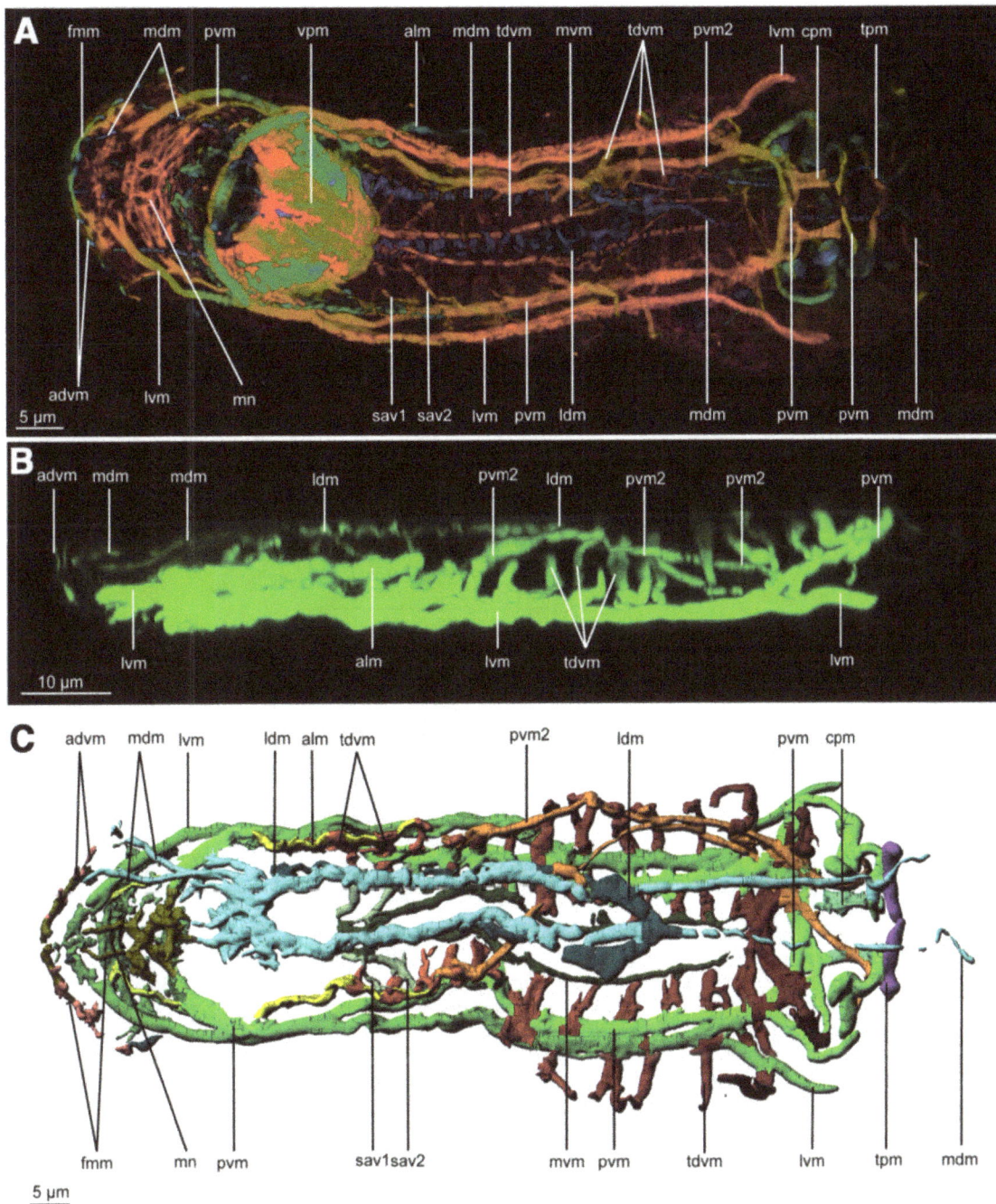

Figure 3 CLSM of phalloidin stained muscle system of *Limnognathia maerski*. Anterior end is positioned left on all pictures. **A**, Ventral view of the maximum depth intensity projection. **B**, lateral view reconstruction of a dorso-ventral Z-stack. Same specimen as Figure 2A,B. **C**, Dorsal view of the isosurface reconstruction of the muscular system. Same specimen as Figure 2A,B. advm, anterior dorso-ventral muscles; alm, anterior lateral muscle; cpm, ciliated pad muscle; fmm, front margin muscles; ldm, lateral dorsal muscle; lvm, Lateral ventral muscle; mdm, medio-dorsal muscle; mn, muscle network; mvm, medio-ventral muscle; pvm, paramedian ventral muscle; pvm2, posterior lateral muscle; sav1,2, small anterior ventral longitudinal muscles; tdvm, trunk dorso-ventral muscles; tpm, transversal posterior muscle; vpm, ventral pharyngeal muscles.

Longitudinal musculature

The longitudinal musculature of the trunk consists of seven pairs of main muscles (*three ventral, two lateral, two dorsal*) as well as two short anterior pairs of muscles and two short posterior pairs of muscles.

Ventral muscles

The three ventral main muscles extend the body length aiding the body contraction and extension (Figures 2A, 3 and 4). The longitudinal ventral muscles are implicated in longitudinal contractions and ventral bending.

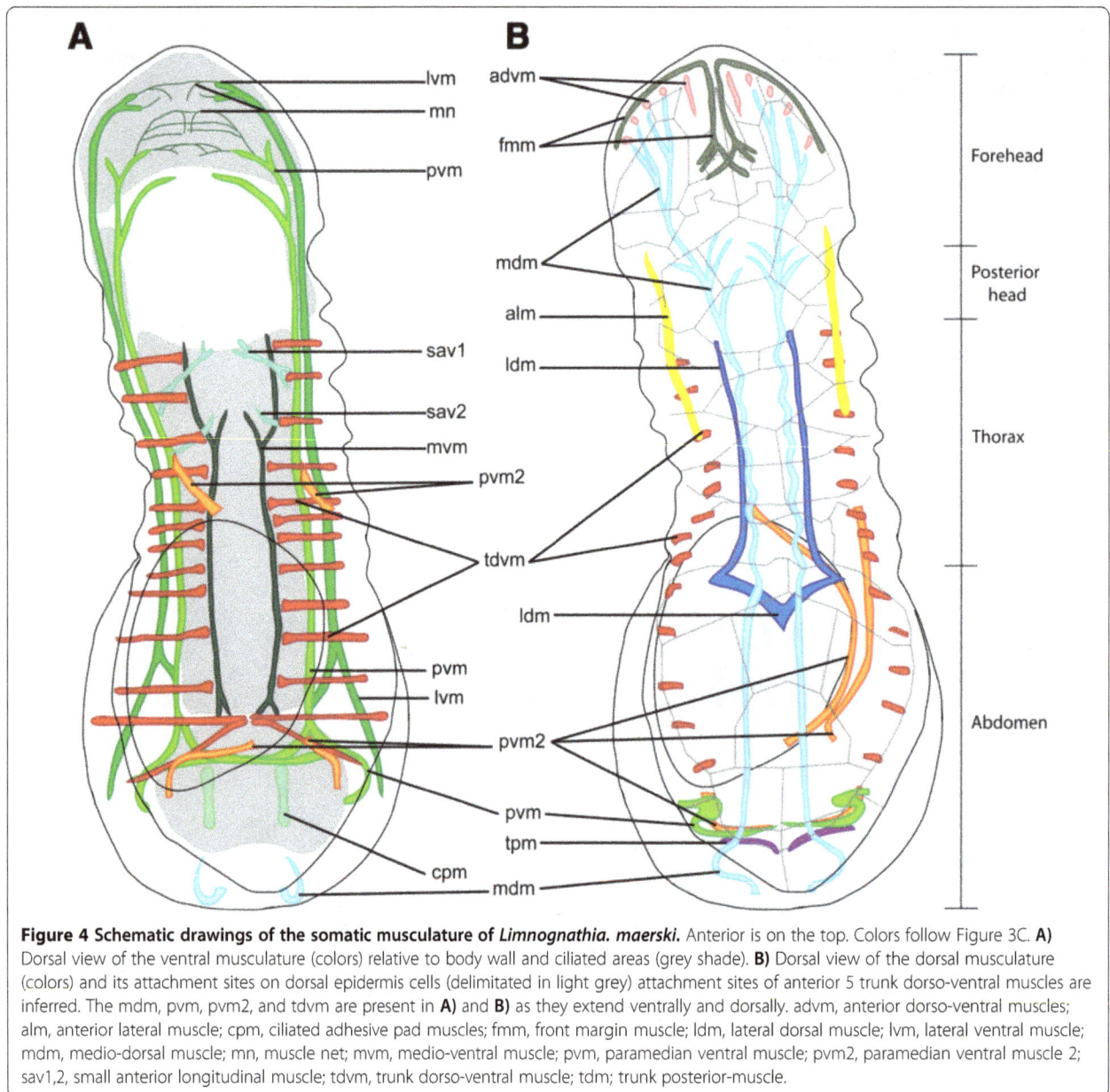

Figure 4 Schematic drawings of the somatic musculature of *Limnognathia. maerski.* Anterior is on the top. Colors follow Figure 3C. **A)** Dorsal view of the ventral musculature (colors) relative to body wall and ciliated areas (grey shade). **B)** Dorsal view of the dorsal musculature (colors) and its attachment sites on dorsal epidermis cells (delimitated in light grey) attachment sites of anterior 5 trunk dorso-ventral muscles are inferred. The mdm, pvm, pvm2, and tdvm are present in **A)** and **B)** as they extend ventrally and dorsally. advm, anterior dorso-ventral muscles; alm, anterior lateral muscle; cpm, ciliated adhesive pad muscles; fmm, front margin muscle; ldm, lateral dorsal muscle; lvm, lateral ventral muscle; mdm, medio-dorsal muscle; mn, muscle net; mvm, medio-ventral muscle; pvm, paramedian ventral muscle; pvm2, paramedian ventral muscle 2; sav1,2, small anterior longitudinal muscle; tdvm, trunk dorso-ventral muscle; tdm; trunk posterior-muscle.

The paired medio-ventral muscles (mvm, Figures 2A, 3A,C, 4 and 5A,B) consist of two muscle fibres that form bundles originating directly posterior to the ventral pharyngeal muscle and extend along the ventral wall of the gut (mvm: Figure 5A). At its posterior extremity, each medio-ventral muscle separates into two very short muscle fibers that each extends four micrometers before inserting into the epidermis that is anterior of the adhesive ciliated pad (mvm: Figure 4).

Medially, two pairs of small anterior ventral longitudinal muscles (sav1, sav2, Figures 2A, 3A,C and 4) supply the anterior part of the thorax, each originating from the midline directly posterior to the ventral pharyngeal muscle. The anteriormost muscle pair (sav1) is bifurcated at both

ends: the anterior bifurcation inserts medially just behind the pharynx, while the posterior bifurcation originates in a more lateral region close to the paramedian ventral muscle (sav1: Figure 4). The posteriormost muscle pair (sav2) inserts medially at the level of mvm and extends laterally toward (and originates close to) the paramedian ventral muscle (described below, sav2: Figure 4).

Latero-anterior to the pharynx are three muscles that come together to form the paramedian ventral muscle (pvm, Figures 2A, 3A,B,C, 4 and 5A,B); consequently, the paramedian ventral muscle is trifurcated at its anterior insertion but extends posteriorly as a single muscle bundle. The paramedian ventral muscle follows the course of the trunk and abdomen, where it eventually

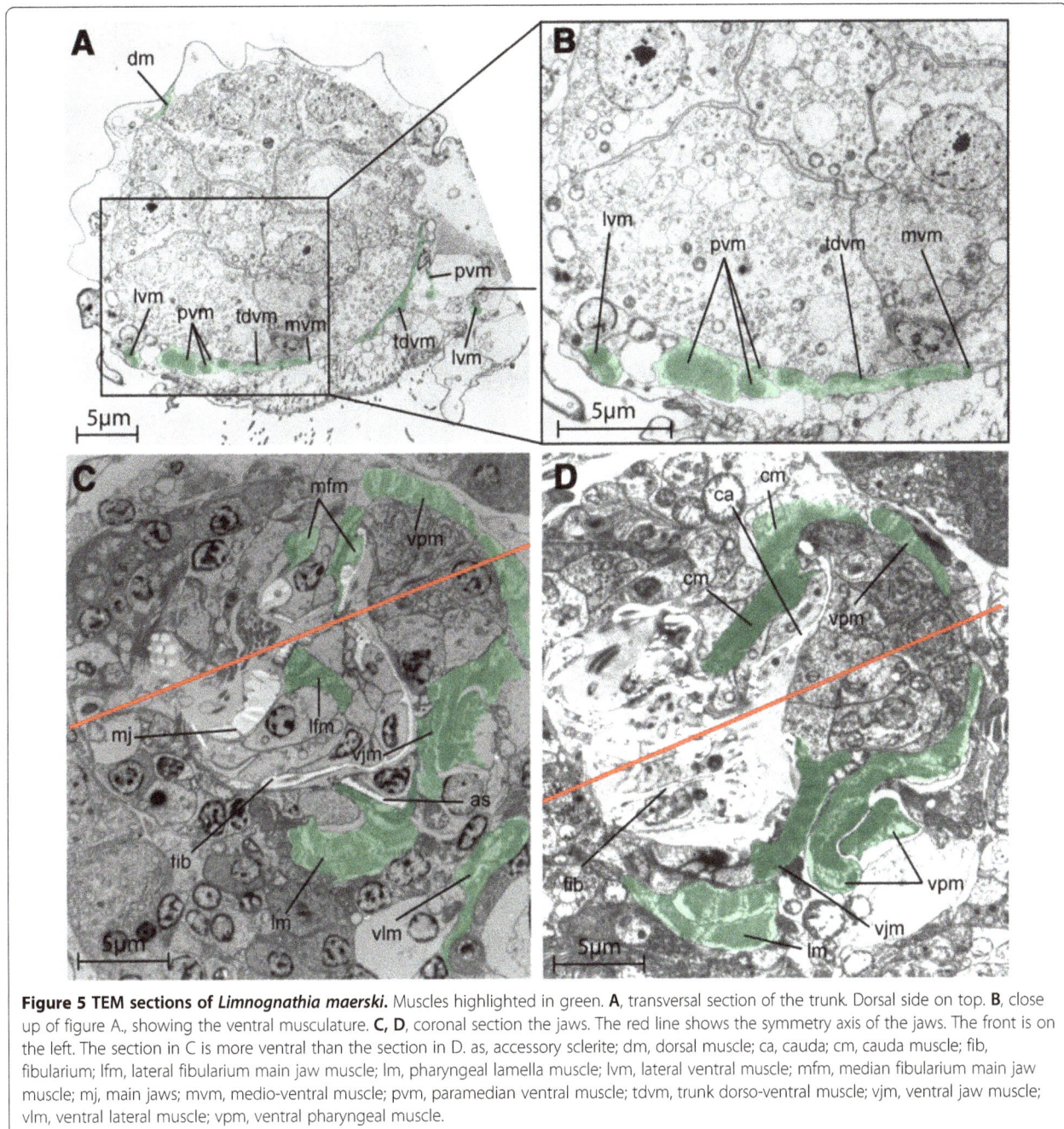

Figure 5 TEM sections of *Limnognathia maerski*. Muscles highlighted in green. **A**, transversal section of the trunk. Dorsal side on top. **B**, close up of figure A., showing the ventral musculature. **C, D**, coronal section the jaws. The red line shows the symmetry axis of the jaws. The front is on the left. The section in C is more ventral than the section in D. as, accessory sclerite; dm, dorsal muscle; ca, cauda; cm, cauda muscle; fib, fibularium; lfm, lateral fibularium main jaw muscle; lm, pharyngeal lamella muscle; lvm, lateral ventral muscle; mfm, median fibularium main jaw muscle; mj, main jaws; mvm, medio-ventral muscle; pvm, paramedian ventral muscle; tdvm, trunk dorso-ventral muscle; vjm, ventral jaw muscle; vlm, ventral lateral muscle; vpm, ventral pharyngeal muscle.

bifurcates into two separate bundles. The ipsilateral muscle bundle extends dorsally where it joins the paramedian ventral muscle 2 on the same side of the abdomen, while the contralateral muscle extends to the opposite side of the body and joins the contralateral last dorso-ventral muscle. Thus, each of the last dorso-ventral muscle bundles consists of three separate muscles: a dorso-ventral muscle, an ipsilateral branch of the paramedian ventral muscle and a contralateral branch of the paramedian ventral muscle from the opposite side of the body. (pvm: Figures 2A, 3A, 4 and 5). The paramedian longitudinal muscle follows the outline of the ventral ciliated area and contractions may change the direction during swimming or crawling (pvm: Figure 4).

Each of the two lateral ventral muscles (lvm; Figures 2A, 3A,B,C and 5A,B) inserts anterior of the mouth where they each bifurcate into two smaller branches. Posteriorly, each lateral ventral muscle extends along the trunk and abdomen as a single bundle that eventually bifurcates again. The inner branch joins the paramedian ventral muscle, while the lateral branch inserts in the region of the large posterior gland.

A pair of ciliated adhesive pad muscles (cpm: Figures 2A, 3A,C and 4), which are present as short longitudinal bands, extend from an anterior zone of the ciliated pad (just posterior of the paramedian ventral muscle midline) to a posterior zone of the ciliated pad (cpm: Figure 4). The adhesive ciliated pad muscle is probably involved in the adhesive ciliated pad area contractions. Contraction of the adhesive ciliated pad muscles could contract this area and allow the animal to release from the substratum.

Lateral muscles

Two pairs of lateral muscles are present in the trunk. The pair of anterior lateral muscles (alm: Figures 2A, 3B, C and 4B) originates anterior of the mouth, probably bifurcating from the paramedian longitudinal muscle, and continues two thirds into the abdomen, appearing to attach to the lateral epidermal cells. They are positioned at a mid dorso-ventral level. The paired paramedian ventral muscles 2 (pvm2: Figures 2A,B and 3A,B,C) originate ventrally to the paramedian ventral muscles, separating at the mid-thoracic level. Each muscle reaches the dorsal side along the anterior part of the abdomen (pvm2: Figure 2B), extends ventrally at the level of the adhesive ciliated pad and returns at an antero-dorsal position, joining the very posterior dorsal epidermal cells and the paramedian ventral muscle. From this point, both paramedian ventral muscle 2 muscles join close to the midline at their posteriormost point, at the level of the last dorsal plate. If an egg is present at the level of the abdomen, one of the posterior lateral muscles is pushed by the egg to the contralateral side to return to the ipsilateral side at the level of the adhesive ciliated pad (pvm2: Figures 2A, 3A,C and 4). This muscle extends along the dorsal side of the gut, being probably implicated in dorsal bending of the animal.

Dorsal muscles

Two dorsal pairs of muscles extend through the trunk. The two pairs are close to the midline and extend as two contiguous muscles (Figures 2B, 3A,C and 4).

The medio-dorsal muscle (mdm: Figures 2B, 3A,B,C and 4. dm: Figure 5) is an elongate band that extends from the head to the abdomen and is composed of several thinner muscles that branch off in the forehead and posterior head regions. Anteriorly, the medio-dorsal muscle branches twice (with additional subbranches) that insert close to the frontal margin where the anterior dorso-ventral muscles insert (advm: Figures 2B, 3A-C and 4). In the posterior head region, the medio-dorsal muscle supplies several short muscle branches just dorsal of the pharynx (mdm: Figures 2B, 3C and 4). At the very posterior part of the animal, the medio-dorsal muscle lines the body wall, to insert at the ventro-posterior extremity of the abdomen (mdm: Figures 2B, 3A,C and 4B).

The lateral dorsal muscles (ldm: Figure 2B, 3A,B, and 4. dm: 5) originate as a pair of muscles that both insert at the midline in the trunk region (mdm: Figure 4). Each muscle extends antero-laterally for about 10 micrometers before curving back medially and continuing anteriorly as a strictly longitudinal muscle band that inserts dorsal to the pharynx (Figures 2B, 3A,C and 4B).

Transversal posterior muscles

Additionally, at the very posterior region, a complex of transversal and dorso-ventral muscles is present (Figures 3C and 4). It is partially formed by the longitudinal muscle extending posteriorly, from the ventral to the dorsal side. Posterior of these muscles, two pairs of dorsal small transversal muscles line each side of the body. It is difficult to determine with certitude if these two pairs are the continuity of the posterior lateral muscle. However, the anteriormost pair of lateral muscles seems to be a continuity of the paramedian ventral muscle (pvm: Figures 2A, 3A,C and 4) while the transversal posterior muscle pair seems to be another set of muscles (tpm: Figures 3A,C and 4). Both pairs of transversal posterior muscles are very dorsal and according their anatomical position could be implicated in a possible anus opening. Along with the posterior longitudinal and dorso-ventral musculature, the complex posterior musculature is probably involved in the oviposition, substrate adherence and, eventually, defecation.

Dorso-ventral musculature

The dorso-ventral musculature consists mostly of two sets of muscles: the anterior dorso-ventral muscles and the trunk dorso-ventral muscles (Figure 3C and 4). The posteriormost dorso-ventral complex is the continuation of the paramedian muscle and the paramedian ventral muscle 2 when they fold in the posterior region, and is not serially homologous to the trunk dorso-ventral muscles.

Anterior muscles

Five pairs of anterior dorso-ventral muscles (advm: Figures 2B, 3A,B,C and 4) supply the front margin. They appear to support the frontal ciliated sensory region. On each side, the medianmost dorso-ventral head muscle inserts dorsally, at the anterior head margin, close to the mid-line (Figure 2B).

Trunk muscles

Thirteen pairs of oblique trunk dorso-ventral muscles (tdvm: Figures 2A,B, 3A-C; 4; 5A,B) supply the thorax and the abdomen. Each trunk dorso-ventral muscle inserts close to the midline on either side of the medio-ventral muscle, extends laterally dorsal to the paramedian ventral muscle and the lateral ventral muscle, and then curves dorsally to insert on epidermal cells (tdvm: Figures 2A; 4; 5A,B). They join the epidermal cells dorsally, extending along the body sides. They line the gut cells very closely, probably functioning as body-wall musculature as well as gut musculature (tdvm: Figure 5A,B). Five pairs supply the thoracic region and eight supply the abdomen region (tdvm: Figures 2A,B; 3A-C; 4). The penultimate and the last pair of dorso-ventral muscles insert ventrally at the midline where the medio-ventral muscle inserts as well, forming a very muscular zone five micrometres anterior of the adhesive pad. A few micrometres posteriorly, the two paramedian muscles cross transversally, forming with the two last dorso-ventral trunk muscles a triangular set of ventral muscles at the anterior area of the adhesive ciliated pad (tdvm: Figures 2A; 3C; 4).

Forehead musculature

The head musculature is a continuity of the longitudinal body musculature as well as a few specific muscles.

On the frontal margin, the paired frontal margin muscles (fmm: Figures 2A; 3A,C; 4) follow the coronal plan supplying the anterior ciliated region. The median extremity of each muscle is dorsal and bends posteriorly to continue dorsally as two longitudinal median head muscles. At the distal extremities, the front margin muscles are more ventral and supply the frontal ciliated zone. The five pairs of anterior dorso-ventral muscles also supply the frontal ciliated area. The anterior dorso-ventral muscles extend dorsally and quite close to the frontal margin muscle, thus appearing to be in contact with it. In front of the pharynx, dorsally, a cross like complex of small muscles consists of the front margin muscles continuing as a longitudinal median head muscle and trifurcates as two lateral small bundles and one median bundle. The bundles of the front margin muscles of each side join the midline with other contralateral front margin muscle (fmm: Figures 3C; 4).

Ventro-anteriorly, in front of the mouth opening the continuity of the lateral ventral muscle and the paramedian ventral muscle form a thin muscle network (mn: Figure 2A; 3A,C; 4), probably implicated in some anterior glands or changes of the shape of the head.

Pharynx musculature

The pharynx musculature includes the major ventral pharyngeal muscle and several paired and unpaired muscles articulating the jaws. Jaw muscles have a non-epidermal origin, with each muscle being connected to an epidermal cell associated to a sclerite (Figure 1D). Thus, the musculature of the jaws is probably of

mesodermal origin. The function of the musculature is interpreted according to previous studies on feeding behaviour and live observations.

Ventral of the trophi, lining the fibularium, several longitudinal fibres form a large ventral pharyngeal muscle plate (vpm, Figures 2A; 3A; 5C,D; 6A-C) and continues anteriorly as two small lateral muscle fibres. This ventral pharyngeal muscle plate is formed by 8-10 longitudinal cross striated muscle fibres (Figures 1B; 5C, D; 6A,C). The longest median muscle filament presents 8 z-bands (Figure 6A-C). However, even though the ventral pharyngeal muscle plate mostly underlies the fibularium, the ventral pharyngeal muscle is shifted more posteriorly compared to the fibularium. The plate is rounded at the lateral and posterior edges, hereby enveloping the trophi (including the fibularium) laterally and caudally (vpm: Figure 5C,D).

Dorsal to the fibularium, two pairs of muscles extend between the fibularium and the main jaws: one pair of lateral fibularium/main jaw muscles (lfm: Figures 5C; 6D-F), and one pair of median fibularium/main jaw muscles (mfm: Figures 5C; 6D-F). Both of them attach to the fibula caudalis of the fibularium. The lateral fibularium/main jaw muscle originates at the fibula caudalis (of the camera dorsalis 1), and supplies the anterior part of the main jaws. The median fibularium/main jaw muscle originates posterior of the fibula caudalis (of the camera dorsalis 1 and 2), and supplies a less anterior part of the main jaws than the lateral fibularium/main jaw muscle.

One pair of strong caudal muscles lines each cauda of the main jaws (cm: Figures 5D; 6D-F). They are thicker in their posterior parts where they follow the paired caudae of the main jaw. The contraction of this muscle moves the main jaws together.

Two short anterior fibularium/main jaw muscles (afm: Figure 6G-I) attach to the anterior part of the fibula lateralis at the camera lateralis, and link in this way the fibularium with the anterior parts of the main jaws.

Altogether, the anterior fibularium main jaw muscle, the lateral fibularium main jaw muscles, the median fibularium/main jaw muscles and the caudal muscle, are probably responsible for the opening of the main jaws and their previously described backward/forward movements (Kristensen and Funch [3]).

An unpaired very thin striated U-shaped dorsal jaw muscle (djm: Figure 6G-I) attaches at each extremity to the posterior ends of each dorsal jaw.

Lateral to the fibularium, one pair of strong cross striated ventral jaw muscles (vjm: Figures 5C,D; 6G-I) inserts at the posterior part of the accessory sclerite. They extend posterior of the trophi, attaching the sides of the fibularium and inserting posteriorly at the pharynx epithelium.

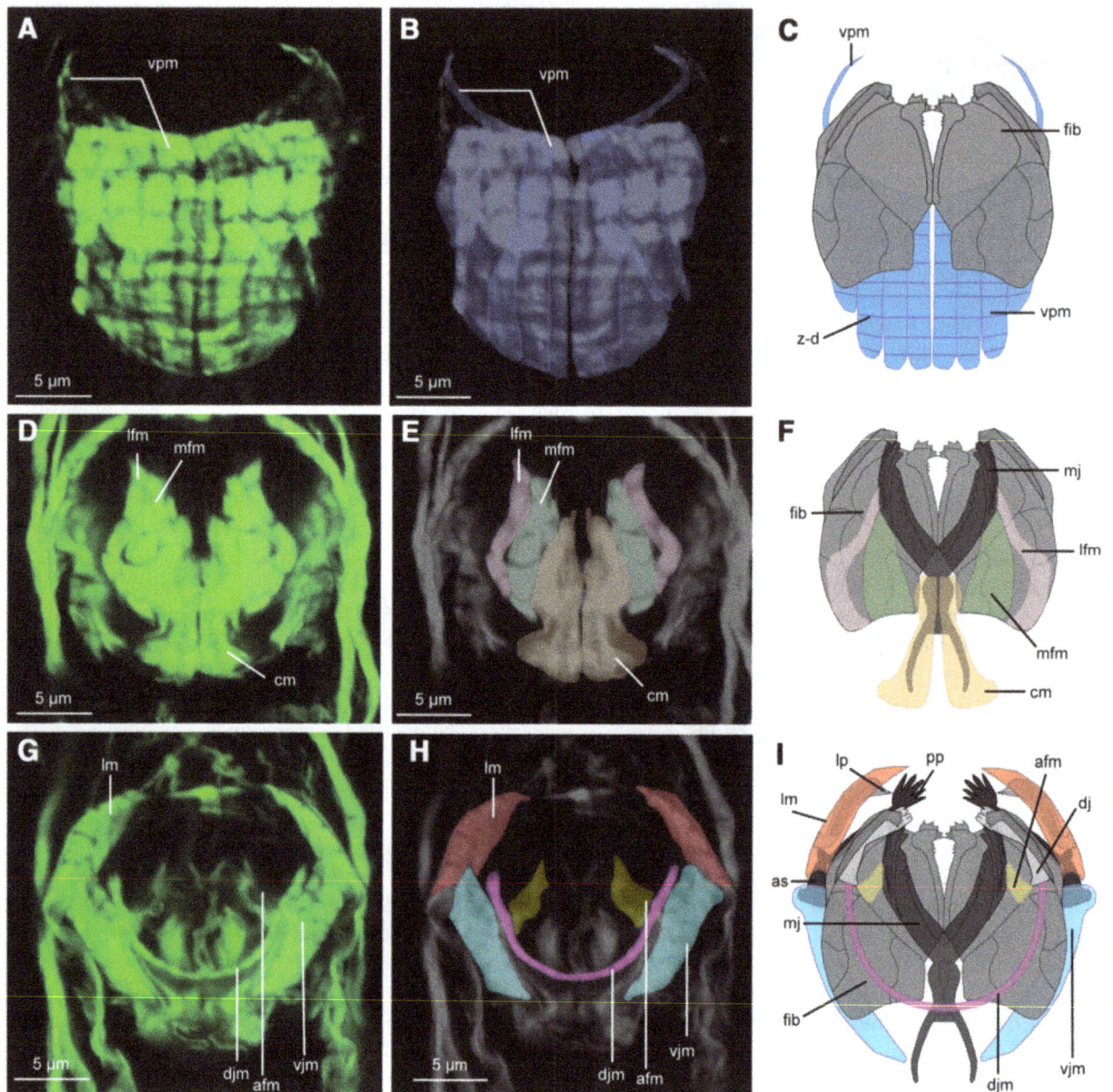

Figure 6 Musculature and reconstruction of the jaw apparatus of *Limnognathia maerski* in dorsal view. Anterior is on the top for all the pictures. **A, B, C**: ventral part of the jaw system. **D, E, F**: median part of the jaw system. **G, H, I**: dorsal part of the jaw system. **A, D, G**: CLSM of phalloidin stained muscle system, dorsal view of a projection of a sub sample of the Z-stack. **B, E, H**: enlightenment of the different muscle systems of the jaws. **C, F, I**: schematic drawing of the dorsal view of the myoanatomy of the jaw system linked to the cuticular elements in greys. Jaw drawing after Sørensen [6]. as: accessory sclerite; afm: anterior fibularium-main jaw muscle; cm: caudal muscle; dj: dorsal jaws; djm: dorsal jaw muscle; fib: fibularium; lm: pharyngeal lamella muscle; lp, pharyngeal lamella; lfm: lateral fibularium-main jaw muscle; mfm: median fibularium-main jaw muscle; mj: main jaws; pp: pseudo-phalangium; vjm: ventral jaw muscle; vpm: ventral pharyngeal muscles; z-b: Z-bands of the cross striated muscles of the ventral pharyngeal muscle.

Anterior to the other parts of the trophi, two strong pharyngeal lamellae muscles (lm: Figures 5C,D; 6G-I) supply the accessory sclerites and the pharyngeal lamellae. The two pharyngeal lamellae muscles are very large and in the continuity of the paramedian ventral muscle and anterior lateral muscle. They enlarge dorso-ventrally at the terminal part. This observation confirms the supposed function of the pharyngeal lamellae (initially lamella oralis) as a supporting structure. This dorso-ventrally enlarged muscle could function in opening and closing the pharyngeal lamellae as a fan, affecting the volume of the pharynx. The ventral jaw muscle is probably functioning together with the pharyngeal lamellae muscle as an antagonist. Indeed, both muscles are connected to the accessory sclerite. When the pharyngeal lamellae muscles are contracted and the ventral jaw muscles relaxed, the pharyngeal lamellae will open and increase the volume of the pharynx cavity, also probably opening the mouth and allowing ventral jaws extrusion.

Anti-Synapsin1 immunoreactivity

Anti-Synapsin 1 immunoreactivity (IR) was tested in ongoing studies of the nervous system (Bekkouche et al. unpublished) and surprisingly yielded a very distinct IR at the borders of the dorsal epidermis cells. This immunoreactivity, which is presented as spots along the borders, resembles the distribution pattern of the unique zip-junctions in *Limnognathia* (equivalents of adherens junctions) (Figure 2C). However this IR interpretation warrants further confirmation. Most importantly, the very distinct cell border signal has been proved useful in the present study for co-localizing the attachment sites of the dorso-ventral muscles. Thereafter it was possible, even in specimens not stained against Synapsin1, to retrieve the borders of the dorsal cells of the epidermis by increasing the brightness of the phalloidin stain (data not shown). The attachment of the last eight trunk dorso-ventral muscles to the dorsal epidermal cells could then be inferred in several specimens (Figure 4B). Furthermore, the synapsin 1 staining clearly shows that *Limnognathia maerski* has cell borders in the epidermis (as opposed to being syncytial) and therefore does not belong to Syndermata (Rotifera and Acanthocephala).

Discussion
Notes on the longitudinal musculature

In *L. maerski* most of the longitudinal musculature extends the entire body length, or at least the entire trunk, yet some muscles are restricted to certain areas, e.g., the adhesive ciliated pad (cpm: Figures 2A; 3A,C; 4A), the thorax (ldm: Figures 3A,C; 4B), the anterior part of the thorax (sav1,2: Figures 2A; 3A,C; 4A), etc. This repartition of the musculature supports functionally the separation of *L. maerski* into a head, a thorax and an abdomen. Similarly, in rotifers, many longitudinal muscles extend a subpart of the body, aiding the retraction of the foot or the corona [11,12]. Contrarily, most of the longitudinal muscles of Gnathostomulida extend the entire body length [9,10].

Is the dorso-ventral musculature of *L. maerski* comparable to circular musculature?

The trunk dorso-ventral musculature of *L. maerski* (tdvm: Figures 2A,B; 3A,B,C; 4; 5A;B) superficially resembles the repeated incomplete circular muscles found in many rotifers. However, as described by Leasi and Ricci [12]: "the muscular system of rotifers generally consists of somatic and splanchnic (visceral) fibers.

Somatic musculature is composed of two layers: an external layer made of separate circular rings and an internal layer of longitudinal muscles". *Limnognathia maerski* lacks splanchnic fibers and the somatic musculature is only composed of longitudinal muscles. However, internal of these are found the dorso-ventral muscles. These are serially repeated along the lateral outline of the gut (tdvm: Figure 5A,B). The median position of the trunk dorso-ventral muscles, relative to the two pairs of lateral and paramedian ventral longitudinal muscles, does not conform to the somatic circular muscles found in rotifers, and homology of these muscles is unlikely. However, they can be functionally compared to those of rotifers: with lack of both outer and inner circular musculature, these dorso-ventral muscles may act both as a splanchnic musculature, aiding the movement of the food throughout the digestive system, as well as somatic dorso-ventral musculature, elongating the body during contraction. In rotifers, the incomplete circular muscles act as antagonists of the longitudinal musculature. When these somatic circular muscles contract, the pressure of the body fluids is redistributed and prompts the extension of the body [12]. The same function is assumed in *L. maerski* for the trunk dorso-ventral muscles. It is interesting to note the medio-ventral longitudinal muscles as they seem to extend at the same level as the ventralmost part of the trunk dorso-ventral muscles (tdvm and mvm: Figures 2A; 3A; 4; 5A,B). This suggests that the medio-ventral longitudinal muscles may specifically work as antagonists of the trunk dorso-ventral muscles in the same way as for rotifers.

Giribet et al. [5] propose, among other hypotheses, a relationship between Micrognathozoa and Cycliophora. In Cycliophora, inner dorso-ventral muscles are also present in the Pandora larva and the dwarf male life stages [33-35]. In the dwarf male, several sets of dorso-ventral muscles are present along the entire body length, while in the Pandora larva, only three pairs of dorso-ventral anterior muscles are present in addition to the incomplete circular muscles repeated through the entire body length. It is, though, difficult to establish any functional comparison with *L. maerski* since there is no gut present in these two cycliophoran stages.

Similar to *L. maerski*, dorso-ventral muscles are found internal of the longitudinal muscles in kinorhynchs [36]. Moreover, in the gastrotrich *Draculiciteria*, two sets of dorso-ventral muscles are found: one inside and one outside the longitudinal musculature, each supposed to be derived from splanchnic and somatic circular muscles, respectively [37]. The organization found in kinorhynchs can be compared to the attachment of the trunk dorso-ventral muscles to the epidermal cells containing the dorsal plates in *L. maerski*, even though the two conditions obviously are analogous. Additionally, in both

Table 1 Previously proposed homologies of *Limnognathia maerski* jaw parts and Rotifera jaw parts

Jaw elements in *Limnognathia maerski*	Proposed homologies with rotifer trophi according to the authors		
	Kristensen and Funch [3]	De Smet [1]	Sørensen [6]
Basal plates		Basal platelet (epipharynx)	Autapomorphy
Fibularium	Ramus	Manubrium + uncus	Autapomorphy
Ventral jaws	Uncus	Pseudomalleus (epipharynx)	Uncus
Accessory sclerites	Manubrium	Pseudomanubrium (epipharynx)	Manubrium
Main jaws dentarium	Ramus	Ramus	Ramus
Main jaws articularium	Fulcrum	Fulcrum	Fulcrum
Lamellae pharyngea	Epipharynx	Oral lamellae (epipharynx)	Epipharynx
Dorsal jaws	Autapomorphy	Pleural rod	Autapomorphy

kinorhynchs and *Draculiciteria*, as well as in rotifers, the contraction of the dorso-ventral musculature is supposed to be involved in the body extension [36,37].

This comparison between small sized pseudoceolomate or acoelomate animals, leads to the supposition that dorso-ventral muscles play a similar role as circular muscles, aiding the fluid circulation in the body and in *L. maerski*, possibly also changing the shape of the relatively large cells of the endodermis. Thus, the dorso-ventral muscle contractions possibly aid the movement of food particles in the gut, the vomit behavior, and the yet non-observed defecation.

Functional considerations of the pharynx musculature
Considerations on the jaw musculature of L. maerski
Six paired main elements are described in the jaws of *L. maerski*: i) The median basal plates ii) the large ventral fibularia, extending dorso-laterally, iii) the lateral-most ventral jaws, iv) the medio-dorsal main jaws, with posteriorly projecting caudae, v) the dorso-lateral dorsal jaws confined to the fibularium area, vi) and the antero-lateral pharyngeal lamellae [6]. For comparison we refer to the Table 1 that summarizes the various jaw homology hypotheses proposed in the literature between the Rotifera and *L. maerski*. A general consensus appears to exist for the homologies between the articularium and cauda of Gnathostomulida, the ramus and fulcrum of Rotifera and the main jaws and caudae of *L. maerski* [1,3,6,38].

No separate musculature associated to the basal plate in *L. maerski* has been found. Moreover, detailed examination of the ventral view of the SEM images of the jaws of *L. maerski* does not show any clear separation between the basal plates and the fibularium [1,6], suggesting that the basal plate could be an integrated part of the fibularium.

The dorsal jaw muscle apparently only connects the two dorsal jaws and is not attached to the pharyngeal wall. In Sørensen [6], the dorsal jaws are described as caudally attached to the internal side of the fibularia, possibly by a flexible ligament on each side, positioning the jaws in a 90° angle to the main jaws. A contraction of the dorsal jaw muscles would then pull apart the tips of the dorsal jaws, turning the jaws about 45° from their resting position.

The fibularium, as the most conspicuous jaw structure, is involved in the attachment of three out of eight jaw muscles systems, suggesting that the fibularium acts primarily as a supporting structure for the jaws and the pharynx, rather than an element directly implicated in the mastication. This assumption is consistent with the strong ventral pharyngeal muscle underlying the fibularium.

Comparison of the pharyngeal musculature of L. maerski with those of other animals
The ventral jaws and accessory sclerites of *L. maerski* make up as a functional unit that has been considered homologous with either the rotifer mallei [3,6] or the rotifer epipharynx [1] (see also Table 1). The ventral jaws can be moved independently and extruded through the mouth opening during foraging while the rest of the jaws are not. In rotifers, the different sclerites are more closely connected through ligaments, and the mallei cannot be fully protruded without also protruding parts of the incus as well (e.g., in *Bryceella stylata* [31] and *Dicranophorus forcipatus* [39]). In *L. maerski* no ligamentous connections exist between the ventral jaws and either the fibularium or main jaws, which allow the ventral jaws to move more independently from the other main elements of the jaw apparatus.

The ventral jaw muscle of *L. maerski* (vjm: Figure 5C, D; 6G-I) can be compared to the musculus circumglandulis of Rotifera. This muscle connects the rami with other parts of the mallei [31,39,40]. Its ventral position, connection with the ramus and conspicuous shape, resembles the ventral pharyngeal muscle (conspicuous muscle made of several bundle) or the ventral jaw muscle (connection and position) in *L. maerski*. However, in rotifers this muscle is assumed to perform the spreading of the rami and eventually also the compression of the salivary glands [31], and such functions are not likely for the ventral jaw muscles in *L. maerski*. Hence, no equivalent of the ventral jaw muscle of *L. maerski* is found in Rotifera.

Underlying the fibularium, the conspicuous plate of the ventral pharyngeal muscle is present (vpm, Figures 2A; 3A; 5C,D; 6A-C). Composed of several longitudinal parallel muscles fibers, this structure is found neither in gnathostomulids nor rotifers. In Gnathostomulida though, a pharyngeal capsule is found, but it is formed by circular muscles enveloping the pharynx [27], which is structurally different from *L. maerski*. However, a strikingly similar ventral set of longitudinal muscles, encompassing two fanlike muscles forming a similar bowl, is found in the microscopic worm *Diurodrilus* (Spiralia *incertae sedis*) [30]. In *Diurodrilus*, this pharyngeal bowl also lines the pharynx ventrally, whereas its posterior part extends further dorsally compared to what is apparent in *L. maerski*. In L. maerski, the configuration of the muscle plate indicates that it is implicated in the extrusion and sinking movements of the fibularium and possibly causes changes in the volume of the pharyngeal cavity.

Functionally, this muscle could also be similar to the mastax receptor retractor found in the rotifer *Pleurotrocha petromyzon* as well as other rotifers with virgate mastax [40], aiding the total movement of the mastax by changing the shape of the pharynx cavity. However, the rotifer mastax receptor retractors are located dorsal to the jaw, which makes an actual homology with the micrognathozoan ventral pharyngeal muscle unlikely. We assume a similar function of the ventral pharyngeal muscle in *L. maerski*, which when contracting seems to move the entire jaws system, during the so-called vomit behavior. Morphologically, the similarity of the plate-bowl-shaped ventral pharyngeal muscle of *L. maerski* and *Diurodrilus* is striking [30] and not found in Rotifera and Gnathostomulida.

The main jaws represent the central element of the micrognathozoan jaw apparatus, and there is a consensus about homologizing the main jaws with the rotifer incus [1,3,6] (see also Table 1). Two different sets of main jaw muscles connect the main jaws with other sclerites or with the pharyngeal wall. The first set, related to the fibularium, is a "lateral connection" created by the anterior fibularium main jaw muscle, the lateral fibularium main jaw muscle and the median fibularium main jaw muscle. The second one, independent of the fibularium, is a "posterior connection" created by the caudal muscle. In *L. maerski*, the "lateral connection" is the most prominent in the main jaws and it is operated by 3 sets of muscles (anterior fibularium main jaw muscle, lateral fibularium main jaw muscle, median fibularium main jaw muscle, respectively afm, lfm, mfm:

Figure 6D-I). In Gnathostomulida, the lateral connection is also dominant, realized by the diductor muscles [9,26] which do not connect to a lateral sclerite but to the dorsal wall of the pharynx. In *L. maerski*, the fibularium has the function of attaching the muscles involved in the lateral connection. Among rotifers, sparse examples of lateral connections can be found. The only muscle having this arrangement is the musculus ramo-manubricus found in *Filinia longiseta* [41] and *Trichocerca rattus* [33], both having very peculiar trophi (respectively malleoramate and asymmetrical virgate). In Rotifera, though, the posterior connection is well documented in the abundant work of the series of confocal and TEM studies by the Ahlrichs Group [31,32,39-41], who refers to this muscle as the musculus fulcro ramicus. Furthermore, Riemann and Ahlrichs, emphasize the wide repartition of this muscle within Rotifera, suggesting the homology of this muscle across the taxon [39]. Then, the cauda muscle of *L. maerski* (cm: Figure 6D-F) could also be homologous to the musculus fulcro ramicus of Rotifera. A difference between those two muscles is that the cauda muscle seems to embed, or at least extend closely the cauda, while the musculus fulcro ramicus is more diagonal in its orientation. Additionally, the cauda muscle goes more posterior and seems to insert in the pharyngeal wall, while the musculus fulcro-ramicus is posteriorly restricted to the fulcrum.

Only muscles functionally implicated in the opening of the main jaws (not in the closing) have been found in *L. maerski*. As proposed for Rotifera and Gnathostomulida, we assume that the kinetic energy release of the cuticular parts provokes a passive closing of the pincer like sclerites in *L. maerski* [26,27,39].

Conclusions

Due to its simplicity, the longitudinal musculature of *L. maerski* is only roughly comparable to the musculature of other groups. However, the dorso-ventral musculature shows a functional similarity to the semi-circular muscles of the closely related Rotifera and other meiofaunal animals.

With regards to the pharyngeal musculature, only one specific homology between the cauda muscle of *L. maerski* and the musculus fulcro ramicus of rotifers can be hypothesized. However, the functional and morphological comparisons of the jaw musculature among gnathiferans aid the understanding of how such small complex systems can be moved. Two different "strategies" can be observed in the jaw apparatus of Rotifera versus Gnathostomulida: in rotifers, sclerites are moved by muscles connected to other jaw parts whereas in gnathostomulids the less complex jaws are moved by muscles connected directly to the pharyngeal wall. It is not surprising considering the complexity of the jaws of

L. maerski that the jaw musculature and function are more comparable to that of Rotifera. However, the independence of the ventral jaw of *L. maerski* relative to the rest of the trophi is an interesting difference between *L. maerski* and Rotifera. Additionally, the striking similarity between the ventral pharyngeal muscle of Micrognathozoa and the pharyngeal bowl-shaped muscle of *Diurodrilus* is interesting in relation to the debated close relationship between the jaw-less *Diurodrilus* and Micrognathozoa [3,30].

Several functional analogies and common patterns could be shown between *L. maerski* and other Gnathifera or small sized animals, but the systematic value of the musculature of *L. maerski* still appears quite limited. However, further studies are needed in Gnathifera. De Smet [1] emphasizes the poor knowledge of the epipharynx of Rotifera. For example, Riemann and Ahlrichs [39], in their study on *Dicranophorus forcipatus* cannot assign any clear function to the hypopharyngeal elements. Furthermore, no complete detailed studies of the musculature and function of trophi of the Seisonidae, Bdelloidea (both Rotifera) and Filospermoidea (Gnathostomulida) have been done so far. Nevertheless, a systematic comparison will still be challenging since the trophi of Bdelloidea and Seisonidea are very modified, and the jaws of Filospermoidea have a relatively simple pincer-like structure, such as in *Haplognathia*.

Material and methods
Collection of specimens
Specimens used for TEM were part of the original material that were collected at the type locality in the Isunngua Spring on Disko Island, West Greenland, 69°43'N 51° 56'W, and used for the description of Micrognathozoa [3]. Specimens for CLSM were collected in July-August 2010 and 2013 at the same locality.

Transmission electron microscopy
Specimens were fixed in trialdehyde 8% (after Kalt and Tandler [42] and Lake, [43], without acrolein) and postfixed in 1% osmium-tetroxide with 0.1M sodium cacodylate buffer for 1 hour (h) at 20°C. Specimens were then dehydrated through an ethanol series, transferred to propylene oxide, and embedded in epoxy resin type TAAB 812®. Ultrathin serial sections were stained with uranyl acetate and lead citrate [44]. TEM examinations were performed with a JEOL JEM 100SX transmission electron microscope.

Cytochemistry and CLSM
Specimens of *L. maerski* were fixed for 2 h at room temperature (or overnight at 4°C) in 2% paraformaldehyde in 0.15M phosphate buffered saline (PBS), pH 7.4, rinsed and stored in PBS plus 0.05% NaN_3. Entire specimens were preincubated two hours in PTA (PBS with 0.5% Triton-X, 0.05% NaN_3, 0.25% bovine serum albumin (BSA) and 5% sucrose) and afterwards incubated for 2h at room temperature in 0.34 µM Alexa fluor 488 phalloidin (Invitrogen, A12379) in PTA and finally mounted in Vectashield® (Vector Laboratories, Burlingame, CA) containing DAPI. For immunostaining against synapsin1, specimens were preincubated two hours in PTA and incubated for 12h at room temperature with antibodies anti synapsin1 raised in Rabbit (ENZO life Sciences, ADI-VAS-SV061-E). Then the specimens were rinsed in PBS, pre-incubated 2h in PTA and incubated 12h at room temperature with the secondary antibody anti-rabbit, conjugated with the fluorophore FITC (SIGMA, prod. num. f0382). Finally the specimens were rinsed in PBS and mounted in Vectashield®. Preparations were analyzed with an Olympus Fluoview FV1000 CLSM or a Leica TCS SP5 CLSM. The specificity of the antibodies was tested by examining specimens where each of the primary and secondary antibodies were omitted.

Image treatment
Z-stacks or parts of them of CLSM files were projected into 2D-images (MIP images = maximum intensity pixel images) and 3D iso-surface reconstructed in Imaris v7 (Bitplane AG, Zürich, Switzerland). Depth coded Z-stack images of F-actin staining are also presented (Leica imaging software), were the depth-gradient follows the area of the spectral light with the uppermost structures appearing red, and the more distant one blue. Free hand drawings and plate setups were done with Adobe Illustrator CS6 and Image modification done with Adobe Photoshop CS6.

Competing interests
The authors declare that they have no competing interests.

Authors' contributions
AH, KW, MVS, NB, RMK collected the animals. RMK made the transmission electron micrographs. AH, KW, NB stained the specimens for phalloidin and scanned specimens for CLSM. KW, MVS, NB, RMK, coordinated and participated in the analysis. KW and NB conceptualized, drafted the manuscript and designed the study. RMK, MVS, AH revised the manuscript. All authors read and approved the final manuscript.

Acknowledgements
The Arctic Station of Qeqertarsuaq, University of Copenhagen provided an excellent working platform with cooling container and we are greatly indebted to the crew of the station as well as R/V Porsild. The fieldwork on Greenland was supported by the Carlsberg Foundation (Grant no. 2009_01_0053), (Grant no. 2012_01_0123), (Grant no. 2010_01_0802) and the Villum foundation (Grant no. 102544). The lab cost and the salary of the first author was supported by the Carlsberg foundation (Grant no. 2010_01_0802) and the Villum foundation (Grant no. 102544). We especially thank Prof. Rick Hochberg for his constructive critical reading and detailed comments on the manuscript.

Author details
[1] Marine Biological Section, Department of Biology, University of Copenhagen, Universitetsparken 4, 2100 Copenhagen Ø, Denmark. [2] Natural History Museum of Denmark, Universitetsparken 15, 2100 Copenhagen Ø, Denmark. [3] Sars International Centre for Marine Molecular Biology, University of Bergen, Thormøhlensgate 55, Bergen N-5008, Norway. [4] Natural History Museum of Denmark, Øster Voldgade 5-7, 1350 Copenhagen K, Denmark.

References

1. De Smet WH: **A new record of** *Limnognathia maerski* **Kristensen & Funch, 2000 (Micrognathozoa) from the subantarctic Crozet Islands, with redescription of the trophi.** *J Zool* 2002, 258:381–393.

2. Kristensen RM: **An introduction to Loricifera, Cycliophora, and Micrognathozoa.** *Integr Comp Biol* 2002, 42:641–651.

3. Kristensen RM, Funch P: **Micrognathozoa: a new class with complicated jaws like those of Rotifera and Gnathostomulida.** *J Zool* 2000, 246:1–49.

4. Funch P, Kristensen RM: **Coda: The Micrognathozoa—a new class or phylum of freshwater meiofauna?** In *Freshwater meiofauna: Biology and ecology.* Edited by Rundle SD, Robertson AL, Schmid-Araya JM. Leiden, The Netherlands: Backhuys Publishers; 2002.

5. Giribet G, Sørensen MV, Funch P, Kristensen RM, Sterrer W: **Investigations into the phylogenetic position of Micrognathozoa using four molecular loci.** *Cladistics* 2004, 20:1–3.

6. Sørensen MV: **Further structures in the jaw apparatus of** *Limnognathia maerski* **(Micrognathozoa), with notes on the phylogeny of the gnathifera.** *J Morphol* 2003, 255:131–145.

7. Sørensen MV, Kristensen RM: **Micrognathozoa.** In *Handbook of Zoology, Gastrotricha, Cycloneuralia and Gnathifera.* Edited by Schmidt-Rhaesa A. Berlin, Boston: Walter De Gruyter GmbH; 2014. In press.

8. Lammert V: **Gnathostomulida.** In *Microscopic Anatomy of Invetebratres, Volume 4 Aschelminthes.* Edited by Harrison FW, Ruppert EE. New York, Chichester, Brisbane, Toronto, Singapore: John Wiley & Sons edition; 1991.

9. Müller MCM, Sterrer W: **Musculature and nervous system of** *Gnathostomula peregrina* **(Gnathostomulida) shown by phalloidin labeling, immunohistochemistry, and cLSM, and their phylogenetic significance.** *Zoomorphology* 2004, 123:169–177.

10. Tyler S, Hooge MD: **Musculature of** *Gnathostomula armata* **Riedl 1971 and its ecological significance.** *Mar Ecol-P S Z N I* 2001, 22:71–83.

11. Leasi F, Neves RC, Worsaae K, Sørensen MV: **Musculature of** *Seison nebaliae* **Grube, 1861 and** *Paraseison annulatus* **(Claus, 1876) revealed with CLSM: a comparative study of the gnathiferan key taxon Seisonacea (Rotifera).** *Zoomorphology* 2012, 131:185–195.

12. Leasi F, Ricci C: **Musculature of two bdelloid rotifers,** *Adineta ricciae* **and** *Macrotrachela quadricornifera*: **organization in a functional and evolutionary perspective.** *J Zoo Syst Evol Res* 2010, 48:33–39.

13. Sørensen MV: **Musculature in three species of Proales (Monogononta, Rotifera) stained with phalloidin-labeled fluorescent dye.** *Zoomorphology* 2005, 124:47–55.

14. Hochberg R, Litvaitis MK: **Functional morphology of the muscles in** *Philodina* **sp. (Rotifera : Bdelloidea).** *Hydrobiologia* 2000, 432:57–64.

15. Wilts EF, Ahlrichs WH, Arbizu PM: **The somatic musculature of** *Bryceella stylata* **(Milne, 1886) (Rotifera: Proalidae) as revealed by confocal laser scanning microscopy with additional new data on its trophi and overall morphology.** *Zool Anz* 2009, 248:161–175.

16. Hochberg R, Lilley G: **Neuromuscular organization of the freshwater colonial rotifer,** *Sinantherina socialis*, **and its implications for understanding the evolution of coloniality in Rotifera.** *Zoomorphology* 2010, 129:153–162.

17. Santo N, Fontaneto D, Fascio U, Melone G, Caprioli M: **External morphology and muscle arrangement of** *Brachionus urceolaris*, *Floscularia ringens*, *Hexarthra mira* **and** *Notommata glyphura* **(Rotifera, Monogononta).** *Hydrobiologia* 2005, 546:223–229.

18. Riemann O, Wilts EF, Ahlrichs WH, Kieneke A: **Body musculature of** *Beauchampiella eudactylota* **(Gosse, 1886) (Rotifera: Euchlanidae) with additional new data on its trophi and overall morphology.** *Acta Zool-Stockholm* 2009, 90:265–274.

19. Kotikova EA, Raikova OI, Flyatchinskaya LP, Reuter M, Gustafsson MKS: **Rotifer muscles as revealed by phalloidin-TRITC staining and confocal scanning laser microscopy.** *Acta Zool-Stockholm* 2001, 82:1–9.

20. Sørensen MV: **An SEM study of the jaws of** *Haplognathia rosea* **and** *Rastrognathia macrostoma* **(Gnathostomulida), with a preliminary comparison with the rotiferan trophi.** *Acta Zool-Stockholm* 2000, 81:9–16.

21. Sørensen MV, Sterrer W: **New characters in the gnathostomulid mouth parts revealed by scanning electron microscopy.** *J Morphol* 2002, 253:310–334.

22. Sterrer W: **Systematics and Evolution within Gnathostomulida.** *Syst Zool* 1972, 21:151–173.

23. Clément P, Wurdak E: **Rotifera.** In *Microscopic Anatomy of Invetebratres, Volume 4, Aschelminthes.* Edited by Harrison FW, Ruppert EE. New York, Chichester, Brisbane, Toronto, Singapore: John Wiley & Sons edition; 1991.

24. Sørensen MV: **On the evolution and morphology of the rotiferan trophi, with a cladistic analysis of Rotifera.** *J Zoo Syst Evol Res* 2002, 40:129–154.

25. Wallace RL, Snell TL, Nogrady T: **Rotifera: Volume 1. Biology, Ecology and Systematics.** In *Guides to the Identification of Microinvertebrates of the Continental Waters of the World.* Edited by Dumont HJ. Leiden: Kenobi Productions, Ghent and Backhyus Publishing; 2006.

26. Sørensen MV, Tyler S, Hooge MD, Funch P: **Organization of pharyngeal hard parts and musculature in** *Gnathostomula armata* **(Gnathostomulida: Gnathostomulidae).** *Can J Zool* 2003, 81:1463–1470.

27. Herlyn H, Ehlers U: **Ultrastructure and function of the pharynx of** *Gnathostomula paradoxa* **(Gnathostomulida).** *Zoomorphology* 1997, 117:135–145.

28. Leasi F, Todaro MA: **The muscular system of** *Musellifer delamarei* **(Renaud-Mornant, 1968) and other chaetonotidans with implications for the phylogeny and systematization of the Paucitubulatina (Gastrotricha).** *Biol J Linn Soc* 2008, 94:379–398.

29. Neves RC, Bailly X, Leasi F, Reichert H, Sørensen MV, Kristensen RM: **A complete three-dimensional reconstruction of the myoanatomy of Loricifera: comparative morphology of an adult and a Higgins larva stage.** *Front Zool* 2013, 10:19.

30. Worsaae K, Rouse GW: **Is** *Diurodrilus* **an Annelid?** *J Morphol* 2008, 269:1426–1455.

31. Wilts EF, Wulfken D, Ahlrichs WH: **Combining confocal laser scanning and transmission electron microscopy for revealing the mastax musculature in** *Bryceella stylata* **(Milne, 1886) (Rotifera: Monogononta).** *Zool Anz* 2010, 248:285–298.

32. Wilts EF, Wulfken D, Ahlrichs WH, Arbizu PM: **The musculature of** *Squatinella rostrum* **(Milne, 1886) (Rotifera: Lepadellidae) as revealed by confocal laser scanning microscopy with additional new data on its trophi and overall morphology.** *Acta Zool-Stockholm* 2012, 93:14–27.

33. Clément P: **Movements in rotifers: correlations of ultrastructure and behavior.** *Hydrobiologia* 1987, 147:339–359.

34. Neves RC, Cunha MR, Funch P, Kristensen RM, Wanninger A: **Comparative myoanatomy of cycliophoran life cycle stages.** *J Morphol* 2010, 271:596–611.

35. Neves RC, Kristensen RM, Wanninger A: **Three-dimensional reconstruction of the musculature of various life cycle stages of the cycliophoran** *Symbion americanus*. *J Morphol* 2009, 271:257–270.

36. Neuhaus B, Higgins RP: **Ultrastructure, biology, and phylogenetic relationships of Kinorhyncha.** *Integr Comp Biol* 2002, 42:619–632.

37. Hochberg R, Litvaitis MK: **The musculature of** *Draculiciteria tessalata* **(Chaetonotida, Paucitubulatina): implications for the evolution of dorsoventral muscles in Gastrotricha.** *Hydrobiologia* 2001, 452:155–161.

38. Sørensen MV: **Phylogeny and jaw evolution in Gnathostomulida, with a cladistic analysis of the genera.** *Zool Scr* 2002, 31:461–480.

39. Riemann O, Ahlrichs WH: **Ultrastructure and function of the mastax in** *Dicranophorus forcipatus* **(Rotifera : Monogononta).** *J Morphol* 2008, 269:698–712.

40. Wulfken D, Wilts EF, Martinez-Arbizu P, Ahlrichs WH: **Comparative analysis of the mastax musculature of the rotifer species** *Pleurotrocha petromyzon* **(Notommatidae) and** *Proales tillyensis* **(Proalidae) with notes on the virgate mastax type.** *Zool Anz* 2010, 249:181–194.

41. Wulfken D, Ahlrichs WH: **The ultrastructure of the mastax of** *Filinia longiseta* **(Flosculariaceae, Rotifera): Informational value of the trophi structure and mastax musculature.** *Zool Anz* 2012, 251:270–278.

42. Kalt MR, Tandler B: **Study of Fixation of Early Amphibian Embryos for Electron Microscopy.** *J Ultra Mol Struct R* 1971, 36:633–645.

43. Lake PS: **Trialdehyde fixation of crustacean tissue for electron microscopy.** *Crustaceana* 1973, 24:244–246.

44. Reynolds ES: **The use of lead citrate at high pH as an electron-opaque stain in electron microscopy.** *J Cell Biol* 1963, 17:208–212.

4

Sympatric prey responses to lethal top-predator control: predator manipulation experiments

Benjamin L Allen[1,2]*, Lee R Allen[2], Richard M Engeman[3] and Luke K-P Leung[1]

Abstract

Introduction: Many prey species around the world are suffering declines due to a variety of interacting causes such as land use change, climate change, invasive species and novel disease. Recent studies on the ecological roles of top-predators have suggested that lethal top-predator control by humans (typically undertaken to protect livestock or managed game from predation) is an indirect additional cause of prey declines through trophic cascade effects. Such studies have prompted calls to prohibit lethal top-predator control with the expectation that doing so will result in widespread benefits for biodiversity at all trophic levels. However, applied experiments investigating *in situ* responses of prey populations to contemporary top-predator management practices are few and none have previously been conducted on the eclectic suite of native and exotic mammalian, reptilian, avian and amphibian predator and prey taxa we simultaneously assess. We conducted a series of landscape-scale, multi-year, manipulative experiments at nine sites spanning five ecosystem types across the Australian continental rangelands to investigate the responses of sympatric prey populations to contemporary poison-baiting programs intended to control top-predators (dingoes) for livestock protection.

Results: Prey populations were almost always in similar or greater abundances in baited areas. Short-term prey responses to baiting were seldom apparent. Longer-term prey population trends fluctuated independently of baiting for every prey species at all sites, and divergence or convergence of prey population trends occurred rarely. Top-predator population trends fluctuated independently of baiting in all cases, and never did diverge or converge. Mesopredator population trends likewise fluctuated independently of baiting in almost all cases, but did diverge or converge in a few instances.

Conclusions: These results demonstrate that Australian populations of prey fauna at lower trophic levels are typically unaffected by top-predator control because top-predator populations are not substantially affected by contemporary control practices, thus averting a trophic cascade. We conclude that alteration of current top-predator management practices is probably unnecessary for enhancing fauna recovery in the Australian rangelands. More generally, our results suggest that theoretical and observational studies advancing the idea that lethal control of top-predators induces trophic cascades may not be as universal as previously supposed.

Keywords: *Canis lupus dingo*, Carnivore conservation, Fauna recovery planning, Ground-dwelling birds, Kangaroo, Poison baiting, Small mammals, Threatened species

* Correspondence: benjamin.allen@daff.qld.gov.au
[1]School of Agriculture and Food Sciences, The University of Queensland, Warrego Highway, Gatton, QLD 4343, Australia
[2]Robert Wicks Pest Animal Research Centre, Biosecurity Queensland, Tor Street, Toowoomba, QLD 4350, Australia
Full list of author information is available at the end of the article

Introduction

Many prey species around the world are threatened or suffering declines in many parts of their ranges due to a variety of interacting biotic, abiotic and anthropogenic causes such as land use change, climatic change, invasive species and novel disease [1-3]. Unbalanced ecosystems with disproportionately high densities of some fauna (e.g. herbivores and mid-sized or mesopredators) can exacerbate the rate of species declines in some cases [4,5]. Apex or top-predators such as lions (*Panthera leo*), bears (*Ursus* spp.) or grey wolves (*Canis lupus*) are expected to stabilise or recalibrate ecosystems by reducing populations of such overabundant species and allowing threatened prey at lower trophic levels to recover [6-8]. Moreover, many top-predators are themselves threatened, in decline or presently absent from large portions of their former ranges, and for this reason alone are worthy of conservation and restoration [9]. Great interest surrounds the recovery and potential use of top-predators as 'natural' and low-cost biodiversity conservation tools [10,11]. Consequently, predator management strategies known or perceived to have negative effects on top-predator populations are expected by some to produce outcomes ultimately detrimental to prey species and even vegetation communities at lower trophic levels [12,13]. Humans are not detached from these processes given their (often unacknowledged) role as the ultimate 'top-predator' or manipulator of species and ecosystems [14-18].

Though the important role that terrestrial top-predators can sometimes play in structuring food webs and ecosystems through their consumptive (e.g. predation) and non-consumptive (e.g. fear, competition) effects on sympatric mesopredator and herbivore species is well known [7-9], top-predators are often lethally controlled to protect livestock, managed game and some threatened fauna from top-predator predation (e.g. [19-22]). Lethal control of top-predators is typically achieved through trapping, shooting and/or poisoning in different parts of the world. Lethal control of rare or threatened top-predators is often unacceptable in many cases (e.g. [23]), and knowledge of or expectations about the ecological roles of top-predators is often used to justify calls to cease lethal control of these threatened top-predators (e.g. [8,24]). But not all top-predators are rare or in decline. In places where top-predator populations are very common, robust and resilient to control (such as Australia), their strategic lethal control (or periodic, temporary suppression) might facilitate profitable livestock or game production while retaining their important functional roles in limiting, suppressing or regulating over-abundant species [25].

Introduced to Australia about 5,000 years ago, dingoes (*Canis lupus dingo* and hybrids) are a relatively small (typically 12–17 kg) but now common and widespread canid top-predator, extant across ~85% of the continent [26,27]. However, some specific dingo genotypes are in decline and worthy of conservation [28-30]. Genetic issues aside, dingoes' distribution and densities are naturally increasing (back into the few remaining areas, <15% of Australia, where they were formerly exterminated) despite often being subject to periodic lethal control programs for the protection of livestock and some threatened fauna [19,27,31]. Faunal biodiversity conservation is expected by some to be compromised by lethal dingo control through its perceived indirect positive effects on mesopredators and their cascading negative effects on prey (e.g. [32-34]). Snapshot, observational, correlative or desktop studies have sometimes reported negative relationships between dingoes and mesopredators or positive relationships between dingoes and some threatened fauna (reviewed in [35,36]). In contrast, long-term and/or experimental studies on the subject have consistently reported that mesopredator populations fluctuate independent of dingoes and dingo control over time ([25,37-39]; see also [40]). Investigation of predator–prey relationships have been a pillar of ecological studies for decades [41], but applied-science studies investigating the indirect *in situ* responses of prey populations to contemporary top-predator management practices are few [12,18,22,42,43].

The prey response to top-predator control is one of the most important variables of interest where threatened prey persist and extant predators of concern can only be managed through lethal control [42]. In these situations, reliably determining causal factors for changes in prey abundance can only be achieved through carefully designed manipulative experiments conducted at spatial and temporal scales relevant to management [35,44,45]. We therefore used a series of predator manipulation experiments – those with the highest level of inference logistically achievable in open rangeland areas [35,46] – to determine (1) whether or not sympatric prey abundances were different between areas that were or were not exposed to top-predator control, (2) whether or not sympatric prey activity levels decreased immediately after top-predator control, and (3) whether or not sympatric prey abundance trends were influenced by top-predator control over time. There are six primary relationships between top-predator control and prey fauna (Figure 1). The relationships (or lack thereof) between mesopredators and dingoes or dingo control (R1, R2 and R4 in Figure 1) were previously reported in Allen et al. [25], and the present study is best understood in conjunction with those results. As an extension to that work, the primary aim of the present study was not to investigate the relationships between dingoes and prey (R5 in Figure 1). Rather, we experimentally assessed whether or not ground-dwelling

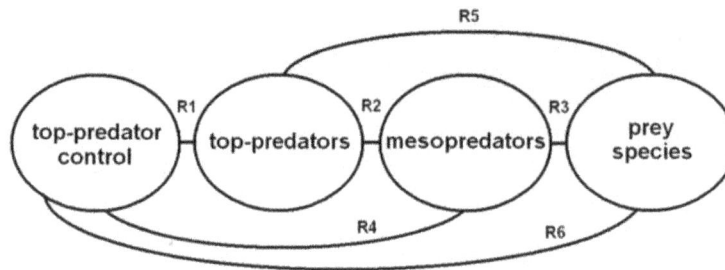

Figure 1 Schematic representation of the six primary relationships of interest (R1–R6) between top-predator control and prey species at lower trophic levels (see [42]).

mammalian, avian, reptilian and amphibian prey populations were influenced by contemporary poison-baiting programs aimed at controlling dingoes (R6 in Figure 1). Comparisons were made between a series of paired poison-baited and unbaited areas monitored over time both before and after multiple baiting events using passive tracking indices (PTI; see Methods for details of study sites and design, prey population monitoring techniques and analytical approaches). These experiments were conducted at nine sites spanning the breadth of the beef-cattle rangelands of Australia, comprising one of the largest geographic scale predator manipulation experiments conducted on any species anywhere in the world [43].

Results

Step 1 – Overall patterns in prey abundance

Linear mixed model analyses revealed a significant interaction between baiting history (i.e. consistent history, historically baited in both treatments before cessation of baiting in the unbaited area, or historically unbaited in both treatments before baiting commenced in the baited area) and treatment (baited or unbaited) for dingoes, but not for any other predator or prey species (Table 1). This is unsurprising given that baiting programs target dingoes, and as expected, baiting history and treatment were similarly significant as univariate factors influencing dingo PTI (Table 1). Overall mean and median macropod PTI was also different between different baiting histories, but not between baited and unbaited treatment areas. No other predator or prey species or species group showed an interaction, nor did any show a difference in PTI between baited and unbaited areas using this approach (Table 1). However, results from this analysis may obscure true prey responses to baiting given the unique combination of experimental design, sampling effort, land system, treatment size, baiting history, baiting context, baiting frequency, fauna assemblage, rainfall conditions and climate trend effects potentially influencing observed fauna responses to baiting at each site. Thus, results for individual 'site x species' responses to top-predator control are described

hereafter to explicitly identify any species- and site-specific responses to baiting.

Repeated measures ANOVA yielded no indication that PTI values for prey were consistently lower in areas exposed to periodic poison-baiting for dingoes (Table 2). Demonstrable differences in PTI between treatments were found in only 20 of the 67 'site x prey species' combinations with sufficient data; in only 11 of these (16% of all cases) was prey PTI lower in baited areas. These 11 cases occurred at different sites for a range of mammals and birds (Table 2). Stratifying the data by season likewise indicated no consistently lower prey PTI values in baited areas (Table 3). Demonstrable differences in PTI between treatments were found in only 29 of the 193 'site x season x prey species' combinations with sufficient data; in only 13 of these (7% of all cases) was prey PTI lower in baited areas. These 13 cases occurred in all seasons except summer, for birds, reptiles and mammals at some sites (Table 3).

Step 2 – Short-term behavioural responses of prey

A total of 25 baiting events from all sites included post-baiting surveys conducted within four months of baiting from both treatments (mean number of days since baiting = 51). Assessing changes in prey PTI between surveys conducted just prior and subsequent to baiting showed little indication of short-term responses of prey at Mt Owen (N = 8 events), Quinyambie (N = 4 events), Strathmore (N = 5 events) or Todmorden (N = 5 events) (Figure 2). An insufficient number of pre- and post-baiting pairs were available to reliably run this analysis for the other sites. Using this approach, demonstrable changes were only found for birds at Strathmore, where PTI values were lower subsequent to baiting (Figure 2).

Step 3 – Longer-term prey abundance trends

Correlations in longer-term PTI trends between baited and unbaited areas were determined for 62 possible 'site x prey species' combinations (Table 4). Of these correlations, 33 (53% of all cases) were demonstrably positive and the remainder were indistinguishable from zero; no

Table 1 P values obtained from linear mixed model analyses investigating the effects of baiting history (consistent, historically baited in both treatments before cessation of baiting in the unbaited area, or historically unbaited in both treatments before commencement of baiting in the baited area), treatment (baited or unbaited) and their interaction on overall mean (Mn) and overall median (Med) predator and prey PTI at nine sites across Australia

	Source	Baiting history (H)	Treatment (T)	T*H
Dingoes~	Mn	0.0242	0.0001	0.0004
	Med	0.0283	0.0001	0.0001
Foxes~	Mn	0.1200	0.1200	0.1400
	Med	0.1000	0.2100	0.1700
Cats~	Mn	0.9500	1.0000	0.1800
	Med	0.8000	0.6100	0.3600
Goannas~	Mn	0.8500	1.0000	0.1800
	Med	0.7400	0.6100	0.9300
Birds~	Mn	0.5200	0.4400	0.2800
	Med	0.5300	0.6700	0.1200
Rabbits~	Mn	0.1900	0.5800	0.0481
	Med	0.1800	0.7600	0.3000
Possums~	Mn	0.4600	0.3800	0.4100
	Med	0.4500	0.3600	0.4200
Small mammals~	Mn	0.3100	0.4100	0.6600
	Med	0.3900	0.3200	0.7700
Macropods~	Mn	0.0185	0.9100	0.1300
	Med	0.0336	0.7000	0.2100
Pigs~	Mn	0.5000	0.4700	0.3600
	Med	0.4400	0.4300	0.3800
Echidnas~	Mn	0.1300	0.3600	0.4200
	Med	0.4200	0.3600	0.4200
Frogs~	Mn	0.5100	0.5700	0.2700
	Med	0.4200	0.3600	0.4200
Hopping-mice	Mn*	0.4400	0.5100	0.5000
	Med*	0.4200	0.4600	0.5100
Reptiles	Mn^	0.5400	0.4700	0.2900
	Med*	0.8800	0.7000	0.1700

Note: ~df values for History = 2, Site(H) = 6, Treatment = 1, T*H = 2, T*S(H) = 6; *df values for History = 1, Site(H) = 2, Treatment = 1, T*H = 1, T*S(H) = 2; ^df values for History = 1, Site(H) = n/a, Treatment = 1, T*H = 1, T*S(H) = 4; insufficient data for koalas and toads.

demonstrably negative correlations were observed. For example, birds were positively correlated between treatments at all sites except Quinyambie and Cordillo. Small mammals were positively correlated between treatments at all sites except Strathmore. Macropods were positively correlated between treatments at Barcaldine, Blackall, Lambina, Mt Owen and Strathmore, but not Cordillo,

Quinyambie, Tambo or Todmorden. No demonstrable correlations between treatments were found for echidnas (*Tachyglossus aculeatus*) at any site. Of the 33 demonstrable and positive correlations observed, 25 (or 76% of cases) showed r values exceeding 0.75, indicating that the positive correlations observed were typically very strong (Table 4). For example, r approached 1.0 and p = <0.001 in most correlations for hopping-mice (*Notomys* spp.) and other small mammals.

Divergence or convergence of PTI trends was assessable for 65 'site x prey species' combinations. Of these, six suggested a demonstrable increase in PTI in baited areas over time, and only three (birds at Quinyambie, and possums at Mt Owen and Strathmore; <5% of all cases) suggested a demonstrable decrease in baited areas over time (Figures 3, 4, 5, 6, 7, 8, 9, 10, 11, 12, 13, 14, 15, 16, 17, 18, 19, 20). No demonstrable changes in PTI differences over time were detected for all other 'site x prey species' combinations. These data demonstrate that longer-term prey PTI trends were typically unaffected by dingo control, instead fluctuating synchronously in baited and unbaited areas over time (Figures 3, 4, 5, 6, 7, 8, 9, 10, 11, 12, 13, 14, 15, 16, 17, 18, 19, 20).

Predator responses to baiting

We previously showed that mesopredator suppression by dingoes was not apparent given that European fox (*Vulpes vulpes*), feral cat (*Felis catus*) and goanna (*Varanus* spp.) trends were not negatively correlated with dingoes over time [25]. However, correlations in longer-term predator PTI trends between baited and unbaited areas were determined here for 33 possible 'site x predator species' combinations (Table 5; see also Figure Two in [25]). Of these correlations, nine were demonstrably positive (27% of all cases) and the remainder were all indistinguishable from zero. No negative correlations between treatments were detected for any predator at any site (Table 5). The nine demonstrably positive correlations were detected for dingoes, foxes, cats and goannas at different sites (Table 5). Divergence or convergence of predator PTI trends was also assessed here for 35 'site x predator species' combinations. Of these, cat PTI apparently increased in baited areas at Tambo and Todmorden yet decreased in baited areas at Quinyambie, and fox PTI apparently decreased in baited areas at Todmorden (Figures 12, 13, 14, 15, 16, 17, 18, 19, 20). However, each of these four outcomes seem artificial given that very few cats or foxes were ever observed at these sites (Table 6 and Figure Two in [25]). Regardless, divergence or convergence of trends was not detected for dingoes (or goannas) at any site (Figures 12, 13, 14, 15, 16, 17, 18, 19, 20), demonstrating that (1) dingo PTI trends were unaffected by dingo control over time and (2) observed convergence or divergence of mesopredator PTI

Table 2 P values obtained from repeated measures ANOVA assessing differences in overall mean prey PTI (all surveys pooled) between baited and unbaited areas at nine sites across Australia (see Table 3 for seasonal breakdown)

Site (N surveys/error df)	Birds	Rabbits	Possums	Small mammals	Macropods	Pigs	Echidnas	Toads	Frogs	Hopping-mice	Reptile
Barcaldine (23/22)	0.0589*	<0.0001*	0.0184*	0.2000	0.6600	0.0081*	0.0194*	0.2100	0.1800	NP	RE
Blackall (21/20)	0.7300	0.5700	X	0.5000	<0.0001*	0.7600	0.6800	X	0.2400	NP	RE
Cordillo (7/6)	0.0186^	0.0019^	NP	0.2100	0.6800	0.2500	X	NP	0.3000	0.1100	0.1300
Lambina (6/5)#	0.2800	0.9000	NP	0.9100	0.5900	NP	X	Np	0.1600	0.6000	0.1300
Mt Owen (19/18)	0.224^	0.0002^	<0.0001^	0.0040^	0.0152*	0.5800	X	NR	NR	NP	RE
Quinyambie (14/13)	0.8100	0.0493^	NP	0.1200	0.2200	ND	0.6700	NP	0.1600	0.0025^	0.7900
Strathmore (9/8)	0.0421^	ND	0.3100	0.1400	0.0001^	0.0003^	X	NR	NR	NP	RE
Tambo (16/15)	0.0024*	0.7600	0.3000	0.5400	0.0928	0.3500	0.3300	0.2200	0.2200	NP	RE
Todmorden (11/10)	0.9100	0.0001*	NP	0.9700	0.4100	NP	X	NP	0.1100	0.2400	0.0778
Greater PTI in Baited areas	2 of 9	2 of 8	1 of 5	0 of 9	2 of 9	1 of 6	1 of 4	0 of 2	0 of 7	0 of 4	0 of 4
Greater PTI in Unbaited areas	3 of 9	3 of 8	1 of 5	1 of 9	1 of 9	1 of 6	0 of 4	0 of 2	0 of 7	1 of 4	0 of 4
Similar PTI between treatment areas	4 of 9	3 of 8	3 of 5	8 of 9	6 of 9	4 of 6	3 of 4	2 of 2	7 of 7	3 of 4	4 of 4

* = Greater in baited areas; ^ = greater PTI in unbaited areas; NP = not present; ND = known to be present but not detected on tracking plots; NR = present and detected but not recorded; RE = mostly *Varanus* spp. and reported in [25]; X = insufficient data to calculate p; #N surveys for hopping-mice at Lambina = 5, error df = 4 for hopping-mice.

trends in these four cases could not be related to changes in dingo PTI trends.

Discussion

Evidence for top-predator control-induced decline of prey fauna

Our results provide demonstrable experimental evidence that the prey populations we monitored are very rarely affected negatively by contemporary dingo control practices in the beef cattle rangelands of Australia. Baiting history was important only to macropods (Table 1). Overall mean prey PTI was seldom lower in baited areas than in paired unbaited areas (Tables 2 and 3). Short-term declines in prey PTI in baited areas (relative to unbaited areas) also occurred rarely (Figure 2). Longer-term prey PTI trends fluctuated similarly in baited and unbaited areas in each case (Table 4, Figures 3, 4, 5, 6, 7, 8, 9, 10, 11). Divergence or convergence of prey PTI trends was seldom observed (Figures 12, 13, 14, 15, 16, 17, 18, 19, 20). These non-effects of baiting were consistent across sites and site histories, environmental contexts, and across assemblages of different ground-dwelling exotic or native and small or large mammalian, avian, reptilian and amphibian prey assessed. Indeed, the few 'significant' differences observed in Steps 1, 2 or 3 of our analyses occurred infrequently and sporadically enough across sites and taxa that they may well have occurred simply by chance. If contemporary dingo control practices truly had detrimental effects on prey abundances,

through either numerical and/or functional changes in predator populations, then: (1) prey PTI should have been lower in baited areas and/or (2) should have declined immediately after baiting and/or (3) should have been negatively correlated between baited and unbaited areas and/or (4) should have shown evidence of decreasing PTI trends in baited areas over time. Rarely did any of these occur for any prey species at any site, and never did our results of Step 3 show evidence of baiting-induced PTI decline for any threatened prey species or species group, such as hopping-mice or other small mammals (Figures 3, 4, 5, 6, 7, 8, 9, 10, 11, 12, 13, 14, 15, 16, 17, 18, 19, 20).

Perhaps our best evidence of baiting-induced changes in prey populations comes from Mt Owen, where a unique combination of baiting history, baiting context, land system, mammal assemblage, rainfall and climate trend suggested that some prey species were affected by dingo control in that given context. Our experiment began at Mt Owen during a period of drought and continued through a period of repeated above-average seasonal conditions when several predator and prey species showed evidence of somewhat linear and bottom-up driven PTI increases in response to rainfall (compare Figures 7, 16 and 21 and Figure Two in [25]). This heterogeneous and structurally complex dry woodland site also supported a relatively high diversity of mammalian prey species of various sizes, including several macropod species [47-49]. The relative abundance of dingoes, cats and goannas increased in both baited and unbaited areas

Table 3 P values obtained from repeated measures ANOVA assessing seasonal differences in mean prey PTI (surveys pooled by season) between baited and unbaited areas at nine sites across Australia

Autumn (March - May)

Site (N surveys/error df)	Birds	Rabbits	Possums	Small mammals	Macropods	Pigs	Echidnas	Toads	Frogs	Hopping-mice	Reptile
Barcaldine (8/7)	0.5604	0.0032*	0.1705	0.2981	0.4162	0.3506	0.1395	0.1114	0.2718	NP	RE
Blackall (9/8)	0.8598	0.8312	1.0000	0.5553	0.0336*	0.1690	0.7287	ND	0.4354	NP	RE
Cordillo (1/0)	X	X	NP	X	X	ND	ND	ND	X	X	X
Lambina (1/0)	X	ND	NP	X	X	NP	ND	NP	X	X	X
Mt Owen (7/6)	0.0304^	0.1624	0.0030^	0.0253^	0.0272*	ND	ND	NR	NR	NP	RE
Quinyambie (4/3)	0.4972	0.0788	NP	0.2963	1.0000	ND	ND	NP	0.3910	0.0184^	0.4444
Strathmore (1/0)	X	ND	X	X	X	X	ND	NR	NR	NP	RE
Tambo (6/5)	0.0358*	0.2752	0.6109	0.8215	0.7518	0.8670	ND	0.2586	0.3632	NP	RE
Todmorden (3/2)	0.0785	0.0196*	NP	0.9531	0.6667	NP	ND	NP	0.4226	0.6892	0.0691

Winter (June - August)

Site (N surveys/error df)	Birds	Rabbits	Possums	Small mammals	Macropods	Pigs	Echidnas	Toads	Frogs	Hopping-mice	Reptile
Barcaldine (6/5)	0.1467	0.0089*	0.3144	0.9490	0.4187	0.2031	0.4206	0.3632	ND	NP	RE
Blackall (6/3)	0.4169	0.8671	X	0.3199	0.0045*	0.4918	0.8240	ND	0.3910	NP	RE
Cordillo (2/1)	0.1869	0.4097	NP	0.5529	0.1257	ND	ND	ND	ND	0.0424*	0.0424^
Lambina (3/2)#	0.2747	0.2697	NP	0.3801	0.5093	NP	ND	NP	0.4226	0.4795	1.0000
Mt Owen (5/4)	0.2927	0.0010^	0.0433^	0.4827	0.1192	0.3739	ND	NR	NR	NP	RE
Quinyambie (6/5)	0.5071	0.0983	NP	0.4130	0.3278	ND	0.3632	NP	ND	0.1059	0.2543
Strathmore (6/5)	0.0677	ND	0.2856	0.4743	0.0011^	0.0023	ND	NR	NR	NP	RE
Tambo (4/3)	0.1064	0.1737	0.3910	0.4564	0.1802	0.8240	0.3910	ND	ND	NP	RE
Todmorden (2/1)	0.7863	0.0577*	NP	0.2284	0.1257	NP	ND	NP	ND	0.2048	0.0903

Spring (September - November)

Site (N surveys/error df)	Birds	Rabbits	Possums	Small mammals	Macropods	Pigs	Echidnas	Toads	Frogs	Hopping-mice	Reptile
Barcaldine (6/5)	0.7361	0.0050*	0.1019	0.2557	1.0000	0.0583*	0.0041	ND	0.3632	NP	RE
Blackall (6/5)	0.8789	0.0578*	ND	0.5583	0.0116*	ND	0.8717	ND	0.3632	NP	RE
Cordillo (3/2)	0.1428	0.0533^	NP	0.3036	0.0198*	0.4226	ND	ND	0.4226	0.9387	0.4400
Lambina (0/0)	NS	NS	NP	NS	NS	NP	NS	NP	NS	NS	NS
Mt Owen (6/5)	0.1927	0.0405^	0.0008^	0.0988	0.5910	1.0000	ND	NR	NR	NP	RE
Quinyambie (3/2)	0.6734	0.9830	NP	0.0991	0.2254	ND	0.4226	NP	0.4226	0.1061	0.3399
Strathmore (2/1)	0.2952	ND	1.000	0.5000	0.3017	0.3529	ND	NR	NR	NP	RE
Tambo (4/3)	0.0904	0.0021^	0.6042	0.0663	0.5720	0.1411	1.0000	ND	ND	NP	RE
Todmorden (2/1)	0.0997	0.3375	NP	0.8688	X	NP	ND	NP	0.5000	0.2857	0.1526

Summer (December - February)

Site (N surveys/error df)	Birds	Rabbits	Possums	Small mammals	Macropods	Pigs	Echidnas	Toads	Frogs	Hopping-mice	Reptile
Barcaldine (3/2)	0.1493	0.3249	0.2254	0.4045	0.3101	0.2254	1.0000	0.4226	ND	NP	RE
Blackall (2/1)	0.5303	0.9097	ND	0.1131	0.2389	0.5000	ND	ND	0.5000	NP	RE
Cordillo (1/0)	X	X	NP	X	X	X	ND	ND	X	X	X
Lambina (2/1)	0.3712	0.6051	NP	0.7018	0.5000	NP	ND	NP	0.5000	0.0622	0.3375
Mt Owen (1/0)	X	X	X	X	X	ND	ND	NR	NR	NP	RE
Quinyambie (1/0)	X	X	NP	X	ND	ND	ND	NP	ND	X	X
Strathmore (0/0)	NS	NS	NS	NS	NS	NS	NS	NS	NS	NP	NS

Table 3 P values obtained from repeated measures ANOVA assessing seasonal differences in mean prey PTI (surveys pooled by season) between baited and unbaited areas at nine sites across Australia *(Continued)*

| Tambo (2/1) | 0.5718 | 0.5000 | ND | 0.5000 | 0.1772 | 0.5000 | 0.5000 | ND | 0.5000 | NP | RE |
| Todmorden (4/3) | 0.8290 | 0.0632 | NP | 0.3193 | 0.1027 | NP | ND | NP | 0.3910 | 0.8470 | 0.0021* |

NP = not present; ND = known to be present but not detected on tracking plots; NR = present and detected on tracking plots but not recorded; RE = Mostly *Varanus* spp. and reported in [25]; X = insufficient data; *All reptiles except for *Varanus* spp. (i.e. predominately agamidae and skincidae); ^All dasyurids and rodents except for hopping-mice.

over the course of the study there (Table 5, and Figure Two in [25]), and were numerically unaffected by baiting (Figures 7 and 16). In this context, however, macropods and rabbits (*Oryctolagus cuniculus*) increased in the baited area where possums (*Trichosurus vulpecula*) decreased (Figures 7 and 16); all other taxa showed no evidence of baiting-induced changes in PTI trends. Possums (53% occurrence), macropods (29% occurrence) and rabbits (7% occurrence) were the three most frequently occurring prey species in dingo diets at Mt Owen, where dingoes switched seasonally between macropods and possums [47,48]. These observations suggest that baiting-induced changes to dingo populations can occur in some contexts, whereby large macropod prey can become unavailable (or uncatchable) to socially-fractured dingo populations exposed to baiting, which then must switch to alternative prey more easily captured [47,50]. In this case, dingoes exposed to baiting appeared to suppress populations of common possums, but not any other more threatened small mammal species. The historical decline of possums in the Australian rangelands has previously been attributed to dingo predation [51-53]. Whether or not these baiting-induced prey responses are sustained subsequent to a change in the ecological context is unknown, but unlikely,

given that baiting-induced functional changes in dingo movement behaviour [54] or prey selection (B. Allen, unpublished data from [55,56]) did not occur at several other sites where these processes were investigated (Table 4, Figures 3, 4, 5, 6, 8, 9, 10, 11, 12, 13, 14, 15 and 17, 18, 19, 20). These variable results suggest that the few numerical changes we observed in some of the preferred dingo prey species at some sites may be related to context-specific functional changes to dingo populations subjected to baiting, which might sometimes occur.

Although patterns in prey PTI were typically unaffected by dingo control, it is possible that prey behaviour might have been altered – perhaps negatively – through changes to the landscape of fear [12,57,58]. In other words, baiting-induced changes to dingo function (if or when it occurs) might allow mesopredators to forage more freely and then increase predation pressure on prey, negatively affecting prey behaviour and fitness [32,33,59]. Changes to the landscape of fear might occur independently of numerical trends in predator populations. Step 2 of our analyses provided the greatest opportunity to assess the behavioural responses of prey to predator control, yet short-term changes in prey PTI were not apparent in most

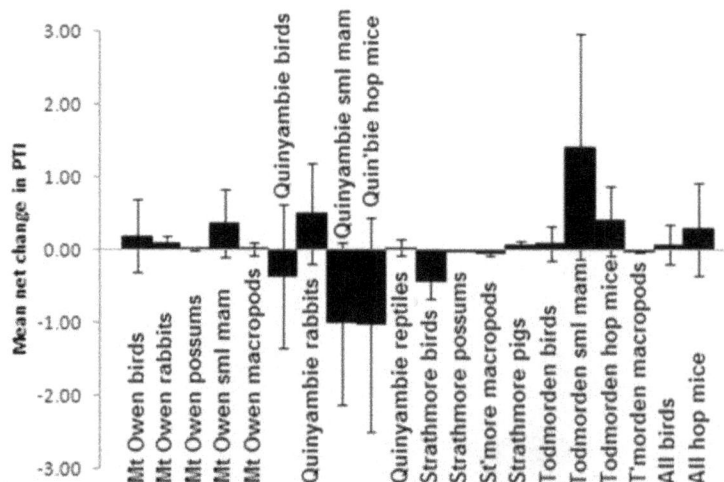

Figure 2 Mean net changes in prey PTI (and 95% confidence intervals) between pre- and post-baiting surveys (conducted within four months of baiting) at Mt Owen (N = 8), Quinyambie (N = 4), Strathmore (N = 5), Todmorden (N = 5) and all sites combined (N = 25), showing little evidence of short-term decreases in prey PTI following dingo control.

Table 4 Correlations (r) and p values (p) for relationships in longer-term prey PTI trends between baited and unbaited areas at nine sites across Australia

Site		Birds	Rabbits	Possums	Small mammals^	Macropods	Pigs	Echidnas	Toads	Frogs	Hopping-mice	Reptiles*
Barcaldine	r	0.7770	0.8134	0.0461	0.8427	0.8297	0.1693	−0.0767	−0.0878	0.6114	NP	RE
	p	<0.0000	<0.0000	0.8346	<0.0000	<0.0000	0.4399	0.7280	0.6905	0.0019		
Blackall	r	0.5024	0.1790	−0.0500	0.5191	0.7382	−0.1062	0.3442	ND	−0.0670	NP	RE
	p	0.0203	0.4376	0.8296	0.0159	0.0001	0.6469	0.1266		0.7730		
Cordillo	r	0.4135	0.4077	NP	0.9100	0.3372	X	ND	NP	X	0.0893	0.3938
	p	0.3564	0.3640		0.0044	0.4595					0.8489	0.3755
Lambina	r	0.8835	0.7018	NP	0.9568	0.8849	NP	ND	NP	0.9931	0.9180	0.9618
	p	0.0196	0.1201		0.0028	0.0191				0.0001	0.0278	0.0022
Mt Owen	r	0.8323	0.0330	0.6828	0.9017	0.7906	−0.0808	X	NR	NR	NP	RE
	p	<0.0000	0.8932	0.0013	<0.0000	0.0001	0.7421					
Quinyambie	r	0.4265	0.8052	NP	0.9834	−0.1040	ND	−0.0769	NP	X	0.9557	0.7755
	p	0.1283	0.0005		<0.0000	0.7235		0.7938			<0.0000	0.0011
Strathmore	r	0.8764	ND	0.5093	0.2370	0.7563	0.8764	X	NR	NR	NP	RE
	p	0.0019		0.1613	0.5392	0.0184	0.0019					
Tambo	r	0.6343	0.4928	0.1108	0.8318	0.4171	−0.1936	−1215	X	0.9932	NP	RE
	p	0.0083	0.0524	0.6829	0.0001	0.1080	0.4726	0.6539	X	<0.0000		
Todmorden	r	0.7226	−0.0273	NP	0.8860	0.8865	NP	ND	NP	X	0.7589	0.1210
	p	0.0120	0.9365		0.0003	0.8865					0.0068	0.7231

NP = not present; X = insufficient data; Goannas represent only *Varanus* spp. at Cordillo, Lambina, Quinyambie and Todmorden, but include a small proportion of other reptiles at the other sites (see also [25]).

cases (Figure 2). Comprehensive reviews of the short-term effects of dingo control on prey concur with our results to show that populations of non-target prey are not negatively affected by dingo control [60,61]. Specifically investigating the behavioural responses of prey to dingo control, Fenner et al. [62] likewise found no change in small mammal prey behaviour following baiting. The predator manipulation experiments conducted by Eldridge et al. [39] similarly show prey populations (such as birds and reptiles) to fluctuate independent of dingo control. Modelling the outcomes of dingo reintroduction and cessation of fox control on prey fauna in forested temperate areas by Dexter et al. [63] also suggests small mammal populations fluctuate largely independent of dingoes. Whereas, the predator exclosure experiments of Kennedy et al. [64] suggest that some small mammal prey of dingoes benefit from dingo exclusion, as predicted by Allen and Leung [55]. The predator exclosure experiment of Moseby et al. [65] showed that some rodents benefited from the removal of rabbits, dingoes and other predators, whereas reptiles, dasyurids and other rodents were largely unaffected by their exclusion. If baiting-induced behaviourally-mediated trophic cascades were occurring at our sites, such changes were not manifest as numerical effects on longer-term prey abundance trends in most cases (Tables 1, 2 and 3, Figures 3, 4, 5, 6, 7, 8, 9, 10, 11).

These results contradict perceptions (reviewed in [35,36]) that (1) prey population abundances are lower in baited areas, (2) prey activity is suppressed shortly after baiting, (3) commencement of baiting produces declines in prey abundances, and (4) cessation of baiting increases prey abundances. Long-term (10–28 years) correlative studies of dingoes, mesopredators and their prey concur with these experimental results (e.g. [37,38]), and "almost all available studies reporting dissimilar results are based on demonstrably confounded predator population sampling methods and/or low-inferential value study designs that simply do not have the capacity to provide reliable evidence for dingo control-induced mesopredator release" ([40], pg. 4). Thus, not only is there a clear absence of reliable evidence for dingo control-induced trophic cascades (e.g. [46,66]), but there is also a strong and growing body of demonstrable experimental evidence that prey populations are usually affected positively (not negatively) by dingo control if prey are affected at all (e.g. [55,59], this study).

Trophic cascade and mesopredator release theory and reality

Trophic cascade and mesopredator release theory predicts that declines of top-predators produce increases of mesopredators and larger herbivores, which then

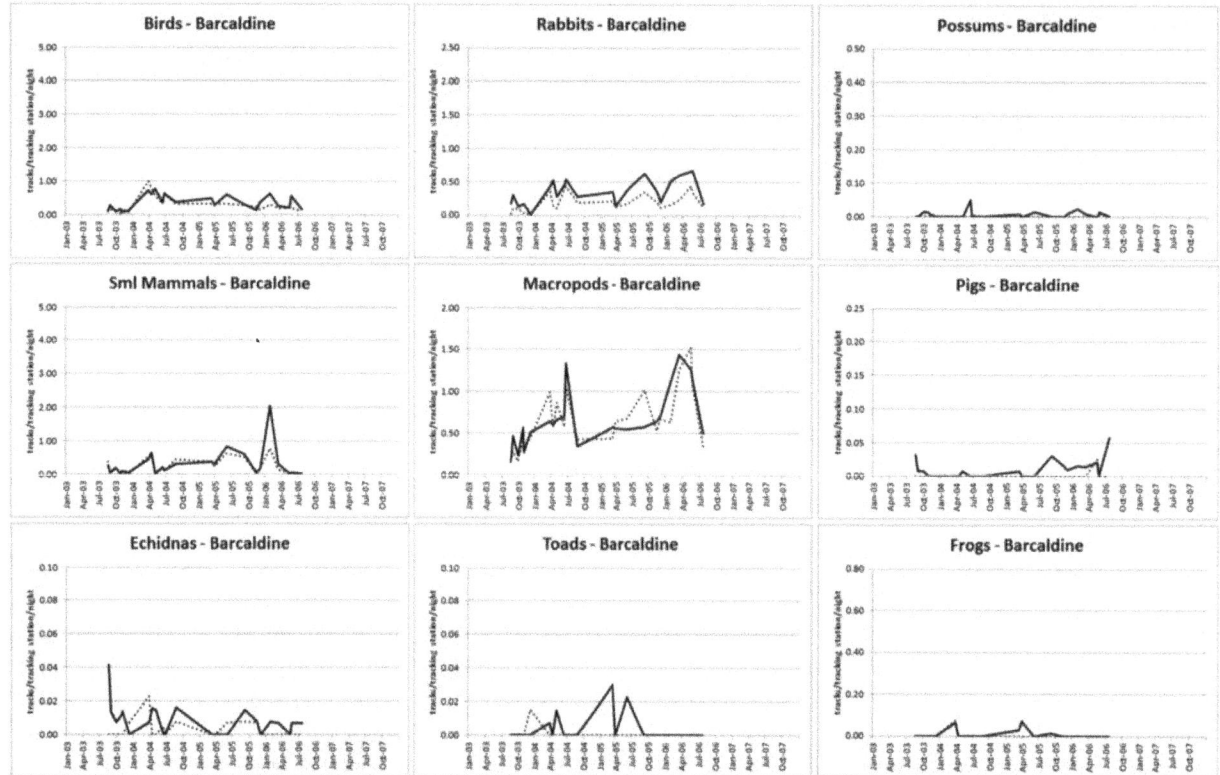

Figure 3 Longer-term prey PTI trends in baited (solid lines) and unbaited (dotted lines) areas at Barcaldine (see Table 4 for associated r and p values).

produce declines in smaller prey, which are often threat-ened [5,7]. The theory appears to work best in reality when food webs are simpler and less complex [67,68]. Deriving their conclusions from desktop studies, snap-shot field studies and/or those conducted on fauna on other continents, some have predicted that dingo control will release feral pigs (*Sus scrofa*), macropods and rabbits from dingo suppression, which will then simultaneously reduce the abundances of hopping-mice and other small mammals, birds and other fauna [13,36,69]. Thus, top-predator management programs that kill, remove or alter the function of top-predators might conceivably produce indirect declines of threatened fauna [8,34]. Despite the potential for substantial and direct negative effects of dingoes on the same threatened fauna through predation [26,51,55,70], such predictions have led some to advocate cessation of dingo control programs with the expectation that doing so will provide widespread net benefits to threatened fauna at lower trophic levels (e.g. [13,32,71]). However, our simultaneous assessment of the effects of dingo control on predator and prey pop-ulations demonstrated that the predicted mesopredator increases do not occur (Table 5 and Figures 12, 13, 14, 15, 16, 17, 18, 19, 20; see also Figure Two in [25])

because contemporary dingo control practices "do not appear to suppress dingo populations to levels low enough and long enough for mesopredators to exploit the situation" ([25], pg. 11). Hence, the consistent ab-sences of negative prey responses to dingo control we found (Tables 1, 2 and 3, Figures 2, 3, 4, 5, 6, 7, 8, 9, 10, 11, 12, 13, 14, 15, 16, 17, 18, 19, 20) should be entirely expected given that the prerequisite first step or trigger for the predicted trophic cascade (i.e. dingo decline) did not occur (Figures 12, 13, 14, 15, 16, 17, 18, 19, 20). What-ever the relationships between dingoes and mesopredators or prey are, they do not appear to be affected by contem-porary dingo control practices to any substantial degree. Alternative dingo control strategies which actually achieve sustained reduction of dingo abundances and/or alteration of dingo function may produce different results that might lend support to popular predictions of dingo control-induced trophic cascades or mesopredator release [25].

Although the occurrence of trophic cascades is well demonstrated [7,8], whether or not they are caused by top-predators or top-predator control is far less certain [46,72,73] and undoubtedly context-specific. For example, results of studies conducted on big cats, bears or wolves in temperate mountainous areas with diverse mammal

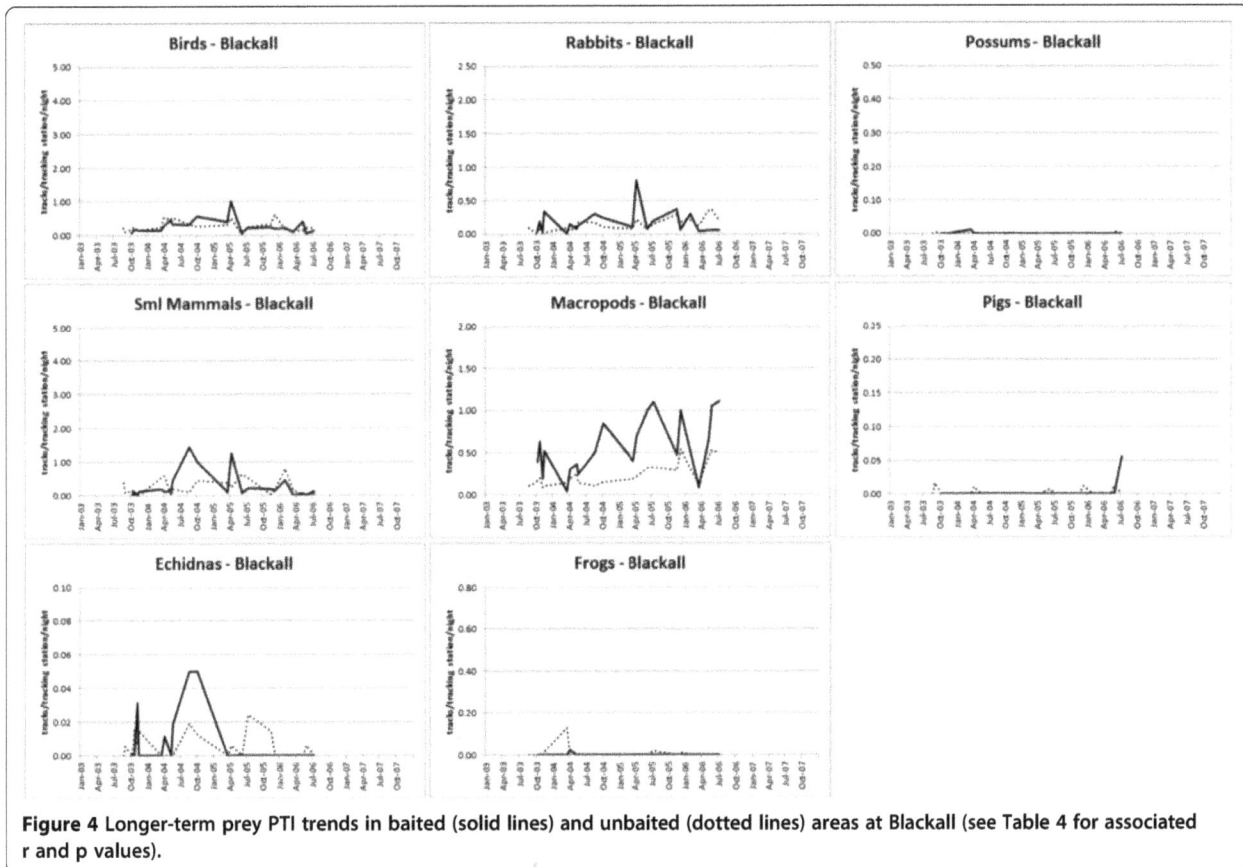

Figure 4 Longer-term prey PTI trends in baited (solid lines) and unbaited (dotted lines) areas at Blackall (see Table 4 for associated r and p values).

assemblages largely untouched by humans are not easily transferable to other predators occupying the severely human-altered areas that dominate the earth's surface [18], such as dingoes and the relatively depauperate mammal assemblages in the beef cattle rangelands of Australia [19]. Moreover, by undertaking applied-science experiments which circumvent investigations of the internal processes at play and instead focus on the actual *in situ* prey responses to top-predator control (R6 in Figure 1) – what Kinnear et al. [74] label the 'black box' approach – our results confirm that prey populations are typically unaffected by contemporary dingo control practices independent of how predators and prey might interact with each other (R2 in Figure 1).

Our findings are in accord with what is known from other predator manipulation experiments worldwide. Fauna at lower trophic levels are unlikely to respond positively to lethal control where (1) multiple predators are removed (i.e. dingoes and foxes are both susceptible to and targeted by baiting [25]) (2) the efficacy of predator removal is low (i.e. where predator populations are resilient to lethal control over time, as in Table 5 or Figures 12, 13, 14, 15, 16, 17, 18, 19, 20; see also [30] or Figure Two in [25]), and/or where (3) the fauna are not the primary

prey species of the predator ([43]; but see [48,55] for information on dingo diets at our sites). Though small and medium-sized mammals (such as rodents, possums and rabbits) are preferred prey for dingoes and other mesopredators alike when available [26,55], fluctuations in the availability of a variety of prey species typically mean that suppression of a given prey species is often only temporary [41,75,76]. Besides targeting multiple predators and producing no lasting changes in predator PTI trends in our experiments (Table 5, Figures 12, 13, 14, 15, 16, 17, 18, 19, 20), the flexible and generalist nature of dingoes', foxes', cats' and goannas' prey preferences may be another reason why we did not detect changes in prey PTI trends following predator baiting.

Factors affecting prey responses to predator control

A large array of factors can influence the outcomes of predator control on prey populations [41,44,77]. The number and type of intraguild predators present, the variety and abundance of available prey, the environmental context in which predator control is undertaken, the responses and resilience of predators to that control, the dietary preferences and habits of predators, and the

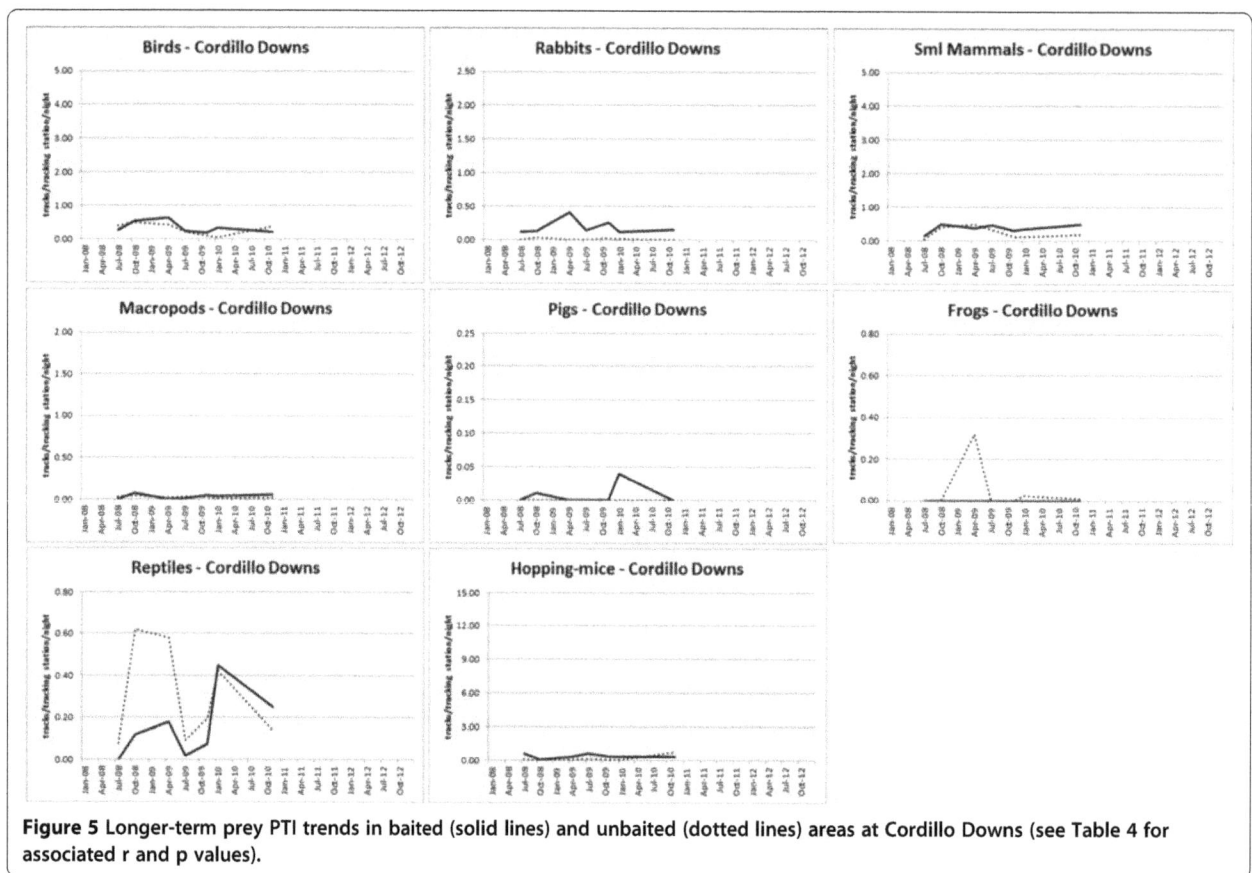

Figure 5 Longer-term prey PTI trends in baited (solid lines) and unbaited (dotted lines) areas at Cordillo Downs (see Table 4 for associated r and p values).

resilience of prey to changes in predator numbers or behaviour are each important factors influencing the responses of prey to predator control [41-43]. Study design and analytical approach also influences the observed outcomes given that 'what you see depends on how you look' (e.g. [40,44,46,56]). Changes in prey abundances following predator control might only be expected where or when predation is actually the limiting factor for prey [51,69]. Many of Australia's threatened fauna are affected to a greater degree by much more than just predator effects [3,70,78-80], suggesting that alteration of predator communities or predator control strategies – in isolation of other, more important drivers of prey decline – might not be universally expected to enhance prey recovery [14,51].

The timeframe over which prey are monitored may also influence the observed prey responses to predator baiting. Snap-shot studies with a single observation conducted at only T_0 (e.g. [81-84]) obviously have no capacity whatsoever to measure a spatial or temporal 'change', 'shift' or 'response' to dingo control [44,45], which is why we conducted multiple surveys over multiple successive years at each site (T_0, T_1, T_2... up to T_{23}; Table 2). Despite conducting our experiments over these timeframes, similar to

most other predator manipulation experiments [43], it is possible that 2–5 years of repeated prey surveys might not be long enough to detect changes in prey abundances following predator removal [65]. However, three lines of evidence suggest that this is not the case for our data.

First, the PTI methodology we applied was sufficient to detect the immediate and longer-term responses of prey (and predators) to the bottom-up effects of rainfall within the timeframe covered (Figures 3, 4, 5, 6, 7, 8, 9, 10, 11, 12, 13, 14, 15, 16, 17, 18, 19, 20, 21; see also [47,85]). Some claim that the top-down effects of dingo control can be stronger than the bottom-up effects of rainfall in the systems we studied [13], so such predicted negative responses of prey to baiting should have been observable. Second, prey responses to dingo control are almost always investigated using observational snap-shot studies or correlative studies of <12 mo duration [46,86], implying that 2–5 years of repeated baiting and population monitoring across spatial scales several orders of magnitude larger should have readily detected both acute and chronic prey declines. Third, prey population trends fluctuated similarly between treatments at Barcaldine, Blackall and Tambo (Table 4, Figures 3, 4, 5, 6, 7, 8, 9, 10, 11, 12, 13, 14, 15, 16, 17, 18, 19, 20) where

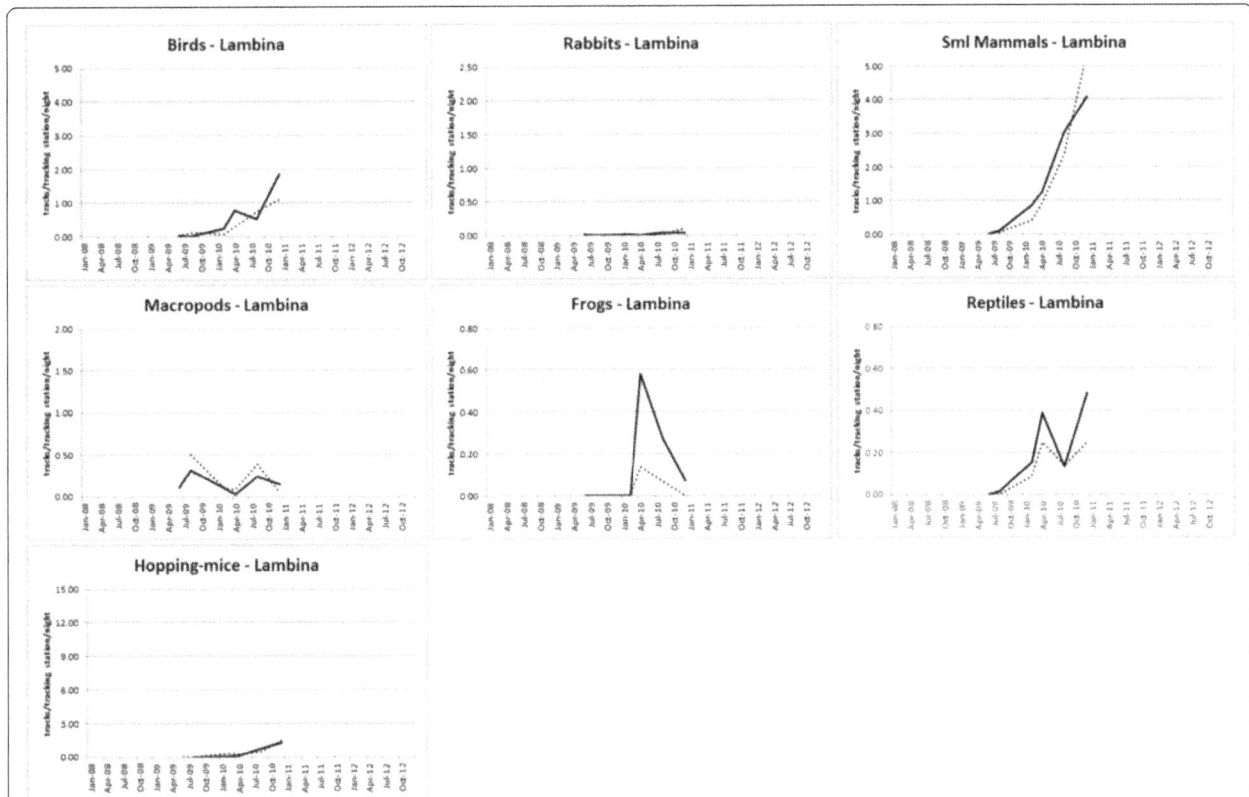

Figure 6 Longer-term prey PTI trends in baited (solid lines) and unbaited (dotted lines) areas at Lambina (see Table 4 for associated r and p values).

baiting had occurred multiple times each year for over 10 years [25]. Viewed in isolation, this latter result might be interpreted to suggest that prey had already declined in baited areas and was now being held below their carrying capacity; however, shorter-term declines were not apparent (Table 4, Figures 3, 4, 5, 6, 7, 8, 9, 10, 11), baiting history was not important to most species (Table 1), overall mean prey PTI was not lower in baited areas for any prey at any of the three Blackall sites (Table 2), and nor were predator PTI trends altered by baiting at these sites

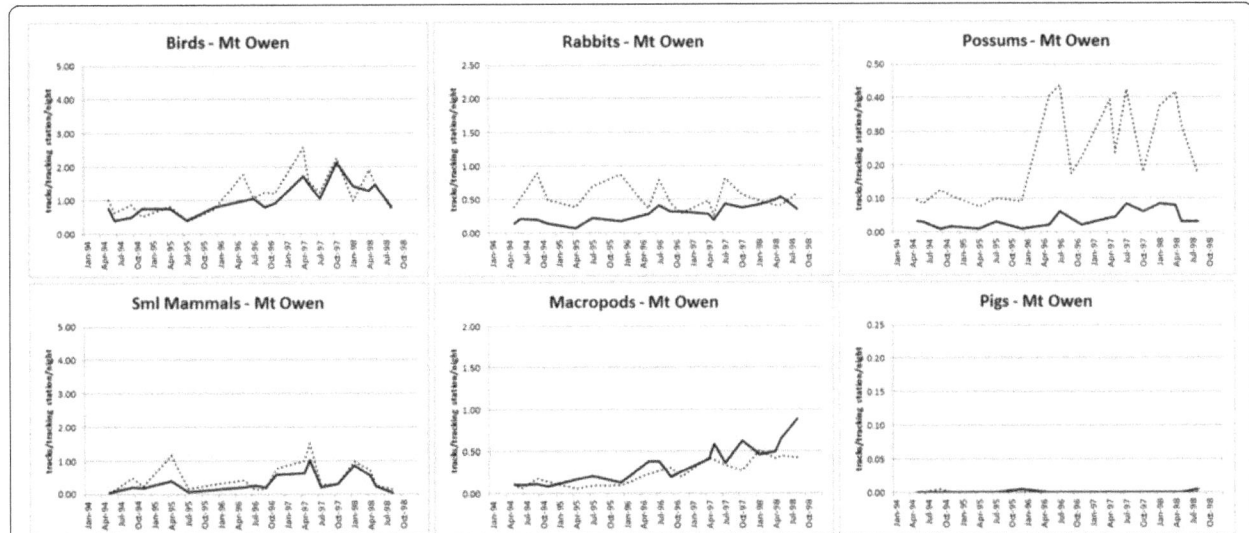

Figure 7 Longer-term prey PTI trends in baited (solid lines) and unbaited (dotted lines) areas at Mt Owen (see Table 4 for associated r and p values).

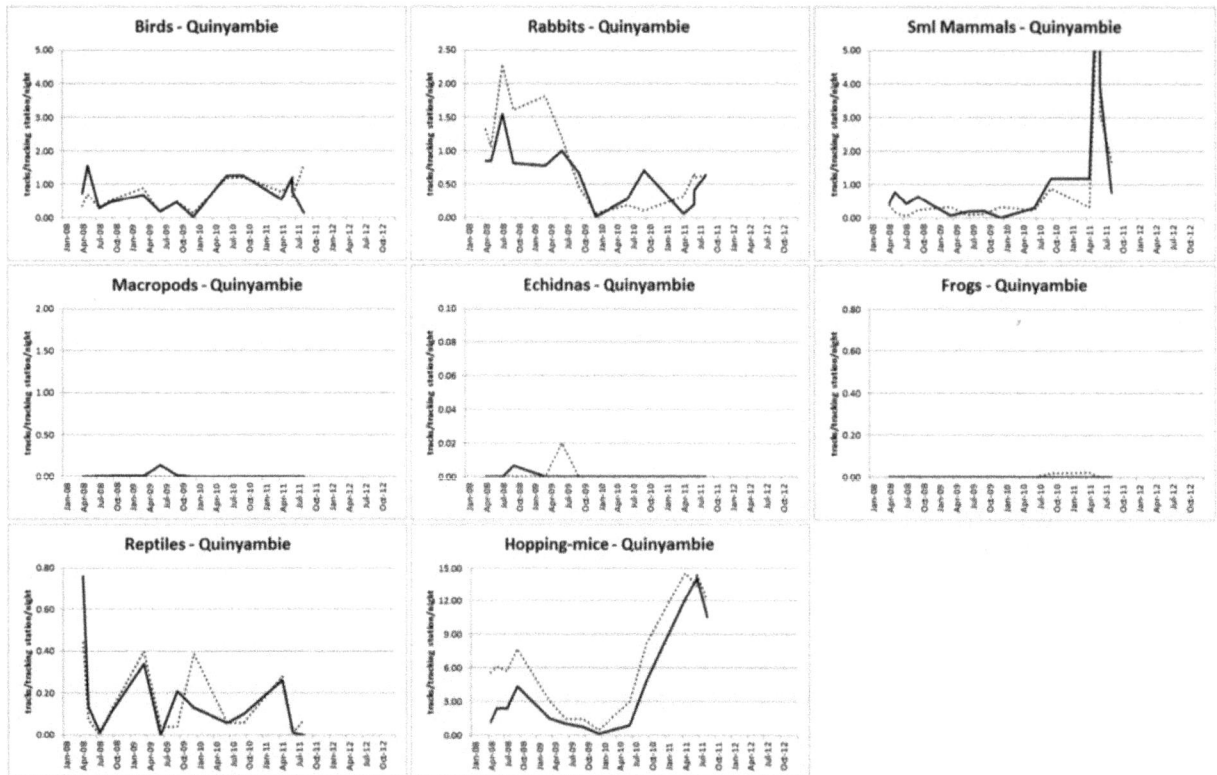

Figure 8 Longer-term prey PTI trends in baited (solid lines) and unbaited (dotted lines) areas at Quinyambie (see Table 4 for associated r and p values).

either (Table 5, Figures 12, 13, 14, 15, 16, 17, 18, 19, 20, and Figure Two in [25]). These lines of evidence indicate that our experimental design was sufficient for detecting baiting-induced changes in prey PTI if they were occurring [44].

The utility of our fauna sampling method (i.e. road-based sand plots) is also likely to vary between species and species groups [87,88]. This may be one reason why the number of tracks observed, and hence PTI values, for some species were low on occasion (Table 6; see also

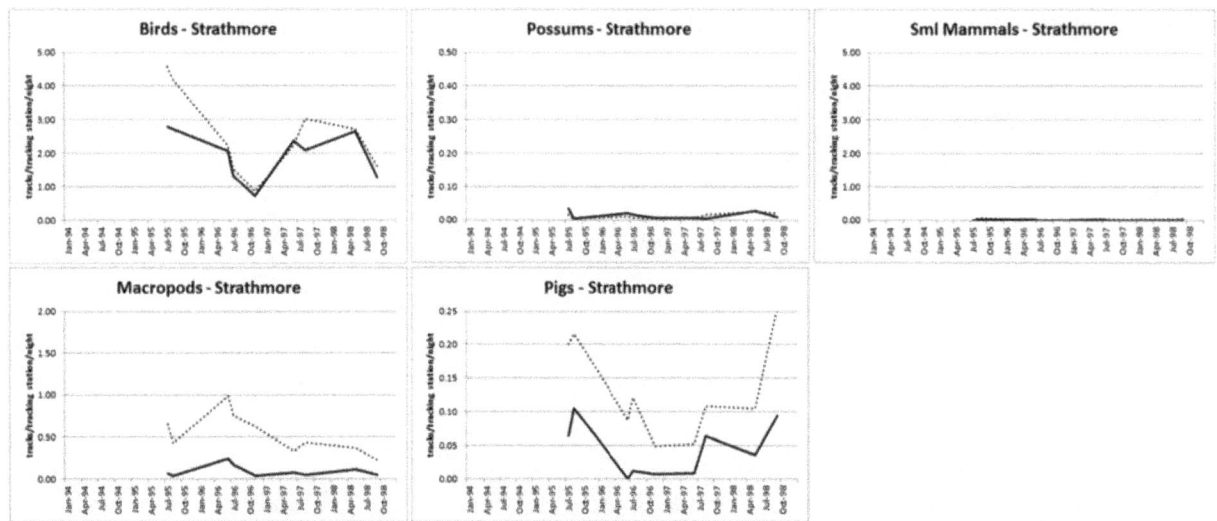

Figure 9 Longer-term prey PTI trends in baited (solid lines) and unbaited (dotted lines) areas at Strathmore (see Table 4 for associated r and p values).

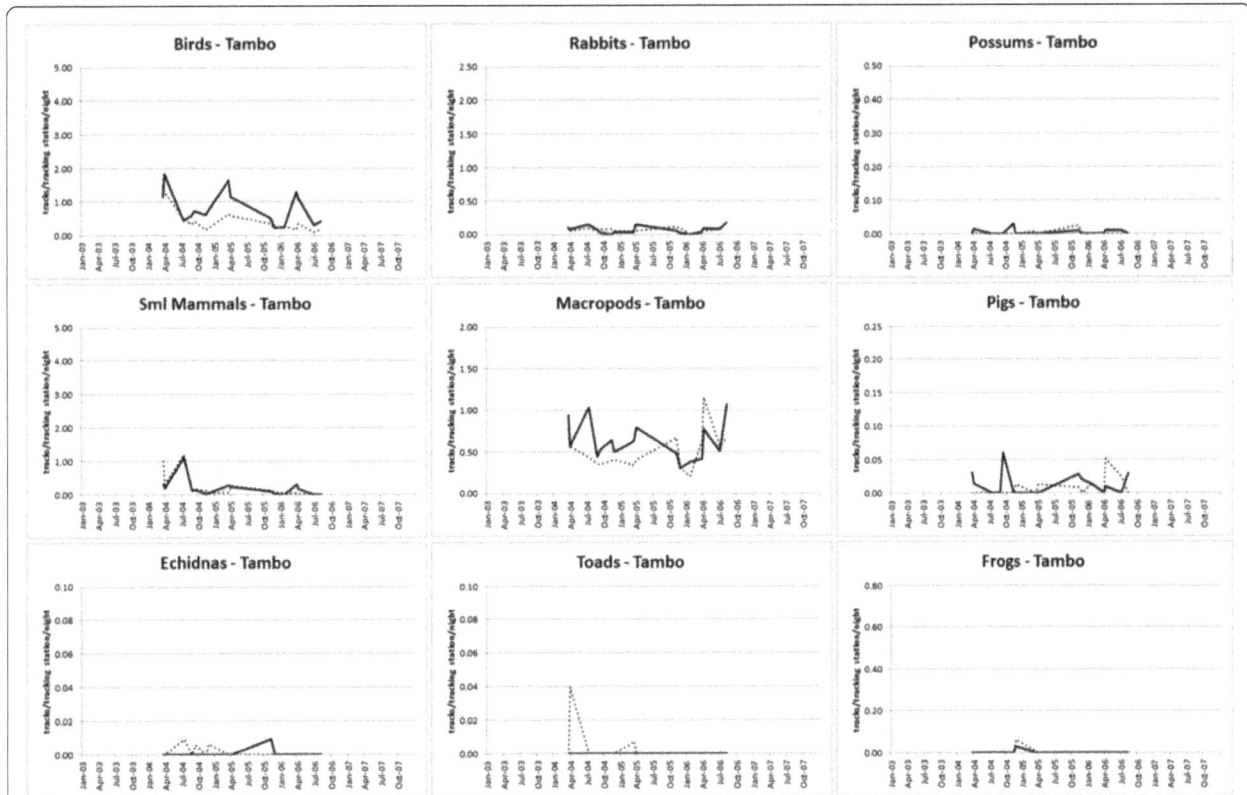

Figure 10 Longer-term prey PTI trends in baited (solid lines) and unbaited (dotted lines) areas at Tambo (see Table 4 for associated r and p values).

Table Six in [25]) and why their analyses yielded no significant responses to baiting (Tables 1, 2, 3, 4 and 5). However, the variable utility of the technique for different species is also of little consequence to our overall conclusions. Far from being a weakness of our study, the observance of few footprints for some species at times (confirming their presence at the study sites) is itself a key result supporting our conclusions given that the number of observations (or PTI values) did not change substantially over time in response to dingo baiting (Figures 3, 4, 5, 6, 7, 8, 9, 10, 11, 12, 13, 14, 15, 16, 17, 18, 19, 20; see also [89]). Our expectation was that if baiting-induced mesopredator releases or prey suppression was occurring (see *Predicted outcomes*, below), then these responses should have been detectable on the 92–166 sand plots interspersed throughout the treatment areas on dirt roads at each site (Note: 'roads' here are simply the two 4WD vehicle wheel tracks that wind throughout the study sites with negligible disturbance or alteration to the extant habitat). To argue that our sampling methodology was unable to detect changes in fauna PTI is to imply that mesopredator releases or prey suppression was occurring elsewhere, or that observed

predator–prey interactions were somehow different on and off the road. This is unlikely given that the activity of almost all the prey species we monitored (e.g. macropods, rabbits, small mammals, birds, reptiles etc.) occurs randomly with respect to roads at our sites, unlike the mammalian predators whose behaviour is influenced by roads [90,91]. Moreover, supplementary studies indicated that baiting did not affect dingo movement behaviour, which was similar both on and off the road just prior and subsequent to baiting [54]. Sampling fauna populations by placing tracking plots on roads is by no means 'insensitive' or of little value just because the approach may produce lower PTI values for some species relative to other tracking plot placements or sampling approaches, providing fauna populations are not below the level of detection by the method. Although some species (e.g. cats, small mammals or reptiles) may have persisted below the level of detectability on roads under certain conditions (e.g. during drought for small mammals or during winter for reptiles), such species were readily detected again on roads when these conditions changed (compare Figures 3, 4, 5, 6, 7, 8, 9, 10, 11 and 21; see also [47]). Thus, Step 3 in our analytical approach (Figures 3, 4, 5, 6, 7, 8, 9, 10, 11, 12,

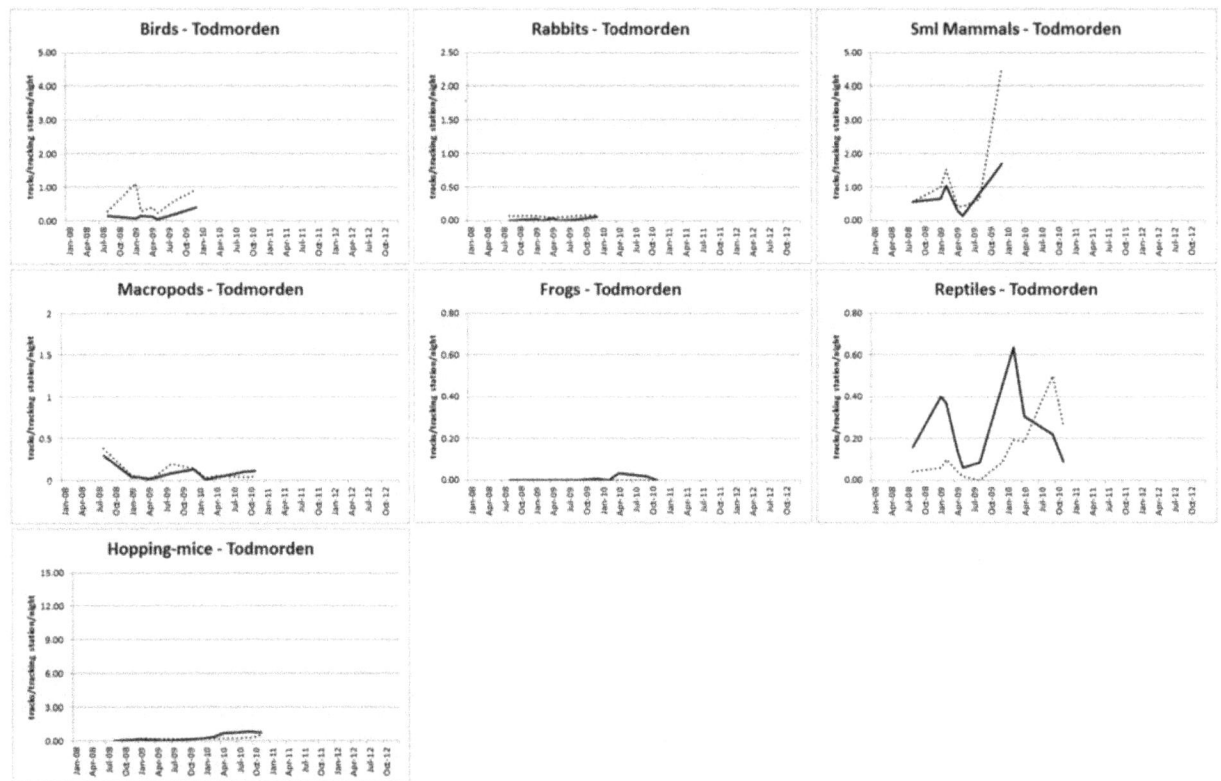

Figure 11 Longer-term prey PTI trends in baited (solid lines) and unbaited (dotted lines) areas at Todmorden (see Table 4 for associated r and p values).

13, 14, 15, 16, 17, 18, 19, 20) should have detected predator and prey PTI responses to baiting if they were occurring, regardless of the variable utility of road-based tracking plots for different species.

Although we undertook our study in an experimental framework inclusive of buffer zones to maximise treatment independence, it is also important to remember that our approach was an evaluation of the overall population-level responses of prey to contemporary top-predator control practices under real-world environmental conditions where predators and prey were each capable of dispersal and migration between treatments over time. In other words, we sought not to compare nil-treatment areas to paired treated areas with 'X% reduction of predators' or 'X density of baits', but with 'contemporary dingo control practices'. This applied-science focus therefore produces results that reflect the *in situ* outcomes of contemporary dingo control practices in the beef-cattle rangelands present across much of the Australian continent. Dingo control strategies that actually achieve complete and sustained dingo removal from the landscape (such as those that include exclusion fencing and eradication) may produce different results, though

such strategies are unlikely to ever occur in the >5.5 million km^2 (or ~75%) of Australia where sheep (*Ovis aries*) and goats (*Capra hircus*) are not commercially farmed [27].

Conclusions and implications

Our results add to the growing body of experimental evidence that prey populations in rangeland Australia are not negatively affected by contemporary dingo control practices through trophic cascade effects. These findings broaden our understanding of the potential outcomes of predator control on prey fauna at lower trophic levels and have important implications for the management of dingoes and threatened fauna. Given the ineffectiveness of contemporary baiting practices at sustainably reducing dingo populations, it might be concluded that dingo control is a pointless waste of time, money and dingoes, which may even be counterproductive to cattle producers at times [50,92]. Importantly however, dingo control is typically undertaken to reduce or avert damage to livestock by dingoes, not to reduce dingo densities per se, and the relationship between dingo density and damage is not well understood

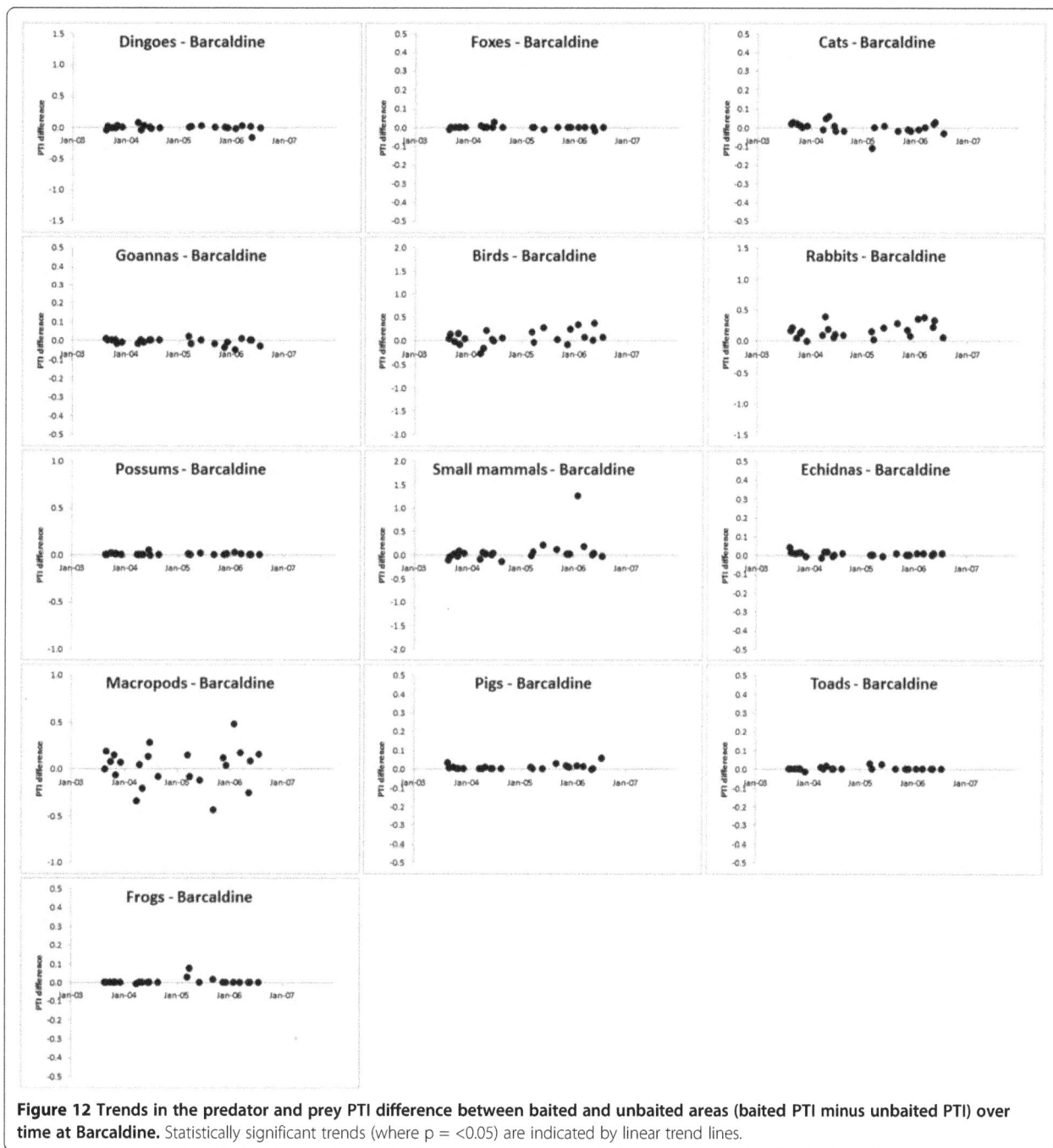

Figure 12 Trends in the predator and prey PTI difference between baited and unbaited areas (baited PTI minus unbaited PTI) over time at Barcaldine. Statistically significant trends (where p = <0.05) are indicated by linear trend lines.

[31,93]. Hence, the 'effectiveness' of dingo control should ultimately be measured in terms of 'damage reduced' or 'losses averted', not in terms of '% reduction in dingo PTI', '% dingoes destroyed', '% people participating in dingo control', or the '% of land area exposed to control' [44,94]. Greater emphasis on measurable damage reduction and/ or mitigation appears warranted in order to ethically justify continued dingo control programs.

Some have also theorised that simply ceasing lethal dingo control is a 'free' or cost-effective strategy able to increase the abundances of threatened prey fauna populations of conservation concern through trophic cascade effects (e.g. [13,32,34,95,96]), but our results demonstrate that such actions do not produce such outcomes. Fauna recovery programs should more carefully consider the factors limiting threatened prey populations

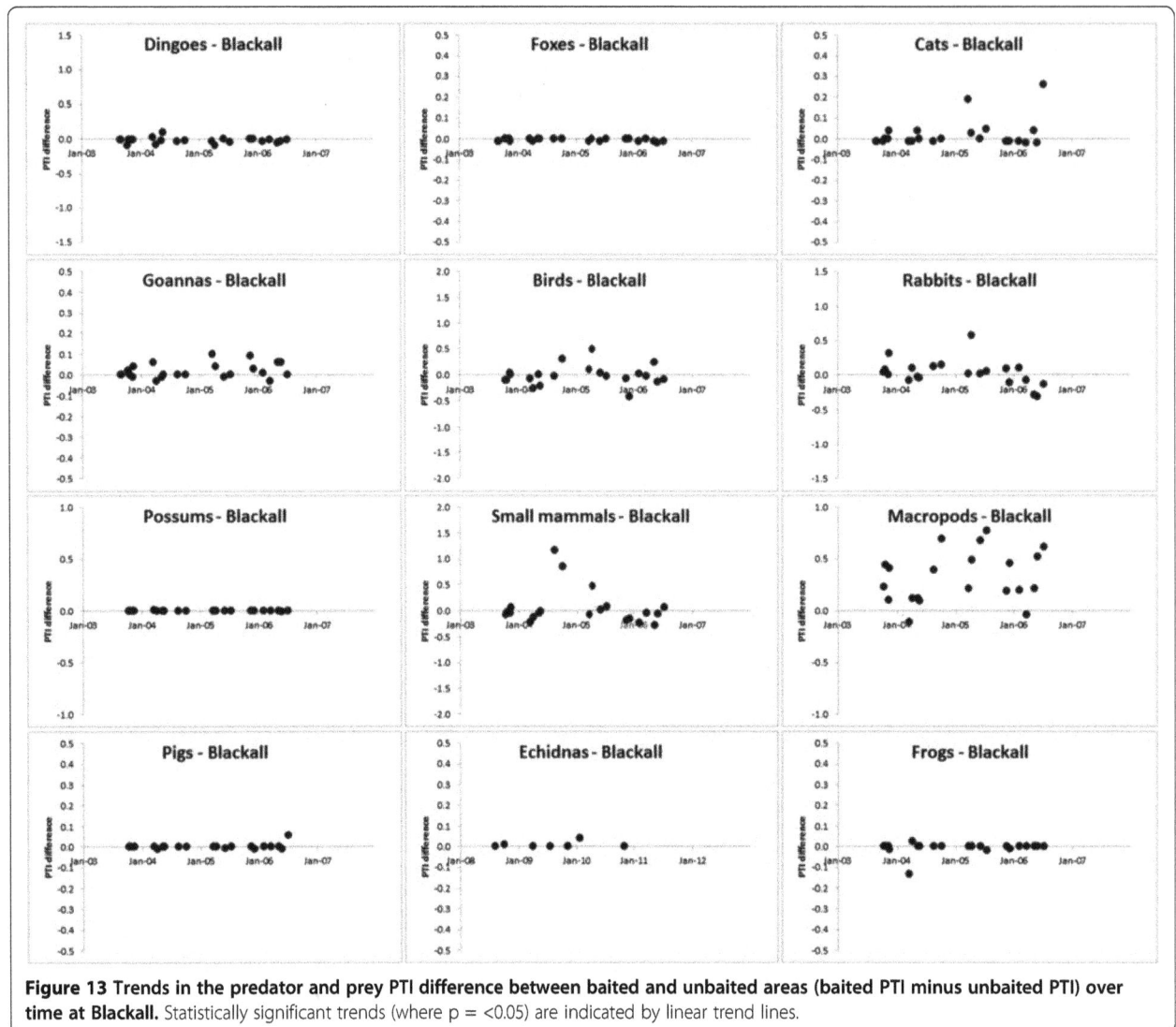

Figure 13 Trends in the predator and prey PTI difference between baited and unbaited areas (baited PTI minus unbaited PTI) over time at Blackall. Statistically significant trends (where p = <0.05) are indicated by linear trend lines.

of interest and the general indifference of predator and prey populations to contemporary dingo control practices before altering current predator control strategies. We conclude, as have others (e.g. [14,59,97]), that proposals to cease dingo control are presently unjustified on grounds that contemporary dingo control somehow harms prey fauna through trophic cascade effects. Our experimental results should be valuable for informing dingo and threatened fauna management plans given that "the majority of work to date has been largely observational and correlative" ([98], pg. 64; see also [35,46]). Future studies might focus on measuring predator control-induced changes in the behaviour of predators (such altered foraging times, prey preferences and space use) and prey (such as selection of non-preferred or safe resources of lesser quality) that may have subtle

effects on prey fitness and long-term population viability not detectable in our experiments.

Methods

Our investigation of the relationships between dingo control and prey (R6 in Figure 1) occurred simultaneously with our investigation of the relationships between dingo control and predators (R1, R2 and R4 in Figure 1), as previously reported in Allen et al. [25]. As such, our description of the methods used in the present study is based heavily on that study. All procedures described were sanctioned by the relevant animal care and welfare authorities for each site (Queensland Department of Natural Resources' Pest Animal Ethics Committee, PAEC 930401 and PAEC 030604; South Australian Department

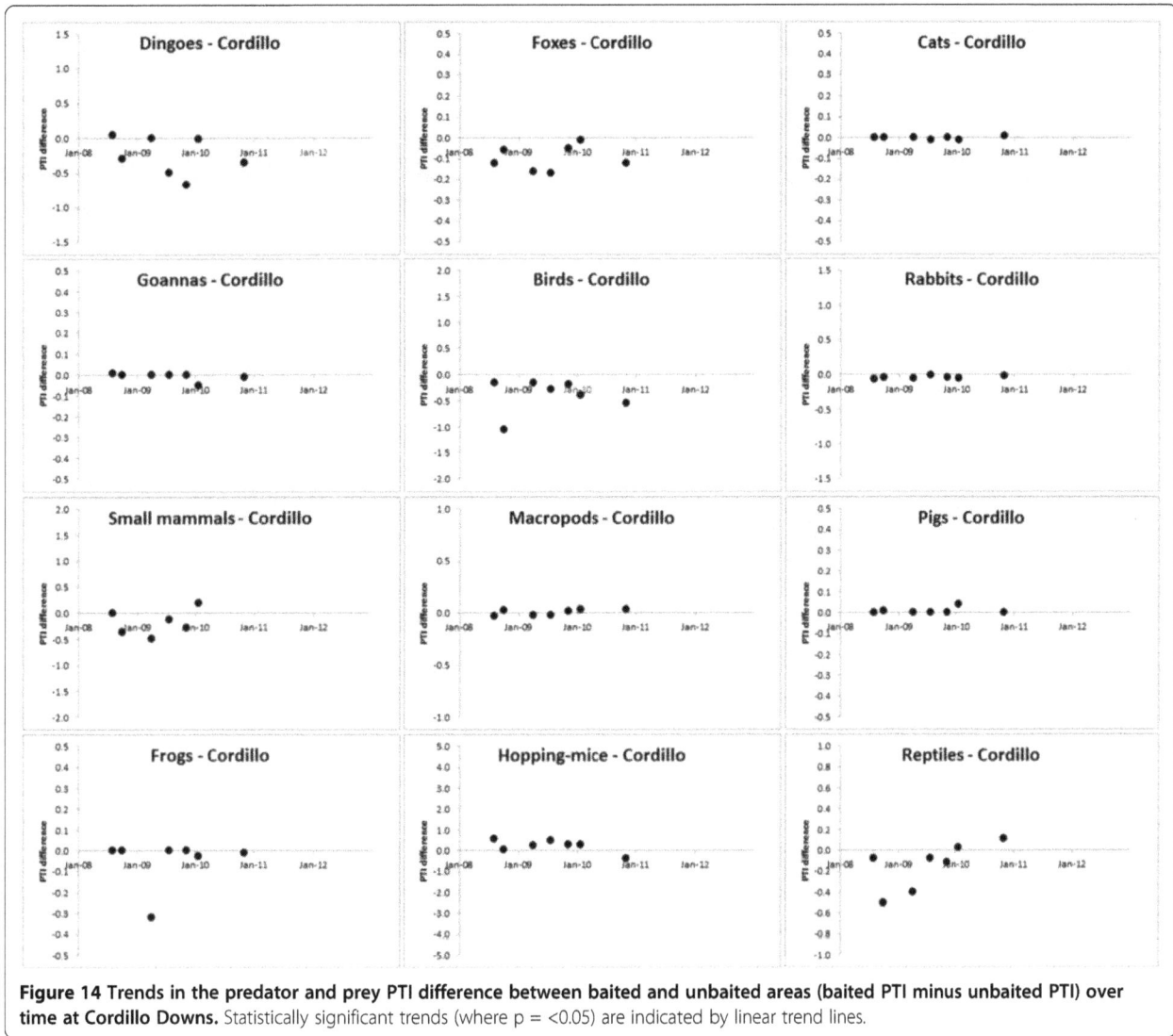

Figure 14 Trends in the predator and prey PTI difference between baited and unbaited areas (baited PTI minus unbaited PTI) over time at Cordillo Downs. Statistically significant trends (where p = <0.05) are indicated by linear trend lines.

of Environment and Heritage's Wildlife Ethics Committee, WEC 16/2008).

Study sites and design

We conducted a series of large-scale, multi-year, predator-manipulation experiments on extensive beef-cattle producing properties in five different land systems representing the breadth of the beef-cattle rangelands of Australia, where mean rainfall varied from 160–772 mm annually, or from arid to tropical areas (see Figure Seven and Table Six in [25]). Seasonal conditions fluctuated between periods of above- and below-average rainfall, or between drought and flush periods at each site during our experiments (Figure 21).

Using paired nil-treatment areas without dingo control for comparison (see Figure EightA in [25]), we examined the relative abundances of prey (and predators) in paired areas subjected to periodic broad-scale poison-baiting for dingoes at six of nine study sites (Strathmore, Mt Owen, Cordillo Downs, Quinyambie, Todmorden and Lambina; see Table Six in [25]), referred to as the six experimental sites. Aerial and/or ground-laid sodium fluoroacetate (or '1080') poison-baits were distributed individually (spaced at least 300 m apart) along landscape features (e.g. drainage lines, ridges, fragment edges etc.) and/or unformed dirt roads or tracks according to local practices and regulations up to five times each year (typically once in spring and again in autumn at the six experimental sites, and every 2–4 months continuously at the other three sites). Baits were distributed over a 1–2 day period to a midway point in the buffer zone between treatments (described below; see also Figure EightA in [25]). Each bait weighed 100–250 g and contained at least 6 mg of 1080,

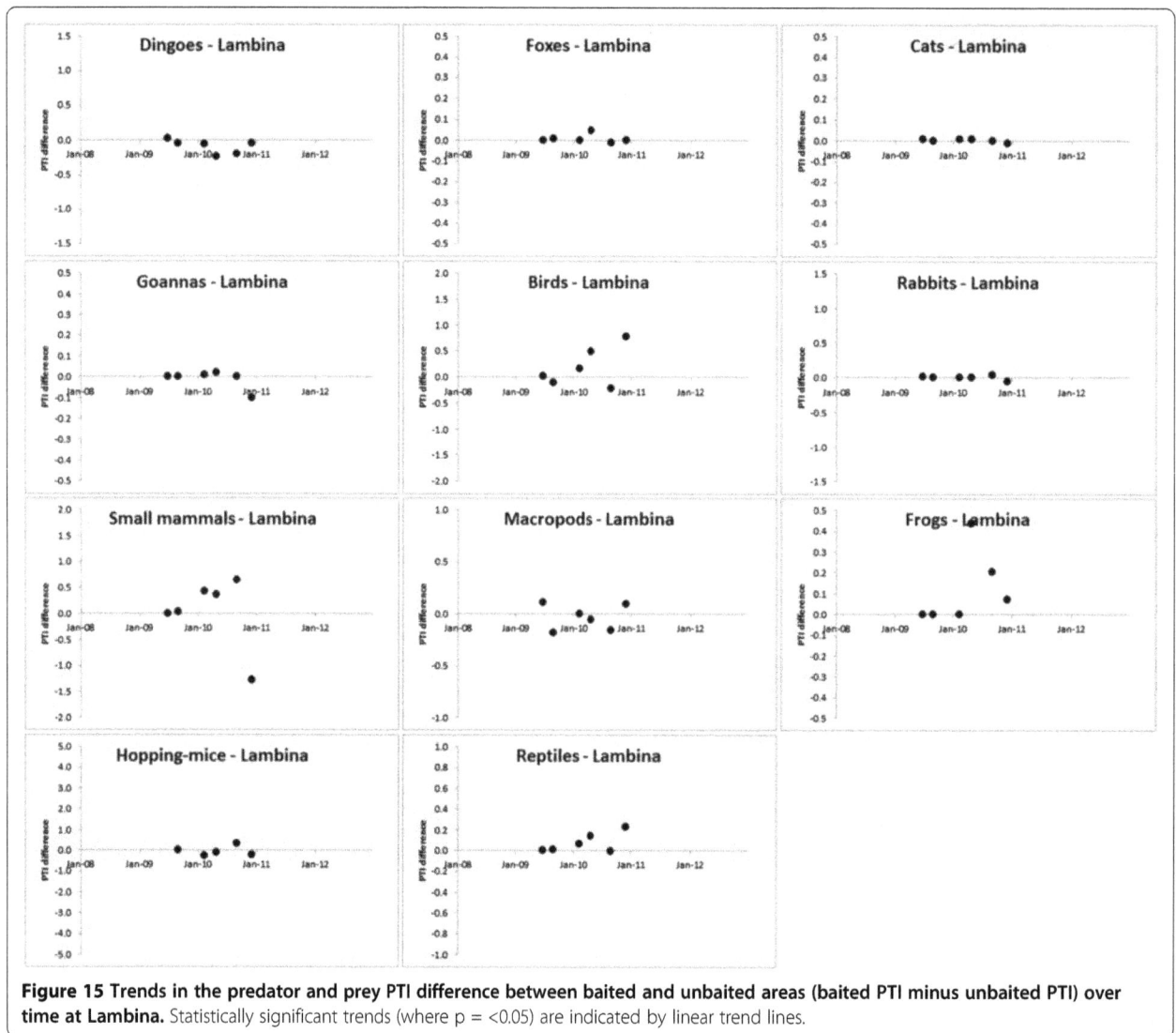

Figure 15 Trends in the predator and prey PTI difference between baited and unbaited areas (baited PTI minus unbaited PTI) over time at Lambina. Statistically significant trends (where p = <0.05) are indicated by linear trend lines.

sufficient to kill adult dingoes, foxes or cats if consumed soon after bait distribution [99]. Such spatially and temporally sporadic baiting practices are common, occur widely across Australia, and are considered the only effective dingo and fox control tool used in rangeland areas [31,100]. Populations of all other extant fauna at our sites are typically not susceptible to such baiting practices because they are either tolerant of the toxin at the low-level doses used in dingo baits and/or rarely consume carrion-like baits, preferring live prey instead (e.g. [60,61,101]).

Experimental treatment (i.e. baited) and nil-treatment (i.e. unbaited) areas were randomly allocated. Treatment and nil-treatment areas were also replicated in some land systems (see Table Six in [25]). Hone [44] defines this study design as an 'unreplicated experiment' or a 'classical experiment' for our site with replication (i.e.

Todmorden and Lambina might be considered a single site with two treatments and two controls). Both treatment and nil-treatment areas at three of the six experimental sites were historically exposed to baiting up until the commencement of the experiment, whereas, both treatment and nil-treatment areas were not historically exposed to baiting at the other three experimental sites (see Table Six in [25]). Such different baiting histories were necessary to investigate the responses of prey to either the commencement or cessation of baiting, or to the 'removal' or 'addition' of predators (i.e. dingoes and foxes were killed at some sites or allowed to increase at others).

The other three sites (Barcaldine, Blackall and Tambo; referred to as the Blackall sites) were monitored for a similar length of time (see Table Six in [25]), but differed

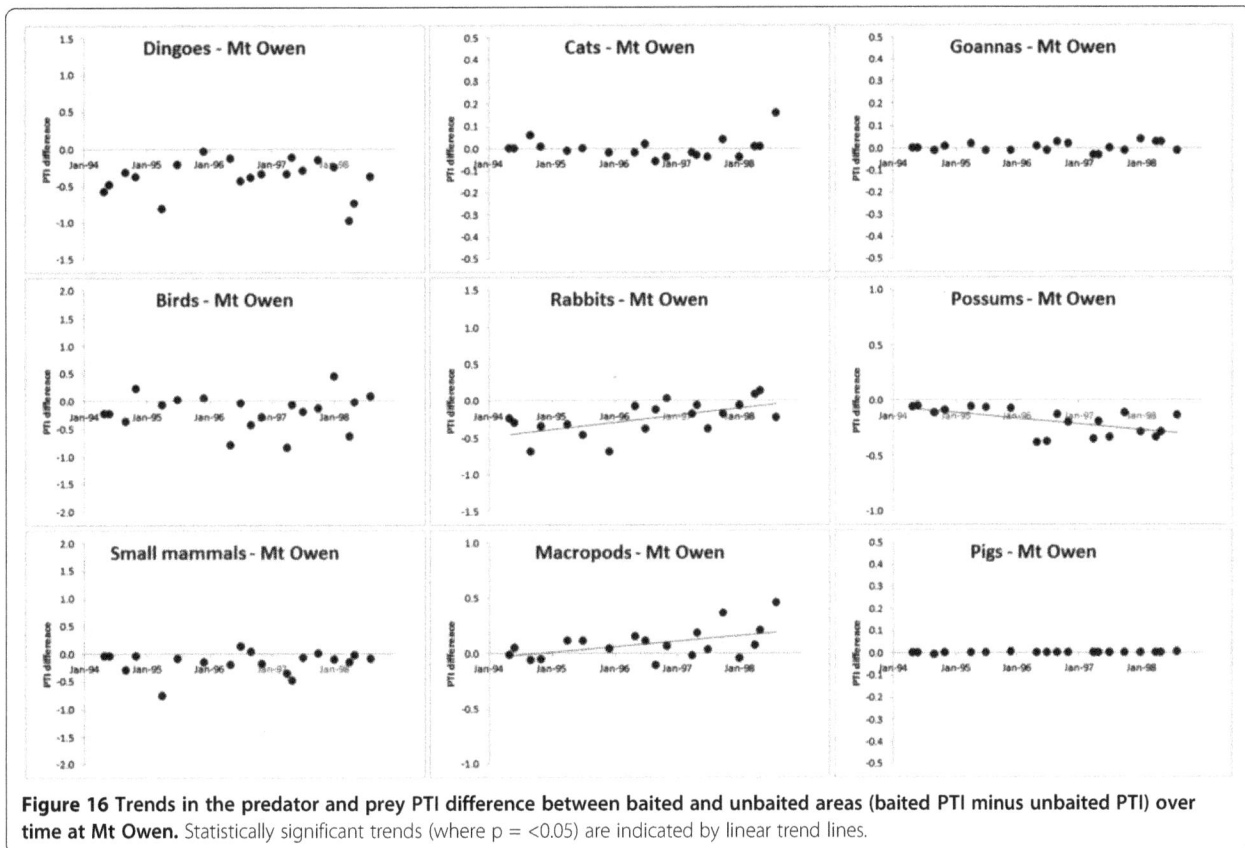

Figure 16 Trends in the predator and prey PTI difference between baited and unbaited areas (baited PTI minus unbaited PTI) over time at Mt Owen. Statistically significant trends (where p = <0.05) are indicated by linear trend lines.

from the six experimental sites in that the treatments and nil-treatments had already been established for over 10 years and they did not have buffer zones between them (see Figure EightB in [25]). This allowed an assessment of the longer-term outcomes of dingo control. Treatment size, independence and baiting practices therefore varied between the nine sites in order to deliver *in situ* tests which reflected contemporary dingo control practices within each bioregion. Experiments were conducted at large spatial scales, where the size of the total treatment and nil-treatment area at each of the nine sites ranged between 800 km^2 and 9,000 km^2, or 45,600 km^2 in total (see Table Six in [25]). The mean property size of properties that bait in north Queensland (where several of our study sites were located) is 400 km^2, and is substantially less elsewhere in Queensland (Queensland Department of Agriculture, Forestry and Fisheries, unpublished data). The size of the baited treatment areas sampled in our experiments ranged from 400 km^2 to 4,000 km^2 (Table Six in [25]). Thus, the sizes of our baited treatment areas represent areas of similar size or up to 10 times larger than those commonly subjected to baiting. Each site was separated by 100–1,500 km, except in the case of Todmorden and

Lambina, which were neighbouring properties (see Figure Seven in [25]).

Prey population monitoring
Prey populations were simultaneously monitored in treatment and nil-treatment areas using passive tracking indices (PTI; [102]), which are commonly used to monitor a variety of ground-dwelling mammals, reptiles and birds both in Australia and elsewhere around the world (e.g. [37,38,88,103-105]. We monitored populations of native and exotic amphibians, reptiles, ground-foraging birds and mammals of various sizes from small rodents (~15 g) to large herbivores such as kangaroos (*Macropus* spp.) and feral pigs using this technique. Larger feral herbivores (e.g. camels *Camelus dromedarius*, donkeys *Equus asinus*, and horses *Equus caballus*) were also recorded on sand plots on a few occasions, but were excluded from analyses because PTI values for these species are confounded by the effects of humans (e.g. culling and harvesting actions), and nor are these very large species likely to be affected by dingoes to any great degree, or vice versa [106].

PTI surveys were conducted several times each year at each site and were repeated at similar times each

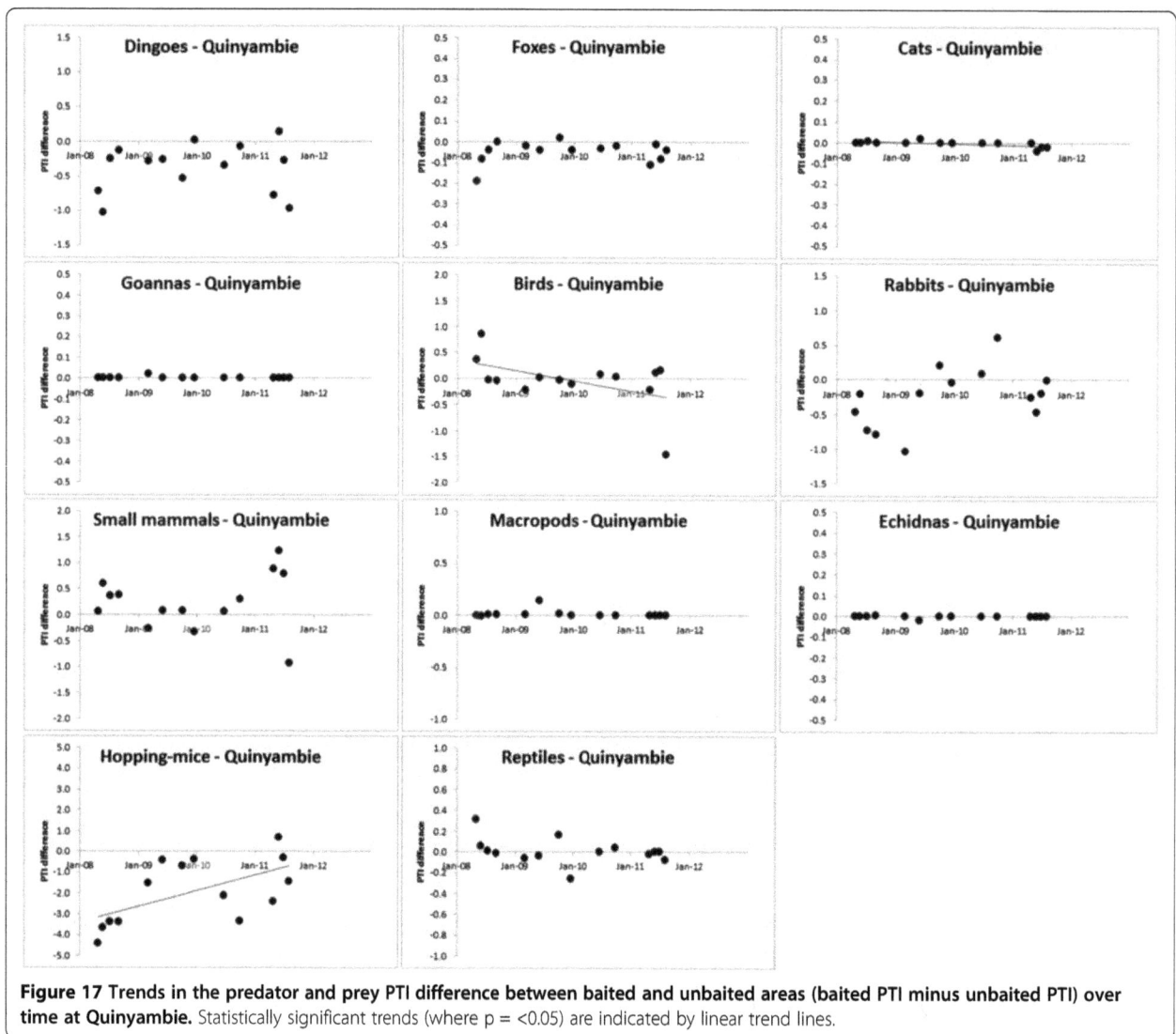

Figure 17 Trends in the predator and prey PTI difference between baited and unbaited areas (baited PTI minus unbaited PTI) over time at Quinyambie. Statistically significant trends (where p = <0.05) are indicated by linear trend lines.

subsequent year over a 2–5 year period (see Table Six in [25]). At the Blackall sites, between 92 and 166 passive tracking plots (or 'sand plots') were spaced at 1 km intervals along unformed vehicle tracks. At the six experimental sites, 50 plots each were similarly established in both the treatment and nil-treatment areas (i.e. 100 plots per site). For any given survey, plots in both treatments were read and refreshed at the same time daily by the same experienced observer and were monitored for up to 10 successive days (usually 2–5). The location of the first tracking plot in each treatment area was randomly allocated and plots were distributed throughout a similar suite of microhabitat types in both treatment areas. Plots rendered unreadable to one or more species by wind, rain or other factors were excluded from analyses. All predator and prey track intrusions were counted (i.e. a continuous measure). However, the tracks of irruptive

small mammals and hopping-mice were limited to a maximum value of 15 tracks per plot per day, which represented saturation of the sand plot with their tracks (i.e. their populations were super-abundant). PTI values for a given survey therefore represented the mean number of prey track intrusions per sand plot tracking station per 24 hr period (i.e. the mean of daily means; [102]). PTIs collected in this way can be reliably interpreted as robust estimates of relative abundance if analysed appropriately (e.g. [66,77,87], but see [107] for an alternative view).

At least one PTI survey was conducted before the imposition of treatments (i.e. before commencement or cessation of baiting in a given treatment) at the six experimental sites to identify any underlying spatial variation in prey population abundances between treatments prior to manipulations. Tracking plot transects at these six sites were separated by a buffer zone 10–50 km wide

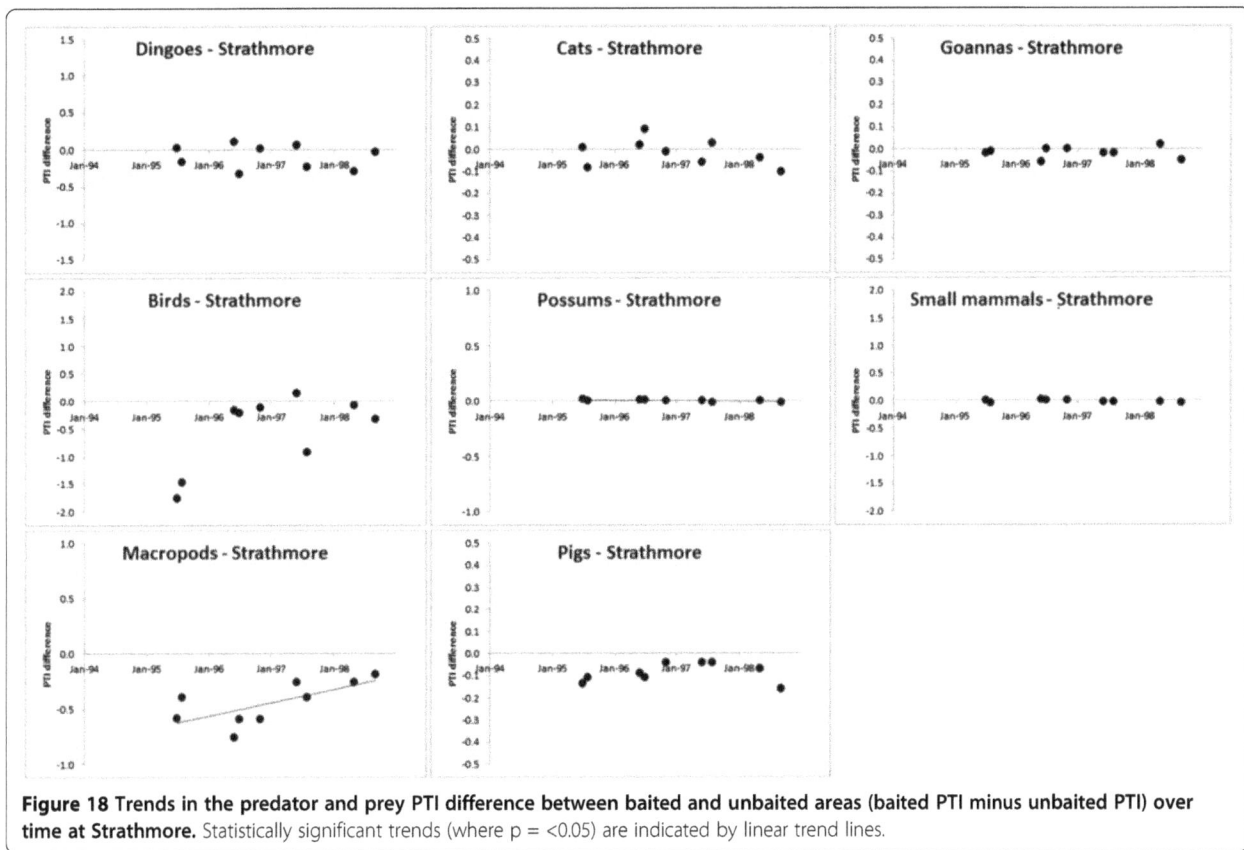

Figure 18 Trends in the predator and prey PTI difference between baited and unbaited areas (baited PTI minus unbaited PTI) over time at Strathmore. Statistically significant trends (where p = <0.05) are indicated by linear trend lines.

to achieve treatment independence during individual surveys (see Figure Eight in [25]). The appropriate width of the buffer zone at each site was based on the width of 1–2 dingo home ranges in the study areas (e.g. [54,91,108]). Tracking plots were located no closer than 5–25 km from the edge of the treatment area (i.e. half the width of the buffer zone) to minimise potential edge effects. Overall, we obtained 35,399 plot-nights of tracking data from 128 surveys conducted over 31 site-years (Table 6; see also Table Six in [25]).

Analytical approaches

Given that each site represented a unique combination of factors including experimental design (experiment or correlation), sampling intensity (N surveys ranged from 6–23), mean annual rainfall (160–772 mm), land system (five different types), treatment size or scale (800–9,000 km^2), baiting context (five different types), baiting frequency (three different types), baiting history (three different types), climate trend during the study period (three different types), and the decade the study was conducted (different sites were sampled up to 30 years apart), reliably assessing their relative effects separately

was not possible for most of these factors. Moreover, a given species (e.g. dingoes) or species group (e.g. small mammals or birds) is also not reliably comparable across sites [87,109]. In the case of species groups, actual PTI values may represent different species (e.g. rodents or dasyurids), which may have completely different life histories [49] and expected responses to predator control. Food web complexity also alters the expected outcomes of predator population changes [67,68], which is why individual species also exist within a unique fauna assemblage that is not equal or comparable across sites. For example, house mouse (*Mus musculus*) populations living at a site with only one common predator and one common rodent competitor (e.g. Quinyambie) are unlikely to respond to predator control in the same way as a house mouse population at a site with three common predators and multiple small mammal competitors (e.g. Mt Owen). Furthermore, prey could not be reliably grouped into functional groups such as 'dingo prey', 'fox prey', 'cat prey' or 'not preyed upon by predators' given that each of these predators are generalists and have extensive dietary overlap (i.e. they each eat and threaten the same prey species; [110-113], but see [51]). For these reasons, different species or species groups cannot be

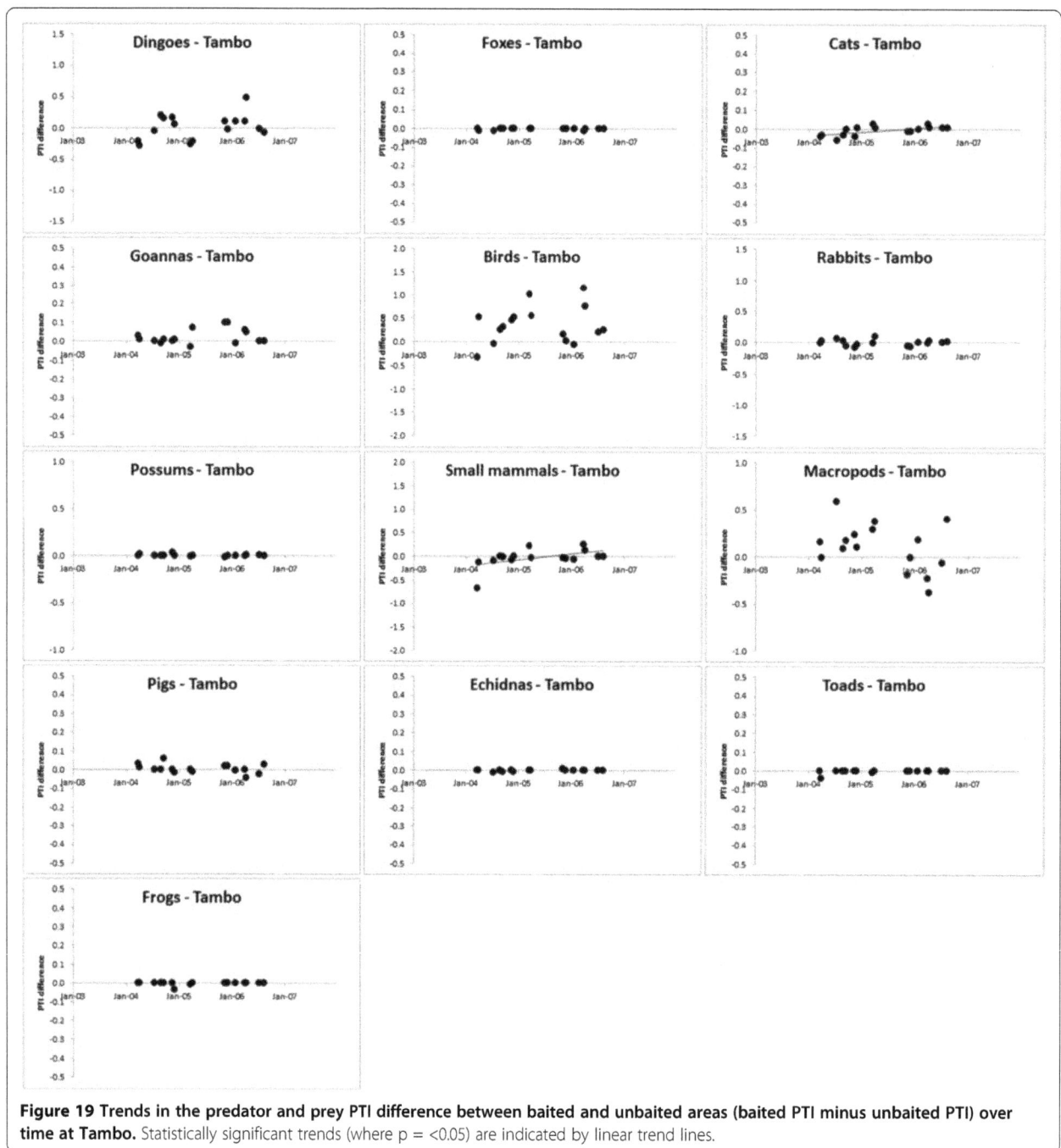

Figure 19 Trends in the predator and prey PTI difference between baited and unbaited areas (baited PTI minus unbaited PTI) over time at Tambo. Statistically significant trends (where p = <0.05) are indicated by linear trend lines.

considered equal between sites and should not be pooled across sites as if they were equal.

These limitations meant that only 'baiting history' and 'treatment' offered reliable variables on which to block or pool data across sites, for a given species or species group. Thus, to avoid a complicated variety of site-, context- and species-specific analytical approaches, we consistently applied a conservative three-step logical

approach to examine the effects of lethal dingo control on sympatric prey populations at each site.

In Step 1, we first used linear mixed model analyses (using SAS PROC MIXED) to investigate the influence of baiting history (i.e. historically baited in both treatments before cessation of baiting in the unbaited area, historically unbaited in both treatments before baiting commenced in the baited area, or consistent baiting

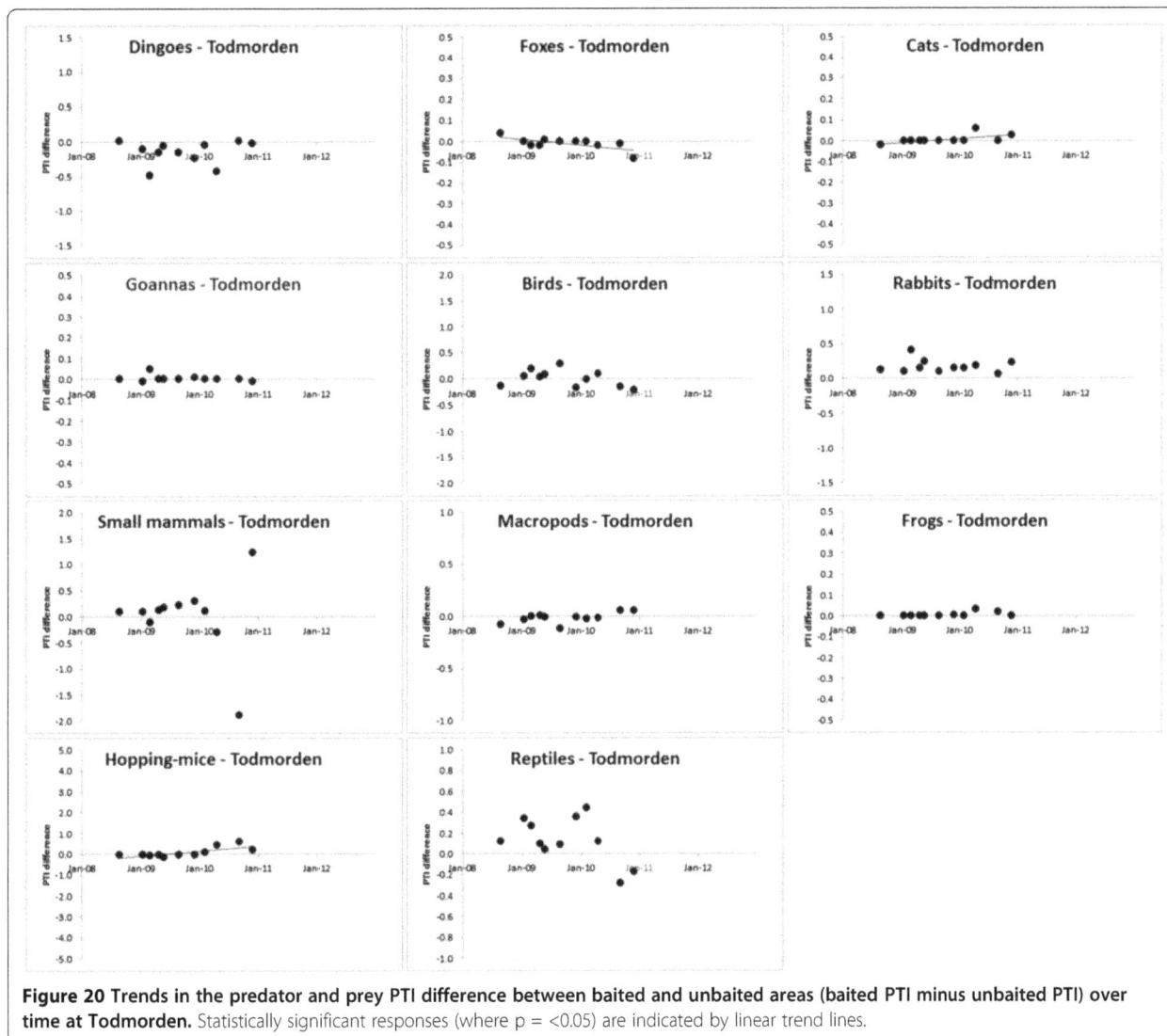

Figure 20 Trends in the predator and prey PTI difference between baited and unbaited areas (baited PTI minus unbaited PTI) over time at Todmorden. Statistically significant responses (where p = <0.05) are indicated by linear trend lines.

histories for over 10 years; N = 3 properties for each), treatment (baited or unbaited) and their interaction on both the overall mean and median PTI for predators and prey. Medians were assessed to address potential issues related to the non-symmetrical distributions of PTI values for some species [89]. Means were assessed for comparative purposes. Results from these analyses yielded little useful information for determining the responses of prey to predator control because the other factors identified above hide or confound any responses that might actually be present. Thus, subsequent analyses focused in detail on individual 'site x species' combinations in order to explicitly identify which (if any) species responded to dingo control at a given site.

We then compared the mean PTI of prey (both overall and also stratified by season) between baited and unbaited

areas at each site using repeated measures ANOVA. That the data are 'approximately normally distributed' is one of the assumptions underlying this approach [114], and given low-detection of some species at times (Table 6, Figures 3, 4, 5, 6, 7, 8, 9, 10, 11), we violated this assumption in some cases [89]. However, repeated measures ANOVA is very robust to deviations from normality, with non-normal distributions seldom affecting the overall outcomes or interpretations [114-116]. Severe deviation from normality can lead to lower p values, or an increased probability of type I errors or false positives. For our study, this simply means that some of the few reported differences in overall mean prey PTI between baited and unbaited areas (Tables 2 and 3) may not be real [89]. This first step determines crude differences in prey PTI between treatments but cannot identify causal factors for any observed differences.

Table 5 Correlations (r, with p values in parentheses) for relationships in longer-term predator PTI trends between baited and unbaited areas at nine sites across Australia (see also Figure Two in [25])

Site	Dingoes	Foxes	Cats	Goannas
Barcaldine	0.6174 (0.0017)	0.1930 (0.3777)	0.2962 (0.1700)	0.3682 (0.0838)
Blackall	0.6409 (0.0017)	0.5137 (0.0172)	0.2660 (0.2437)	−0.3321 (0.1413)
Cordillo	−0.0919 (0.8446)	0.2185 (0.6379)	−0.2582 (0.5761)	0.6285 (0.1306)
Lambina	0.5761 (0.3094)	−0.0199 (0.9746)	0.7845 (0.1162)	−0.3542 (0.5587)
Mt Owen	0.6152 (0.0051)	0.2679 (0.2674)	0.6828 (0.0013)	0.9518 (0.0000)
Quinyambie	0.4028 (0.1533)	0.4199 (0.1350)	−0.1627 (0.5785)	X
Strathmore	0.4313 (0.2464)	NP	0.4393 (0.2368)	0.7577 (0.0180)
Tambo	0.2969 (0.2642)	X	0.9564 (0.0000)	0.2254 (0.4013)
Todmorden	0.1463 (0.6678)	0.0145 (0.9663)	0.3002 (0.3698)	0.6259 (0.0394)

NP = not present; ND = known to be present but not detected on tracking plots; BP = believed present but not detected on tracking plots; NR = present and detected on tracking plots but not recorded; RE = Mostly *Varanus* spp. and reported in [25]; *All reptiles except for *Varanus* spp. (i.e. predominately agamidae and skincidae); ^All dasyurids and rodents except for hopping-mice; N surveys per site are given in Table 2.

In Step 2, we determined short-term changes in prey PTI values between pre- and post-baiting surveys by assessing mean net changes in PTI (i.e. changes in the baited area after accounting for changes in the unbaited area) with one-factor repeated measures analyses, or t-tests. To ensure maximum analytical power, this was done for each site where at least four pre- and post-baiting surveys were conducted. This step identifies any short-term responses to baiting and their cause (i.e. baiting) but cannot determine whether or not these observed responses are sustained over longer timeframes. Greater detail on the resilience of dingoes to lethal control and the factors affecting the efficacy of individual baiting programs can be found elsewhere in [19,47,85].

In Step 3, we assessed (1) temporal correlations between predator and prey abundance trends in baited and unbaited areas and (2) whether or not the difference in species' PTI values between baited and unbaited areas increased, decreased or did not change over time. This third and final step assesses whether or not population trends in baited and unbaited areas fluctuate synchronously, identifies causal factors (i.e. baiting), and determines whether or not predator or prey population trends in baited and unbaited areas are diverging or converging over longer timeframes. Step 3 is the most conclusive of our analyses for determining the responses of predators and prey to baiting.

This three-step analytical approach was designed to assess the outcomes of baiting at each site for each species, and was not designed to assess the relative influence of the many other factors that might also influence predator and prey population dynamics, such as rainfall and those others mentioned above (e.g. [117,118]). Analyses were performed using all available data. However, data were not available for all prey species at all sites

because the distribution of various species does not extend to all sites [49] or because extant species were not detected on tracking plots or recorded (Table 6). For example, though they were present at some sites, no analyses could be performed on koalas (*Phascolarctos cinereus*) due to insufficient data (Table 6). Additional details on the sensitivity and reliability of our methods can be found in Allen et al. [25,89] or Allen [85,119] and Allen [19].

Predicted outcomes (which seldom, if ever, occurred; see Results)

Whether through numerical and/or functional changes to predator and/or prey populations, overall negative effects of lethal top-predator control on prey populations are expected to be manifest as a numerical decline in prey population abundance indices in baited areas (e.g. [13,32,33]). Thus, our three-step analytical approach would detect dingo control-induced prey declines where dingo PTI trends diverge or converge and:

1. Mean overall prey PTI is lower in baited areas (potentially indicative of greater mesopredator abundances and predation pressure on prey in baited areas)
2. Mean net prey PTI significantly decreases shortly after dingo control (potentially indicative of an immediate increase in mesopredator activity, predation pressure on prey or fear-induced behavioural avoidance of predators by prey)
3. (A) Prey PTI trends are negatively correlated over time between treatments and/or (B) divergence or convergence of PTI trends is apparent (potentially indicative of longer-term dingo control-induced prey declines).

Table 6 The total number of prey tracks observed in paired baited and unbaited areas at nine sites in Australia

Site	Treatment	Plotnights	Birds	Rabbits	Possums	Small mammals^	Macropods	Pigs	Echidnas	Koalas	Toads	Frogs	Hopping mice	Reptiles*
Barcaldine	B	3015	998	1005	17	901	1901	29	24	BP	10	24	NP	RE
	UB	3131	921	608	2	703	1969	3	9		2	10		
Blackall	B	809	187	120	1	185	348	1	5	BP	ND	2	NP	RE
	UB	3271	808	473	2	749	762	8	19			19		
Tambo	B	1352	1001	93	5	222	782	17	1	BP	0	2	NP	RE
	UB	2130	901	148	4	448	960	13	3		6	11		
Mt Owen	B	4389	4554	1292	175	1442	1512	2	1	2	NR	NR	NP	RE
	UB	4350	5549	2229	1048	2186	1157	1	1	1				
Strathmore	B	2066	4254	ND	32	21	200	89	3	BP	NR	NR	NP	RE
	UB	2186	5633		26	53	1183	295	0					
Lambina	B	750	380	11	NP	5128	127	NP	ND	NP	NP	106	276	132
	UB	750	275	10		1051	152					24	303	84
Quinyambie	B	1400	1026	922	NP	942	12	ND	1	NP	NP	0	6105	213
	UB	1400	948	1234		554	1		1			3	8883	190
Cordillo Downs	B	900	134	16	NP	593	25	7	ND	NP	NP	0	356	136
	UB	900	459	53		980	20	0				37	130	252
Todmorden	B	1300	496	221	NP	366	89	NP	ND	NP	NP	8	393	377
	UB	1300	470	9		333	108					0	272	173
Grand Total		35399	28995	8444	1312	16857	11308	465	68	3	18	246	16718	1557

* = Greater in baited areas; ^ = greater PTI in unbaited areas; NP = not present; ND = known to be present but not detected on tracking plots; NR = present and detected but not recorded; RE = mostly Varanus spp. and reported in [25]; X = insufficient data to calculate p; #N surveys for hopping-mice at La mbina = 2, error df = 1 for hopping-mice.

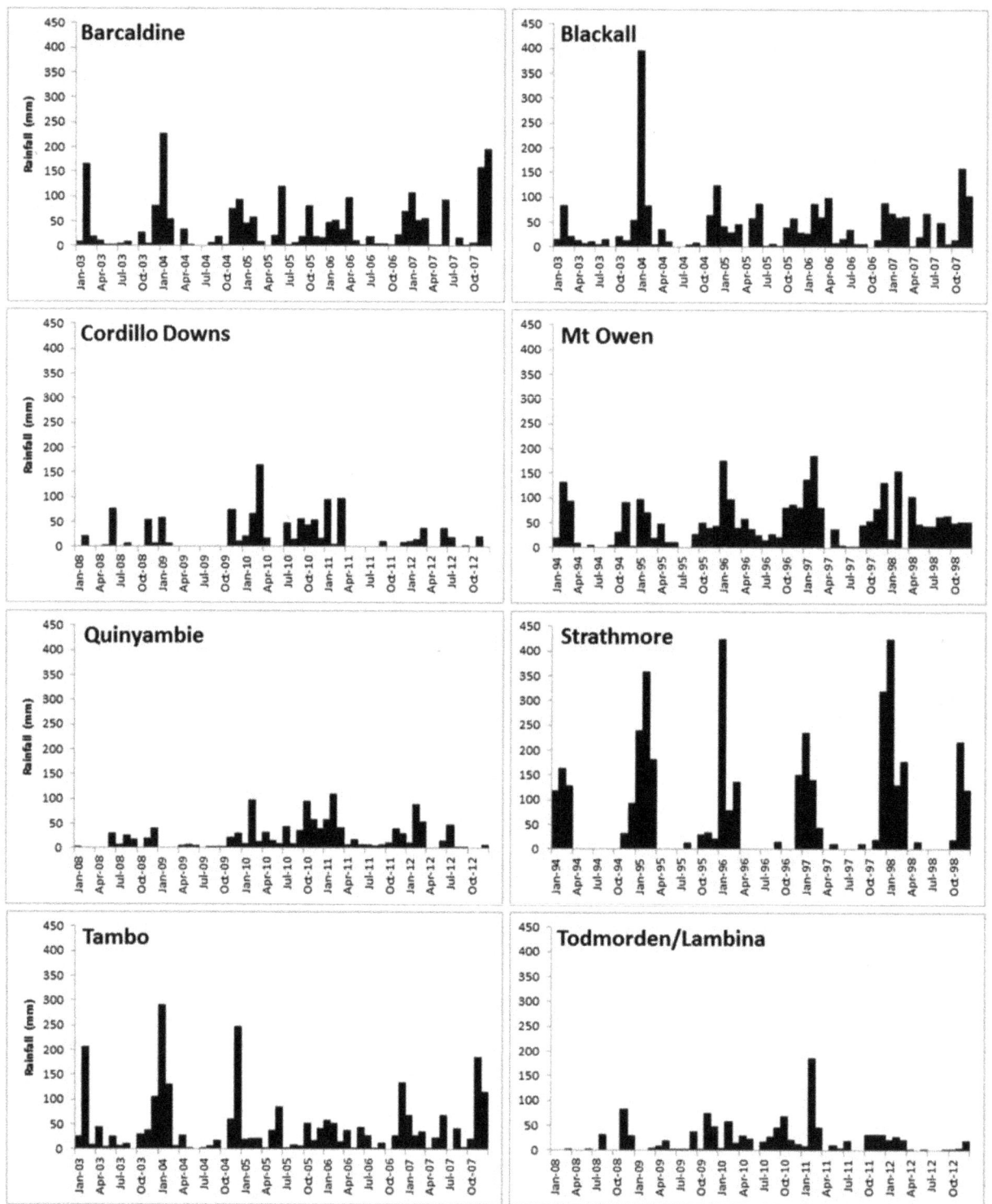

Figure 21 Monthly rainfall trends (mm rain) over the study periods at the nine study sites.

Availability of supporting data

All PTI values obtained during our study are presented in the tables and figures.

Abbreviations

PTI: Passive tracking index; ANOVA: Analysis of variance.

Competing interests

The authors declare that they have no competing interests.

Authors' contributions

LA designed and supervised the study. BA and LA collected field data and performed preliminary analyses. BA performed the remaining analyses, constructed the tables and figures and wrote the majority of the manuscript. RE performed statistical analyses. LA, RE and LL contributed further to the writing of the manuscript. All authors read and approved the final manuscript.

Acknowledgements

Generous in-kind support and hospitality was freely provided by the owners and managers of the beef-cattle properties on which we worked – this research would not have been possible without them. Damien Byrne, Heather Miller, James Speed, Steve Eldridge and Tony Gonzales assisted in the collection of field data. Analytical and editorial advice on earlier drafts of the paper was provided by Joe Scanlan, Matt Gentle and Matt Hayward. Allan Lisle provided additional statistical advice. Al Glen, Camilla Myers, Dane Panetta, Glen Saunders, Greg Campbell, Jim Hone, John Read, Peter Bird, Peter Fleming, Simon Humphrys and Tony Pople provided advice on the associated study of Allen et al. [25], which advice was further used to shape aspects of this related paper as well. Generous funding for components of this study was provided by the following Australian Government funding sources: Bureau of Resource Sciences, Caring for Our Country, and Natural Heritage Trust. These funding bodies had no role in the design, implementation, analysis or publication of this study. Some of this funding was administered by the South Australian Arid Lands Natural Resources Management Board. The production of this report was undertaken as part of work supported by the Invasive Animals Cooperative Research Centre.

Author details

[1]School of Agriculture and Food Sciences, The University of Queensland, Warrego Highway, Gatton, QLD 4343, Australia. [2]Robert Wicks Pest Animal Research Centre, Biosecurity Queensland, Tor Street, Toowoomba, QLD 4350, Australia. [3]National Wildlife Research Centre, US Department of Agriculture, LaPorte Avenue, Fort Collins, CO 80521-2154, USA.

References

1. Tscharntke T, Clough Y, Wanger TC, Jackson L, Motzke I, Perfecto I, Vandermeer J, Whitbread A: Global food security, biodiversity conservation and the future of agricultural intensification. *Biol Conserv* 2012, 151:53–59.
2. Tylianakis JM, Didham RK, Bascompte J, Wardle DA: Global change and species interactions in terrestrial ecosystems. *Ecol Lett* 2008, 11:1351–1363.
3. McKenzie NL, Burbidge AA, Baynes A, Brereton RN, Dickman CR, Gordon G, Gibson LA, Menkhorst PW, Robinson AC, Williams MR, Woinarski JCZ: Analysis of factors implicated in the recent decline of Australia's mammal fauna. *J Biogeogr* 2007, 34:597–611.
4. Ritchie EG, Johnson CN: Predator interactions, mesopredator release and biodiversity conservation. *Ecol Lett* 2009, 12:982–998.
5. Crooks KR, Soulé ME: Mesopredator release and avifaunal extinctions in a fragmented system. *Nature* 1999, 400:563–566.
6. Estes J, Crooks K, Holt RD: Ecological Role of Predators. In *Encyclopedia of Biodiversity*, Volume 6. 2nd edition. Edited by Levin SA. Waltham: Academic; 2013.
7. Terborgh J, Estes JA: *Trophic Cascades: Predator, Prey, and the Changing Dynamics of Nature*. Washington D.C: Island Press; 2010.
8. Ripple WJ, Estes JA, Beschta RL, Wilmers CC, Ritchie EG, Hebblewhite M, Berger J, Elmhagen B, Letnic M, Nelson MP, Schmitz OJ, Smith DW, Wallach AD, Wirsing AJ: Status and ecological effects of the world's largest carnivores. *Science* 2014, 343:151–163.
9. Hayward MW, Somers MJ: *Reintroduction of Top-order Predators*. Oxford: Wiley-Blackwell; 2009.
10. Ritchie EG, Elmhagen B, Glen AS, Letnic M, Ludwig G, McDonald RA: Ecosystem restoration with teeth: what role for predators? *Trends Ecol Evol* 2012, 27:265–271.
11. Sergio F, Caro T, Brown D, Clucas B, Hunter J, Ketchum J, McHugh K, Hiraldo F: Top predators as conservation tools: ecological rationale, assumptions, and efficacy. *Annu Rev Ecol Evol Syst* 2008, 39:1–19.
12. Ordiz A, Bischof R, Swenson JE: Saving large carnivores, but losing the apex predator? *Biol Conserv* 2013, 168:128–133.
13. Wallach AD, Johnson CN, Ritchie EG, O'Neill AJ: Predator control promotes invasive dominated ecological states. *Ecol Lett* 2010, 13:1008–1018.
14. Fleming PJS, Allen BL, Ballard G: Seven considerations about dingoes as biodiversity engineers: the socioecological niches of dogs in Australia. *Australian Mammalogy* 2012, 34:119–131.
15. Muhly TB, Hebblewhite M, Paton D, Pitt JA, Boyce MS, Musiani M: Humans strengthen bottom-up effects and weaken trophic cascades in a terrestrial food web. *PLoS ONE* 2013, 8:e64311.
16. Wright HL, Lake IR, Dolman PM: Agriculture—a key element for conservation in the developing world. *Conserv Lett* 2012, 5:11–19.
17. Phalan B, Balmford A, Green RE, Scharlemann JPW: Minimising the harm to biodiversity of producing more food globally. *Food Policy* 2011, 36(Supplement 1):S62–S71.
18. Linnell JDC: The relative importance of predators and people in structuring and conserving ecosystems. *Conserv Biol* 2011, 25:646–647.
19. Allen BL: The effect of lethal control on the conservation values of *Canis lupus dingo*. In *Wolves: Biology, Conservation, and Management*. Edited by Maia AP, Crussi HF. New York: Nova Publishers; 2012:79–108.
20. Valeix M, Hemson G, Loverage AJ, Mills G, Macdonald DW: Behavioural adjustments of a large carnivore to access secondary prey in a human-dominated landscape. *J Appl Ecol* 2012, 49:73–81.
21. Sidorovich VE, Tikhomirova LL, Jedrzejewska B: Wolf *Canis lupus* numbers, diet and damage to livestock in relation to hunting and ungulate abundance in northeastern Belarus during 1990–2000. *Wildl Biol* 2003, 9:103–111.
22. Treves A: Hunting for large carnivore conservation. *J Appl Ecol* 2009, 46:1350–1356.
23. Chapron G, Miquelle DG, Lambert A, Goodrich JM, Legendre S, Clobert J: The impact on tigers of poaching versus prey depletion. *J Appl Ecol* 2008, 45:1667–1674.
24. Ritchie EG: The world's top predators are in decline, and it's hurting us too. *The Conversation* 2014. 10th January 2014, accessed 11th January 2014: available at http://theconversation.com/the-worlds-top-predators-are-in-decline-and-its-hurting-us-too-21830.
25. Allen BL, Allen LR, Engeman RM, Leung LK-P: Intraguild relationships between sympatric predators exposed to lethal control: predator manipulation experiments. *Frontiers in Zoology* 2013, 10:39.
26. Corbett LK: *The Dingo in Australia and Asia*. 2nd edition. Marleston: J.B. Books, South Australia; 2001.
27. Allen BL, West P: The influence of dingoes on sheep distribution in Australia. *Aust Vet J* 2013, 91:261–267.
28. Stephens D: *The molecular ecology of Australian wild dogs: hybridisation, gene flow and genetic structure at multiple geographic scales*, PhD. Australia, Perth: The University of Western Australia; 2011.
29. Corbett LK: *Canis lupus ssp. dingo*. IUCN 2010. IUCN Red List of Threatened Species. Version 2010.4. www.iucnredlist.org. 2008. Downloaded on 20 April 2011.
30. Allen BL, Higginbottom K, Bracks JH, Davies N, Baxter GS: Balancing dingo conservation with human safety on Fraser Island: the numerical and demographic effects of humane destruction of dingoes. *Australas J Environ Manag*. In press.
31. Fleming PJS, Allen BL, Allen LR, Ballard G, Bengsen AJ, Gentle MN, McLeod LJ, Meek PD, Saunders GR: Management of wild canids in Australia: free-ranging dogs and red foxes. In *Carnivores of Australia: Past, Present and Future*. Edited by Glen AS, Dickman CR. Collingwood: CSIRO Publishing; 2014.
32. Johnson C: *Australia's Mammal Extinctions: A 50 000 Year History*. Melbourne: Cambridge University Press; 2006.
33. Wallach AD, Ritchie EG, Read J, O'Neill AJ: More than mere numbers: the impact of lethal control on the stability of a top-order predator. *PLoS ONE* 2009, 4:e6861.
34. Letnic M, Greenville A, Denny E, Dickman CR, Tischler M, Gordon C, Koch F: Does a top predator suppress the abundance of an invasive mesopredator at a continental scale? *Glob Ecol Biogeogr* 2011, 20:343–353.

35. Glen AS, Dickman CR, Soulé ME, Mackey BG: Evaluating the role of the dingo as a trophic regulator in Australian ecosystems. *Austral Ecology* 2007, **32**:492–501.

36. Letnic M, Ritchie EG, Dickman CR: Top predators as biodiversity regulators: the dingo *Canis lupus dingo* as a case study. *Biol Rev* 2012, **87**:390–413.

37. Arthur AD, Catling PC, Reid A: Relative influence of habitat structure, species interactions and rainfall on the post-fire population dynamics of ground-dwelling vertebrates. *Austral Ecology* 2013, **37**:958–970.

38. Claridge AW, Cunningham RB, Catling PC, Reid AM: Trends in the activity levels of forest-dwelling vertebrate fauna against a background of intensive baiting for foxes. *For Ecol Manag* 2010, **260**:822–832.

39. Eldridge SR, Shakeshaft BJ, Nano TJ: *The Impact of Wild Dog Control on Cattle, Native and Introduced Herbivores and Introduced Predators in Central Australia, Final Report to the Bureau of Rural Sciences*. Alice Springs: Parks and Wildlife Commission of the Northern Territory; 2002.

40. Allen BL, Lundie-Jenkins G, Burrows ND, Engeman RM, Fleming PJS, Leung LK-P: Does lethal control of top-predators release mesopredators? A re-evaluation of three Australian case studies. *Ecol Manag Restor* 2014, **15**:1–5.

41. Barbosa P, Castellanos I: *Ecology of Predator–prey Interactions*. New York: Oxford University Press; 2005.

42. Allen BL, Fleming PJS, Hayward M, Allen LR, Engeman RM, Ballard G, Leung LK-P: Top-predators as biodiversity regulators: contemporary issues affecting knowledge and management of dingoes in Australia. In *Biodiversity Enrichment in a Diverse World*, Volume 4. Edited by Lameed GA. Rijeka, Croatia: InTech Publishing; 2012:85–132.

43. Salo P, Banks PB, Dickman CR, Korpimaki E: Predator manipulation experiments: impacts on populations of terrestrial vertebrate prey. *Ecol Monogr* 2010, **80**:531–546.

44. Hone J: *Wildlife Damage Control*. Collingwood, Victoria: CSIRO Publishing; 2007.

45. Krebs CJ: *Ecology: The Experimental Analysis of Distribution and Abundance*. 6th edition. San Francisco: Benjamin-Cummings Publishing; 2008.

46. Allen BL, Fleming PJS, Allen LR, Engeman RM, Ballard G, Leung LK-P: As clear as mud: a critical review of evidence for the ecological roles of Australian dingoes. *Biol Conserv* 2013, **159**:158–174.

47. Allen LR: *The Impact of Wild Dog Predation and Wild Dog Control on Beef Cattle: Large-scale Manipulative Experiments Examining the Impact of and Response to Lethal Control*. Saarbrucken, Germany: LAP Lambert Academic Publishing; 2013.

48. Allen LR, Goullet M, Palmer R: The diet of the dingo (*Canis lupus dingo* and hybrids) in north-eastern Australia: a supplement to Brook and Kutt. *The Rangeland Journal* 2012, **34**:211–217.

49. Van Dyck S, Strahan R: *The Mammals of Australia*. 3rd edition. Sydney: Reed New Holland; 2008.

50. Allen LR: Wild dog control impacts on calf wastage in extensive beef cattle enterprises. *Anim Prod Sci* 2014, **54**:214–220.

51. Allen BL, Fleming PJS: Reintroducing the dingo: the risk of dingo predation to threatened vertebrates of western New South Wales. *Wildl Res* 2012, **39**:35–50.

52. Foulkes JN: *The ecology and management of the common brushtail possum Trichosurus vulpecula in central Australia*, PhD. Canberra: The University of Canberra, Applied ecology research group; 2001.

53. Kerle JA, Foulkes JN, Kimber RG, Papenfus D: The decline of the brushtail possum, *Trichosurus vulpecula* (Kerr 1798), in arid Australia. *The Rangeland Journal* 1992, **14**:107–127.

54. Allen BL, Engeman RM, Leung LK-P: The short-term effects of a routine poisoning campaign on the movement behaviour and detectability of a social top-predator. *Environ Sci Pollut Res* 2014, **21**:2178–2190.

55. Allen BL, Leung LK-P: Assessing predation risk to threatened fauna from their prevalence in predator scats: dingoes and rodents in arid Australia. *PLoS ONE* 2012, **7**:e36426.

56. Allen BL: Scat happens: spatiotemporal fluctuation in dingo scat collection rates. *Aust J Zool* 2012, **60**:137–140.

57. Haber GC: Biological, conservation, and ethical implications of exploiting and controlling wolves. *Conserv Biol* 1996, **10**:1068–1081.

58. Webb NF, Allen JR, Merrill EH: Demography of a harvested population of wolves (*Canis lupus*) in west-central Alberta, Canada. *Can J Zool* 2011, **89**:744–752.

59. Wang Y, Fisher D: Dingoes affect activity of feral cats, but do not exclude them from the habitat of an endangered macropod. *Wildl Res* 2012, **39**:611–620.

60. APVMA: *Review Findings for Sodium Monofluoroacetate: The Reconsideration of Registrations of Products Containing Sodium Monofluoroacetate and Approvals of their Associated Labels, Environmental Assessment*. Canberra: Australian Pesticides and Veterinary Medicines Authority; 2008.

61. Glen AS, Gentle MN, Dickman CR: Non-target impacts of poison baiting for predator control in Australia. *Mammal Rev* 2007, **37**:191–205.

62. Fenner S, Körtner G, Vernes K: Aerial baiting with 1080 to control wild dogs does not affect the populations of two common small mammal species. *Wildl Res* 2009, **36**:528–532.

63. Dexter N, Ramsey DSL, MacGregor C, Lindenmayer D: Predicting ecosystem wide impacts of wallaby management using a fuzzy cognitive map. *Ecosystems* 2012, **15**:1363–1379.

64. Kennedy MS, Baxter GS, Spencer RJ: Towards a more detailed understanding of habitat: the responses of bush rats to manipulation of food and predation after fire. In *The 10th Biennial Australasian Bushfire Conference: Bushfire 2006; Brisbane*. SEQ Fire & Biodiversity Consortium; 2006.

65. Moseby KE, Hill BM, Read JL: Arid recovery - a comparison of reptile and small mammal populations inside and outside a large rabbit, cat, and fox-proof exclosure in arid South Australia. *Austral Ecology* 2009, **34**:156–169.

66. Allen BL, Engeman RM, Allen LR: Wild dogma I: an examination of recent "evidence" for dingo regulation of invasive mesopredator release in Australia. *Current Zoology* 2011, **57**:568–583.

67. Finke DL, Denno RF: Predator diversity dampens trophic cascades. *Nature* 2004, **429**:407–410.

68. Hairston N, Smith F, Slobodkin L: Community structure, population control and competition. *Am Nat* 1960, **94**:421–425.

69. Choquenot D, Forsyth DM: Exploitation ecosystems and trophic cascades in non-equilibrium systems: pasture – red kangaroo – dingo interactions in arid Australia. *Oikos* 2013, **122**:1292–1306.

70. Allen BL: A comment on the distribution of historical and contemporary livestock grazing across Australia: implications for using dingoes for biodiversity conservation. *Ecological Management & Restoration* 2011, **12**:26–30.

71. Carwardine J, O'Connor T, Legge S, Mackey B, Possingham HP, Martin TG: Prioritizing threat management for biodiversity conservation. *Conserv Lett* 2012, **5**:196–204.

72. Mech LD: Is science in danger of sanctifying the wolf? *Biol Conserv* 2012, **150**:143–149.

73. Marris E: Rethinking predators: legend of the wolf. *Nature* 2014, **507**:158–160.

74. Kinnear JE, Krebs CJ, Pentland C, Orell P, Holme C, Karvinen R: Predator-baiting experiments for the conservation of rock-wallabies in Western Australia: a 25-year review with recent advances. *Wildl Res* 2010, **37**:57–67.

75. Holt RD, Huxel GR: Alternative prey and the dynamics of intraguild predation: theoretical perspectives. *Ecology* 2007, **88**:2706–2712.

76. Corbett L, Newsome AE: The feeding ecology of the dingo. III. Dietary relationships with widely fluctuating prey populations in arid Australia: an hypothesis of alternation of predation. *Oecologia* 1987, **74**:215–227.

77. Caughley G: *Analysis of Vertebrate Populations. Reprinted with Corrections Edn*. Chichester: John Wiley & Sons Ltd; 1980.

78. Burbidge AA, McKenzie NL: Patterns in the modern decline of Western Australia's vertebrate fauna: causes and conservation implications. *Biol Conserv* 1989, **50**:143–198.

79. Lunney D: Causes of the extinction of native mammals of the western division of New South Wales: an ecological interpretation of the nineteenth century historical record. *Rangel J* 2001, **23**:44–70.

80. Smith PJ, Pressey RL, Smith JE: Birds of particular conservation concern in the western division of New South Wales. *Biol Conserv* 1994, **69**:315–338.

81. Brook LA, Johnson CN, Ritchie EG: Effects of predator control on behaviour of an apex predator and indirect consequences for mesopredator suppression. *J Appl Ecol* 2012, **49**:1278–1286.

82. Colman NJ, Gordon CE, Crowther MS, Letnic M: Lethal control of an apex predator has unintended cascading effects on forest mammal assemblages. *Proc R Soc B Biol Sci* 2014, **281**. doi:10.1098/rspb.2013.3094.

83. Letnic M, Koch F, Gordon C, Crowther M, Dickman C: Keystone effects of an alien top-predator stem extinctions of native mammals. *Proc R Soc B Biol Sci* 2009, **276**:3249–3256.

84. Wallach AD, O'Neill AJ: Threatened species indicate hot-spots of top-down regulation. *Anim Biodivers Conserv* 2009, **32**:127–133.

85. Allen LR: *Best-practice Baiting: Evaluation of Large-scale, Community-based 1080 Baiting Campaigns.* Toowoomba: Robert Wicks Pest Animal Research Centre, Department of Primary Industries (Biosecurity Queensland); 2006.

86. Reddiex B, Forsyth DM: **Control of pest mammals for biodiversity protection in Australia. II. Reliability of knowledge.** *Wildl Res* 2006, **33**:711–717.

87. Engeman R: **Indexing principles and a widely applicable paradigm for indexing animal populations.** *Wildl Res* 2005, **32**:202–210.

88. Engeman R, Pipas M, Gruver K, Bourassa J, Allen L: **Plot placement when using a passive tracking index to simultaneously monitor multiple species of animals.** *Wildl Res* 2002, **29**:85–90.

89. Allen BL, Allen LR, Engeman RM, Leung LK-P: **Reply to the criticism by Johnson *et al.* (2014) on the report by Allen *et al.* (2013).** *Frontiers in Zoology* 2014, accessed 1st June 2014: Available at: http://www.frontiersinzoology.com/content/11/11/17/comments#1982699.

90. Mahon PS, Banks PB, Dickman CR: **Population indices for wild carnivores: a critical study in sand-dune habitat, south-western Queensland.** *Wildl Res* 1998, **25**:11–22.

91. Allen LR: **Best practice baiting: dispersal and seasonal movement of wild dogs (*Canis lupus familiaris*).** In *Technical Highlights: Invasive Plant and Animal Research 2008–09.* Brisbane: QLD Department of Employment, Economic Development and Innovation; 2009:61–62.

92. Wicks S, Allen BL: *Returns on Investment in Wild Dog Management: Beef Production in the South Australian Arid Lands.* Canberra: Australian Bureau of Agricultural and Resource Economics and Sciences, Department of Agriculture, Fisheries and Forestry; 2012.

93. Fleming PJS, Allen BL, Ballard G, Allen LR: *Wild Dog Ecology, Impacts and Management in Northern Australian Cattle Enterprises: A Review with Recommendations for R, D&E Investments.* Sydney: Meat and Livestock Australia; 2012.

94. Braysher M: *Managing Vertebrate Pests: Principles and Strategies.* Canberra: Bureau of Rural Sciences, Australian Government Publishing; 1993.

95. Carwardine J, O'Connor T, Legge S, Mackey B, Possingham HP, Martin TG: *Priority Threat Management to Protect Kimberley Wildlife.* Brisbane: CSIRO Ecosystem Sciences; 2011.

96. Dickman C, Glen A, Letnic M: **Reintroducing the dingo: Can Australia's conservation wastelands be restored?** In *Reintroduction of Top-order Predators.* Edited by Hayward MW, Somers MJ. Oxford: Wiley-Blackwell; 2009:238–269. Conservation science and practice series.

97. Glen AS: **Enough dogma: seeking the middle ground on the role of dingoes.** *Current Zoology* 2012, **58**:856–858.

98. Ritchie EG, Dickman CR, Letnic M, Vanak AT: **Dogs as predators and trophic regulators.** In *Free-ranging Dogs and Wildlife Conservation.* Edited by Gompper ME. New York: Oxford University Press; 2014:55–68.

99. McIlroy JC: **The sensitivity of Australian animals to 1080 poison. II. Marsupial and eutherian carnivores.** *Australian Wildlife Research* 1981, **8**:385–399.

100. Allen LR, Fleming PJS: **Review of canid management in Australia for the protection of livestock and wildlife - potential application to coyote management.** *Sheep and Goat Research Journal* 2004, **19**:97–104.

101. McIlroy JC: **The sensitivity of Australian animals to 1080 poison. IX. Comparisons between the major groups of animals and the potential danger non-targets face from 1080-poisoning campaigns.** *Australian Wildlife Research* 1986, **13**:39–48.

102. Allen L, Engeman R, Krupa H: **Evaluation of three relative abundance indices for assessing dingo populations.** *Wildl Res* 1996, **23**:197–206.

103. Engeman R, Evangelista P: **Investigating the feasibility of a tracking index for monitoring wildlife in the Lower Omo Valley, Ethiopia.** *Afr J Ecol* 2006, **45**:184–188.

104. Olifiers N, Loretto D, Rademaker V, Cerqueira R: **Comparing the effectiveness of tracking methods for medium to large-sized mammals of Pantanal.** *Zoologia* 2011, **28**:207–213.

105. Engeman RM, Massei G, Sage M, Gentle MN: **Monitoring wild pig populations: a review of methods.** *Environ Sci Pollut Res* 2013, **20**:8077–8091.

106. Fleming P, Corbett L, Harden R, Thomson P: *Managing the Impacts of Dingoes and Other Wild Dogs.* Canberra: Bureau of Rural Sciences; 2001.

107. Hayward MW, Marlow N: **Will dingoes really conserve wildlife and can our methods tell?** *J Appl Ecol*, **51**:835–838.

108. Allen BL: **Do desert dingoes drink daily? Visitation rates at remote waterpoints in the Strzelecki Desert.** *Australian Mammalogy* 2012, **34**:251–256.

109. Caughley G, Sinclair ARE: *Wildlife Ecology and Management.* Cambridge, Massachusetts: Blackwell Sciences; 1994.

110. Cupples JB, Crowther MS, Story G, Letnic M: **Dietary overlap and prey selectivity among sympatric carnivores: could dingoes suppress foxes through competition for prey?** *J Mammal* 2011, **92**:590–600.

111. Glen AS, Pennay M, Dickman CR, Wintle BA, Firestone KB: **Diets of sympatric native and introduced carnivores in the Barrington Tops, eastern Australia.** *Austral Ecology* 2011, **36**:290–296.

112. Letnic M, Dworjanyn SA: **Does a top predator reduce the predatory impact of an invasive mesopredator on an endangered rodent?** *Ecography* 2011, **34**:827–835.

113. Mitchell BD, Banks PB: **Do wild dogs exclude foxes? Evidence for competition from dietary and spatial overlaps.** *Austral Ecology* 2005, **30**:581–591.

114. Girden ER: *ANOVA: Repeated Measures (Quantitative Applications in the Social Sciences).* London: Sage Publications; 1992.

115. Glass GV, Peckham PD, Sanders JR: **Consequences of failure to meet assumptions underlying the fixed effects analyses of variance and covariance.** *Rev Educ Res* 1972, **42**:237–288.

116. Lindman HR: *Analysis of Variance in Complex Experimental Designs.* San Francisco: W.H. Freeman & Co; 1974.

117. Dickman CR, Letnic M, Mahon PS: **Population dynamics of two species of dragon lizards in arid Australia: the effects of rainfall.** *Oecologia* 1999, **119**:357–366.

118. Dickman CR, Mahon PS, Masters P, Gibson DF: **Long-term dynamics of rodent populations in arid Australia: the influence of rainfall.** *Wildl Res* 1999, **26**:389–403.

119. Allen LR: **The Impact of Wild Dog Predation and Wild Dog Control on Beef Cattle Production.** In *PhD Thesis.* Australia: Department of Zoology, The University of Queensland; 2005.

5

Song characteristics track bill morphology along a gradient of urbanization in house finches (*Haemorhous mexicanus*)

Mathieu Giraudeau[1,2]*, Paul M Nolan[3], Caitlin E Black[4], Stevan R Earl[5], Masaru Hasegawa[6] and Kevin J McGraw[1]

Abstract

Introduction: Urbanization can considerably impact animal ecology, evolution, and behavior. Among the new conditions that animals experience in cities is anthropogenic noise, which can limit the sound space available for animals to communicate using acoustic signals. Some urban bird species increase their song frequencies so that they can be heard above low-frequency background city noise. However, the ability to make such song modifications may be constrained by several morphological factors, including bill gape, size, and shape, thereby limiting the degree to which certain species can vocally adapt to urban settings. We examined the relationship between song characteristics and bill morphology in a species (the house finch, *Haemorhous mexicanus*) where both vocal performance and bill size are known to differ between city and rural animals.

Results: We found that bills were longer and narrower in more disturbed, urban areas. We observed an increase in minimum song frequency of urban birds, and we also found that the upper frequency limit of songs decreased in direct relation to bill morphology.

Conclusions: These findings are consistent with the hypothesis that birds with longer beaks and therefore longer vocal tracts sing songs with lower maximum frequencies because longer tubes have lower-frequency resonances. Thus, for the first time, we reveal dual constraints (one biotic, one abiotic) on the song frequency range of urban animals. Urban foraging pressures may additionally interact with the acoustic environment to shape bill traits and vocal performance.

Keywords: Urban impacts, Bill shape, Singing behavior, Noise pollution, Vocal communication

Introduction

Humans continue to urbanize Earth's landscape and alter wildlife habitat in many ways [1-5]. Many species suffer from anthropogenic disturbance, leading to depleted urban biodiversity, although some populations thrive and expand in cities [3,6]. Factors such as human activity [7], pollution exposure [8], artificial lighting [9], elevated temperature [10], and food and water availability [11] can directly impact wildlife success in urban areas [12].

Among the unique conditions that animals experience in urban habitats, ambient city noise is key because it can limit the sound space available for animals to communicate using acoustic signals [13,14]. Many animal species, and especially birds, use acoustic signals to attract mates and/or communicate with competitors [15,16]. Low-frequency urban noise may overlap with songs of native species, limiting sound reception and ultimately even decreasing fitness and population viability [17].

Recent studies have shown that some bird species adjust their song characteristics – specifically by increasing their minimum frequency – to be heard in a noisy urban environment [13,18-20]. However, the ability to make such song modifications may be constrained by several morphological factors, including bill gape, size, and shape [21-25]. During sound production, the vocal tract (i.e. trachea, larynx, and bill) acts as a resonance chamber for the sound frequencies produced by the syrinx [26-28], and subtle variation in bill size/shape and vocal tract morphology affects sound production [22].

* Correspondence: giraudeau.mathieu@gmail.com
[1]School of Life Sciences, Arizona State University, Tempe, AZ 85287-4501, USA
[2]Present address: School of Biological Sciences A08, University of Sydney, Sydney, NSW 2006, Australia
Full list of author information is available at the end of the article

Generally, birds with longer, deeper, and wider beaks produce songs with significantly lower minimum frequencies, maximum frequencies, and frequency bandwidths [29]. Ultimately, such bill morphological factors may limit the degree to which certain species can vocally adapt to urban settings; however, to our knowledge, the few studies that have investigated covariance between avian bill and song traits [30,31,24] have been done outside of an urbanization context.

Bill shape in birds is also strongly associated with diet (e.g. short and thin in insectivores, deep and hooked in granivores and carnivores; [32]), such that foraging pressures can work either with or against directional selection on bill size/shape for song production [23,24]. Urban environments offer novel foraging opportunities that may shift bill morphology, and in fact city effects on bill morphology have been documented in house finches, *Haemorhous mexicanus* [33], whereby bill size increased in urban birds perhaps as a function of the availability of larger, harder-to-husk seeds at backyard bird feeders. This modification of bill shape/size may strongly impact song characteristics in urban birds. What is now needed is an integrated understanding of the relationship between bill morphology and song output in the context of urbanization.

Here, we examined the relationship between song characteristics and bill morphology in house finches along a gradient of urbanization in the Phoenix (Arizona, USA) metropolitan area. Song is a sexually selected trait in this species [34], with females preferring to mate with males that sing more, longer songs. For the first time in studies examining urban impacts on animal signals, we quantified a series of different metrics of urbanization, including human population density and seven measurements describing land-use patterns within the 1-km radius around each of our trapping sites [35], to assess how these factors may be associated with bill shape/size and song characteristics. Based on previous studies, we predicted that minimum song frequency would be associated with urban background noise [13,18,19] and that bill size (length, width, and height) would increase in urban habitats [32]. In addition, given that the angle between the upper and lower mandibles should decrease with an increase in bill length (considering a similar aperture at the bill tip), urban birds with longer bills should have a proportionally longer orotracheal cavity with a reduced high resonating frequency compared to rural individuals with shorter bills [22]. Thus, the shift in bill morphology in urban birds should be associated with a decrease in the highest song frequency produced in urban compared to desert areas. To summarize, we predicted a reduction in the song frequency range for finches in human-modified areas, due to both an increase in minimum frequency in response to the urban background noise and a decrease of the highest song frequency associated with the longer bills of urban birds.

Results

Habitat description

Using principal component analysis (PCA), urbanization scores were generated using the data for the 8 variables cited below (7 land use variables and human population density). PCA indicated that three PCs captured >84% of habitat variation. PC1 summarized 47% of the variance, while PC2 and PC3 summarized 24% and 13% of the variance, respectively. PC1 correlated negatively and strongly (component loading >94%) with the percentage of land covered by native undisturbed (desert) habitat. PC2 correlated positively and strongly (component loading >81%) with the percentage of land covered by cultivated vegetation. Finally, PC3 was positively correlated with the percentage of land covered by native vegetation (component loading >74%, Table 1).

Urbanization and morphometrics

Tarsus length and body mass were not correlated with bill morphometrics (all P > 0.39). However, tarsus length (rho = 0.76, P = 0.03) was correlated with urbanization PC2 scores, such that birds captured from sites where more land was covered by cultivated vegetation had longer tarsi.

Bill length (rho = 0.76, P = 0.03; Figure 1) was positively correlated with urbanization PC1 scores, while bill width (rho = –0.79, P = 0.02) was negatively correlated with this urbanization metric (PCA 1 vs bill height: rho = –0.52, P = 0.18). Thus, bill length increased and bill width decreased at sites where less land was covered by native undisturbed habitat. None of the bill traits were correlated with urbanization PC2 and PC3 scores (all P > 0.49).

Using site averages, we found a significant negative relationship between bill width and length, (rho = –0.86, P = 0.006; Figure 1) and a positive association between bill width and height (rho = 0.76, P = 0.03). Bill height was not significantly linked to bill length at the population level (rho = –0.55, P = 0.16).

Urbanization and song characteristics

Mean maximum frequency of background noise at each site was negatively correlated with urbanization PC2 scores (rho = –0.83, P = 0.04) but not with urbanization PC1 (rho = 0.03, P = 0.96) or PC3 scores (rho = 0.60, P = 0.21). Thus, background noise frequency was higher at sites with less land covered by vegetation.

Using site averages, minimum song frequency was significantly positively correlated with maximum frequency of background noise (rho = 0.94, P = 0.005; Figure 2). None of the other song characteristics were related to maximum background-noise frequency (all P > 0.6). Minimum song

Table 1 Characteristics of the sites at which we studied house finches in Maricopa County, USA

Capture site	Phoenix downtown	ASU campus	Mesa organic farm	Crossroads district park	Chandler neighborhood	Phoenix zoo	South mountain park	Estrella mountain regional park
City	Phoenix	Tempe	Mesa	Gilbert	Chandler	Phoenix	Phoenix	Goodyear
Geographical coordinates	33°27'N 112°03'W	33°25'N 111°55'W	33° 27'N 111° 49'W	33° 19'N 111°43' W	33° 18'N 111°55' W	33°27'N 111°57'W	33°21'N 112°4'W	33° 25'N 112°25' W
Number of humans living within 1 km of the study site	7291	10385	4600	17175	3948	50	1001	11
Sample size bill measurements	23	20	21	23	22	21	22	20
Sample size for song analyses	9	10	–	11	10	13	–	10
Mean song frequencies: lowest, highest and range in Hz (SE)	2162 (47), 6573 (206), 4411 (234)	2137 (55), 6446 (137), 4310 (188)	–	1844 (43), 5890 (195), 4045 (199)	2008 (60), 6212 (182), 4205 (202)	1968 (24), 6720 (112), 4752 (125)	–	1858 (60), 6806 (182), 4949 (200)
Mean bill size: Length, height, and width in mm (SE)	9.84 (0.076), 8.16 (0.043), 7.26 (0.041)	9.85 (0.062), 8.00 (0.037), 7.17 (0.040)	9.71 (0.072), 8.12 (0.044), 7.22 (0.047)	10.04 (0.072), 8.06 (0.059), 7.09 (0.055)	10.05 (0.083), 8.04 (0.042), 7.07 (0.041)	9.71 (0.056), 8.17 (0.050), 7.28 (0.042)	9.76 (0.057), 8.14 (0.046), 7.28 (0.070)	9.44 (0.073), 8.14 (0.047), 7.29 (0.043)
Habitats (% of land covered by):								
Cultivated vegetation and grass	0.00	0.00	3.44	9.07	0.67	8.05	0.05	1.11
Disturbed (Mesic and Xeric vegetation)	57.55	38.79	48.72	40.28	59.53	9.01	17.26	1.73
Compacted soil (prior agriculture or not)	0.78	1.01	4.5	1.2	5.47	1.99	1.81	0.67
Disturbed (commercial, industrial, asphalt)	30.11	48.38	22.51	21.31	20.82	18.64	7.85	2.87
Undisturbed	10.55	9.56	14.45	12.05	10	57.45	68.89	67.44
Vegetation	1.01	2.25	6.37	3.32	3.47	3.16	4.12	3.62
River gravel and water	0	0	0	3.97	0.04	1.68	0.03	22.55

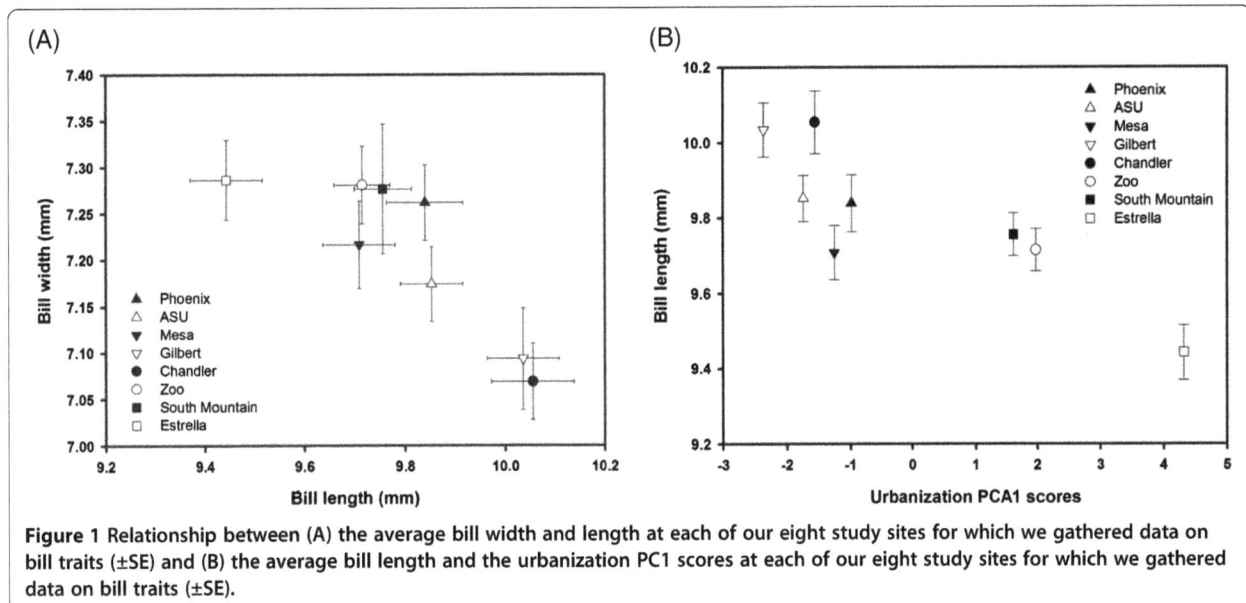

Figure 1 Relationship between (A) the average bill width and length at each of our eight study sites for which we gathered data on bill traits (±SE) and (B) the average bill length and the urbanization PC1 scores at each of our eight study sites for which we gathered data on bill traits (±SE).

frequency was negatively correlated with urbanization PC2 scores (rho = −0.94, P = 0.005), while maximum song frequency (rho = −0.83, P = 0.04) and frequency range (rho = −0.84, P = 0.04) were positively correlated with urbanization PC1 scores. In other words, minimum song frequency increased at sites with less land is covered by vegetation, and maximum frequency and the frequency range decreased at sites with a greater percentage of disturbed habitat.

Intersite covariance between bill shape and song characteristics

Using site averages, bill length and width, but not height (all P > 0.5), were significantly related to song characteristics; finches sang at lower maximum frequencies (rho = −0.94, P = 0.005; Figure 3) and with a decreased

frequency range (rho = −0.94, P = 0.005; Figure 4) at sites where bills were longer and narrower. The lowest song frequencies used by birds were not linked with bill morphology (all P > 0.8). In other words, modifications of bill shape associated with the life in the city were correlated with song maximum frequency and frequency range.

Correlations among bill traits and among song traits within birds and sites are provided in Additional file 1.

Discussion

We examined relationships between song characteristics and bill morphology in house finches along a gradient of urbanization. We found a gradual increase in bill length and decrease in bill width at progressively more disturbed urban areas. Urban and rural finches differ considerably in

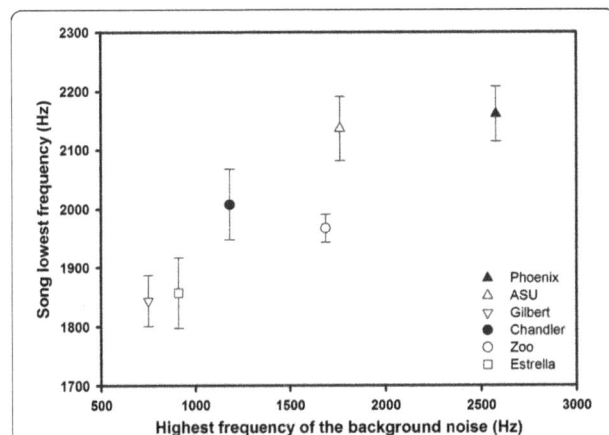

Figure 2 Relationship between the average song lowest frequency and the highest frequency of the background noise at the six study sites for which we gathered song data (±SE).

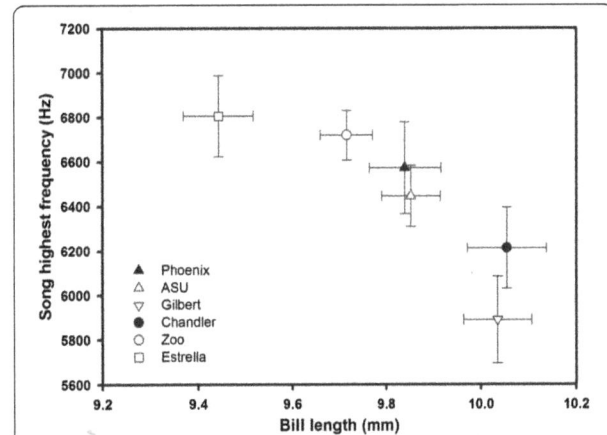

Figure 3 Relationship between the song highest frequency and the average bill length at the six study sites for which we gathered data on both song and bill traits (±SE).

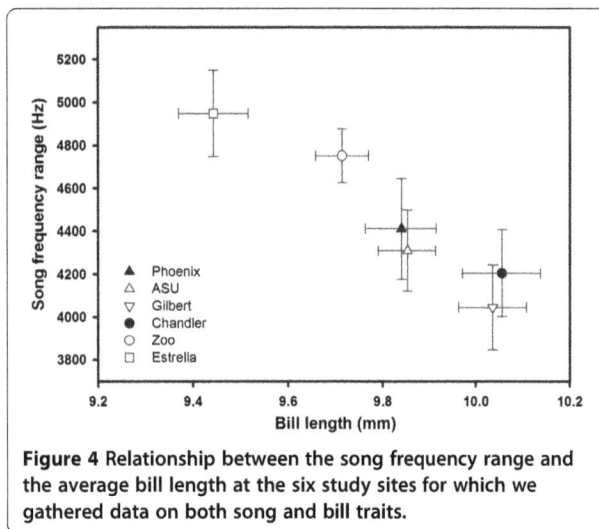

Figure 4 Relationship between the song frequency range and the average bill length at the six study sites for which we gathered data on both song and bill traits.

foraging and food consumption, given the prevalence of bird seed feeders with large seeds (e.g. sunflower, millet) in suburban and urban areas [36-38,6]. Seeds provided at feeders are argued to require greater bite force to open than natural small seeds of cacti and grasses [33], but this cannot explain why city birds in Phoenix had thinner, longer beaks than did rural finches. The most plausible explanation is that birds with longer beaks gain an advantage in handling large seeds at feeders, while shorter beaks are better suited for processing smaller native seeds. In accordance with this hypothesis, Soobramoney and Perrin [39] showed that passerine species with the smallest bills husked the smallest seeds fastest, while species with the largest bills husked the largest seeds fastest.

More proximately, a modification of bill morphology arises through ontogenetic changes in bill tissue proliferation and migration, a process largely regulated by the expression of bone morphogenetic proteins (BMPs) during early development [40-43]. If juvenile birds in more urban areas are exposed to seeds with different characteristics from those in more rural regions, then different levels of mechanical stress during foraging in early life could drive patterns of BMP production and bill growth (up to 2.5 months after hatching in house finches, [44]). Recently, Badyaev et al. [33] showed that the difference of BMP expression between urban and desert house finches may arise even before hatching. Thus, natural selection may even have favored pre-hatching overexpression of BMP proteins in birds from urban areas (with longer, thinner beaks) compared to natural areas.

We also found that differences in bill morphology along the urban gradient were associated with modifications of song characteristics. The increased bill length and decreased bill width observed in finches from more human-disturbed environments were associated with a decrease in maximum song frequency. These results

are in accordance with prior studies showing that larger bills increase the length of the vocal tract, making it more suitable for the production of lower-frequency songs [22,45,29]. However, we did not find a reduction in the minimum song frequency associated with the bill-length increase in urban areas. Conversely, and in accordance with previous studies in house finches and other species, we found a significant positive correlation between the highest frequencies of the background noise and the lowest frequencies of bird song [13,18,19]. Thus, it is likely that urban birds increased their minimum song frequency in order to be heard in a noisy environment, although, with their longer bill, they are probably able to produce lower minimum frequency song than desert birds. Taken together, these results show that the range of song frequencies used by urban birds was drastically reduced (by ca. 20%) compared to those used by rural birds.

In addition to being shaped by static bill morphology, these song patterns may also be due to individual plasticity; house finches have been shown to quickly increase or decrease their minimum song frequencies in response to different experimental noise treatments [46]. However, we are not aware of any studies showing plasticity in the maximum frequencies sung in response to urban noise. Future experiments should examine the role of genetics, development, learning, and vocal-tract plasticity in generating diverse vocal characteristics in an anthropogenic environment.

So what might be the ecological and evolutionary consequences of these song modifications in urban and suburban bird populations? Song traits are key indicators of male quality in many bird species [47]. In the house finch, females showed significant mate preferences based on song characteristics [34]. Therefore, reduction in frequency bandwidth of the signal in response to human activity could have profound reproductive consequences for males. Alternatively, plasticity in female choice could allow receivers to use alternative vocal components that more reliably reflect male quality in the novel environment [48,49]. In line with this hypothesis, Halfwerk et al. (2011, [49]) recently experimentally demonstrated a signaling advantage in male great tits (*Parus major*) for high-frequency songs in noisy conditions, whereas low-frequency songs are likely to be preferred in natural, less noisy environments.

Conclusions
We have shown for the first time a gradual modification of bill morphology and song characteristics along an urban gradient in populations of house finches. These findings demonstrate the extent to which human activities may strongly impact both the morphology of animals and the associated quality of their sexual signals.

Methods

Bill morphology

We used basket traps and Potter traps baited with sunflower seeds to capture 172 adult male house finches at eight sites (two urban, one city park, two suburban, two desert, and one rural; Table 1) in the Phoenix metropolitan area in August-September 2011. At capture, each bird was leg banded with a numbered United States Geological Survey metal ring for individual identification. We also measured body mass (to the nearest 0.1 g with a digital scale), tarsus length (to the nearest 0.1 mm with digital calipers), and bill morphology (to the nearest 0.01 mm with calipers; sensu [33]). Bill length was measured from the anterior end of the nostril to the tip of the upper mandible, bill width was measured at the anterior end of the nostril, and bill height was measured in a vertical plane at the anterior end of the nostrils over both mandibles.

Song measurement and analyses

We recorded songs from adult males during the breeding season (May 2011) from six of the aforementioned study sites (Table 1). We used a Marantz PMD 661 digital recording device (sampling rate: 44,000 Hz; Mahwah, NJ, USA) and a Sennheiser ME 60 directional microphone (Old Lyme, CT, USA) to record the songs. We recorded birds from 0600–1400 hrs. At each location, we listened opportunistically for males to sing, and then recorded them. No playback was used to elicit vocalizations. We approached the birds as closely as possible without disrupting them, and we separated each of our recordings by ≥100 meters to minimize the chance that we inadvertently recorded the same male twice.

Audio recordings were analyzed using Raven Pro 1.4 audio editing software (Cornell University, Ithaca, NY, USA). A song was defined as a set of ≥4 elements with ≥1 second between songs. We isolated 1,165 songs from a total of 79 individuals along the gradient. Using Raven Pro, we generated spectrograms using standardized parameters (Hann window, size =512 samples; DFT size =1024 samples; values below −120 dB were clipped). Recordings of individuals with < 6 songs were discarded (Badyaev et al. [33]). From each song, we extracted the following variables: 1) frequency range, 2) upper frequency and 3) lower frequency. Characteristics were averaged for each bird and then for each study site, to determine the relationship among song characteristics within birds, within sites, and among sites in relation to urbanization (see below) and bill morphology. Once the songs were isolated, we also measured the highest frequency of the background noise during each recording, at a standardized contrast level on the computer display, and calculated the average background noise at each site.

Habitat description

Most studies on urbanization and animal behavior limit the study sampling to single urban vs rural sites (but see [50,51]). However, a more ecologically appropriate sampling scheme is to measure traits at more than one site per habitat type. Moreover, these "urban" and "rural" sites typically vary in many anthropogenic parameters, so it is appropriate to specifically quantify types and degrees of human impact. To assess relationships between finch traits and anthropogenic environmental characteristics, we obtained several urbanization parameters around our eight trapping sites from a local database that is part of the Central Arizona-Phoenix Long-Term Ecological Research program [52-54]: (1) human population density within a radius of 1 km around each trapping site, estimated from the 2010 US Census data; (2) landuse and landcover (LULC, in 2007) variables within the same 1 km radius. From satellite images, we determined percentage of land dedicated to 7 land uses: cultivated vegetation and cultivated grass, river gravel and water, vegetation, disturbed-commercial/industrial and asphalt, undisturbed, disturbed-compacted soil, disturbed-mesic and xeric vegetation residential (see [52-54] for a full description of the LULC types). Using principal component analysis (PCA), urbanization scores were generated using the data for the 8 variables cited above (7 land use variables and human population density, see Results section).

Statistics

All statistical analyses were carried out with Statistica software (StatSoft, Inc. Tulsa, USA) with α set at 0.05. We ran non-parametric Spearman rank correlations between the three urbanization metrics extracted from the PCA and the average values for bill morphology and song characteristics. We also ran non-parametric correlations among the song traits and among the bill traits within birds and for all birds within a site. As recommended by Nakagawa (2004, [55]), we did not use Bonferroni or similar adjustments to correct for multiple comparisons in order to avoid a reduction of power and an increase of Type II error to unacceptable levels. We did not include the date or time of day in our analyses since song was recorded at every site within a one-week time period and always from 0600–1400 hrs.

Additional file

Additional file 1: Supplementary material.

Competing interests
The authors declare that they have no competing interests.

Authors' contributions
MG conceived and designed the study, and drafted the manuscript. MG, PMN, CEB and MH performed the fieldwork. SRE has provided the urbanization data.

MG, PMN, CEB, MH, SE and KJM revised the manuscript. All authors approved the final manuscript.

Acknowledgements
This work was supported by the National Science Foundation (IOS-0923694 to KJM and CAP3: BCS-1026865 to the CAP-LTER program), the citadel foundation and the College of Charleston Office of Undergraduate Research and Creative Activities. We are grateful to Michael Lundgren, Ron Rutowski, the staffs of the Love grows farm, Phoenix zoo, the city of Gilbert, the South Mountain and Estrella mountain regional parks for providing facilities during field work; and to Erick Lundgren for the suggestion to study beak morphology. All research was conducted with approval from the ASU Institutional Animal Care and Use Committee (protocol 09-1054R).

Author details
[1]School of Life Sciences, Arizona State University, Tempe, AZ 85287-4501, USA. [2]Present address: School of Biological Sciences A08, University of Sydney, Sydney, NSW 2006, Australia. [3]Department of Biology, The Citadel, Charleston, SC 29409, USA. [4]Department of Biology, The College of Charleston, Charleston, SC 29424, USA. [5]Global Institute of Sustainability & School of Sustainability, Arizona State University, Tempe, AZ 85287-5402, USA. [6]Graduate School of Life and Environmental Sciences, University of Tsukuba, 1-1-1 Tennoudai, Tsukuba-shi, Ibaraki 305-8572, Japan.

References
1. Sauvajot RM, Buechner M, Kamradt DA, Schonewald CM: Patterns of human disturbance and response by small mammals and birds in chaparral near urban development. Urban Ecosyst 1998, 2:279–297.
2. Grimm NB, Grove JM, Pickett STA, Redman CL: Integrated approaches to long-term studies of urban ecological systems. Bioscience 2000, 50:571–584.
3. Marzluff JM, Bowman R, Donnelly R: Avian ecology and conservation in an urbanizing world. Norwell, MA, USA: Kluwer Academic Publishers; 2001.
4. Milesi C, Elvidge CD, Nemani RR, Running SW: Assessing the impact of urban land development on net primary productivity in the southeastern United States. Remote Sens Environ 2003, 86:401–410.
5. Shochat E, Warren PS, Faeth SH, McIntyre NE, Hope D: From patterns to emerging processes in mechanistic urban ecology. Trends Ecol Evol 2006, 21:186–191.
6. Shochat E, Lerman SB, Katti M, Lewis DB: Linking optimal foraging behavior to bird community structure in an urban-desert landscape: field experiments with artificial food patches. Am Nat 2004, 164:232–243.
7. Fernández-Jurici E, Jimenez MD, Lucas E: Bird Tolerance to Human Disturbance in Urban Parks of Madrid (Spain). Management Implications. In Avian Ecology and Conservation in an Urbanizing World. Edited by Marzluff JM, Bowman R, Donnelly R. Norwell, MA; USA: Kluwer; 2001:261–275.
8. Janssens E, Dauwe T, Bervoets L, Eens M: Heavy metals and selenium in feathers of great tits (Parus major) along a pollution gradient. Environ Toxicol Chem 2001, 20:2815–2820.
9. Longcore T, Rich C: Ecological light pollution. Front Ecol Environ 2004, 2(4):191–198.
10. Luniak M: Urban Ecological Studies in Central and Eastern Europe. Wroclaw, Poland: Polish Academy of Sciences; 1990.
11. Eeva T, Lehikoinen E, Rönkä M: Air pollution fades the plumage of the great tit. Funct Ecol 1998, 12:607–612.
12. Bokony V, Kulcsar A, Liker A: Does urbanization select for weak competitors in house sparrows? Oikos 2010, 119:437–444.
13. Slabbekoorn H, Peet M: Ecology: birds sing at a higher pitch in urban noise. Nature 2003, 424:267.
14. Katti M, Warren PS: Tits, noise and urban bioacoustics. Trends Ecol Evol 2004, 19:109–110.
15. Catchpole CK, Slater PJB: Bird Song: Biological Themes and Variations. Cambridge: Cambridge University Press; 1995.
16. Marler P, Slabbekoorn H: Nature's Music. USA: Elsevier Academic Press; 2004.
17. Klump GM: Bird Communication in the Noisy World. In Ecology and Evolution of Acoustic Communication in Birds. Edited by Kroodsma DE, Miller EH. Ithaca, New York: Cornell University Press; 1996:321–338.
18. Fernandez-Juricic E, Poston R, Collibus KD, Morgan T, Bastain B, Martin C, Jones K, Tremininio R: Microhabitat selection and singing behavior patterns of male house finches (Carpodacus mexicanus) in urban parks in a heavily urbanized landscape in the western U.S. Urban Habitats 2005, 3:49–69.
19. Wood WE, Yezerinac SM: Song sparrow (Melospiza melodia) song varies with urban noise. Auk 2006, 123:650–659.
20. Potvin DA, Parris KM, Mulder RA: Geographically pervasive effects of urban noise on frequency and syllable rate of songs and calls in silvereyes (Zosterops lateralis). Proc R Soc Lond B 2011, 278:2464–2469.
21. Westneat MW, Long JH, Hoese W, Nowicki S: Kinematics of birdsong—functional correlation of cranial movements and acoustic features in sparrows. J Exp Biol 1993, 182:147–171.
22. Palacios MG, Tubaro PL: Does beak size affect acoustic frequencies in woodcreepers? Condor 2000, 102:553–560.
23. Podos J, Nowicki S: Performance Limits on Birdsong Production. In Nature's Musicians: The Science of Birdsong. Edited by Marler P, Slabbekoorn H. New York: Elsevier/Academic Press; 2004:318–341.
24. Podos J, Nowicki S: Beaks, adaptation, and vocal evolution in Darwin's finches. Bioscience 2004, 54:501–510.
25. Nelson BS, Beckers GJ, Suthers RA: Vocal tract filtering and sound radiation in a songbird. J Exp Biol 2005, 208:297–308.
26. Nowicki S: Vocal-tract resonances in oscine bird sound production - evidence from birdsongs in a helium atmosphere. Nature 1987, 325:53–55.
27. Nowicki S, Marler P: How do birds sing? Music Percept 1988, 5:391–426.
28. Beckers GJL, Suthers RA, Cate C: Pure-tone birdsong by resonance filtering of harmonic overtones. Proc Natl Acad Sci U S A 2003, 100:7372–7376.
29. Huber SK, Podos J: Beak morphology and song features covary in a population of Darwin's finches (Geospiza fortis). Biol J Linn Soc 2006, 88:489–498.
30. Bowman RI: Morphological differentiation and adaptation in the galapagos finches. Univ Caliornia Publ Zool 1961, 58:1–302.
31. Slabbekoorn H, Smith TB: Habitat-dependent song divergence in the little greenbul: an analysis of environmental selection pressures on acoustic signals. Evolution 2002, 56:1849–1858.
32. Francis CD, Guralnick RP: Fitting the bill: do different winter food resources influence juniper titmouse (Baeolophus ridgwayi) bill morphology? Biol J Lin Soc 2010, 101:667–679.
33. Badyaev AV, Young RL, Oh KP, Addison C: Evolution on a local scale: developmental, functional, and genetic bases of divergence in bill form and associated changes in song structure between adjacent habitats. Evolution 2008, 62:1951–1962.
34. Nolan PM, Hill GE: Female choice for song characteristics in the house finch. Anim Behav 2004, 67:403–410.
35. Giraudeau M, Mousel M, Earl SE, McGraw KJ: Parasites in the city: degree of urbanization predicts poxvirus and coccidian infection in house finches. PLoS One 2014, 9:e86747.
36. Hensley MM: Ecological relations of breeding bird populations of the desert biome in Arizona. Ecol Monogr 1954, 24:185–208.
37. Mills GS, Dunning JB Jr, Bates JM: Effects of urbanization on breeding bird community structure in southwestern desert habitats. Condor 1989, 91:416–428.
38. Hill GE: Geographic variation in carotenoid plumage pigmentation of house finches. Biol J Lin Soc 1993, 49:63–86.
39. Soobramoney S, Perrin MR: The effect of bill structure on seed selection and handling ability of five species of granivorous birds. Emu 2007, 107:169–176.
40. Hogan BL: Bone morphogenetic proteins in development. Curr Opin Genet Dev 1996, 6:432–438.
41. Urist MR: Bone morphogenetic protein: the molecularization of skeletal system development. J Bone Miner Res 1997, 12:343–346.
42. Tsumaki N, Yoshikawa H: The role of bone morphogenetic proteins in endochondral bone formation. Cytokine Growth 2005, 16:279–285.
43. Young RL, Badyaev AV: Evolution of ontogeny: linking epigenetic remodeling and genetic adaptation in skeletal structures. Integr Comp Biol 2007, 47:234–244.
44. Badyaev AV, Martin TE: Individual variation in growth trajectories: Phenotypic and genetic correlations in ontogeny of the house finch (Carpodacus mexicanus). J Evol Biol 2000, 13:290–302.
45. Podos J: Correlated evolution of morphology and vocal signal structure in Darwin's finches. Nature 2001, 409:185–188.

46. Bermúdez-Cuamatzin E, Ríos-Chelén AA, Gil D, Macías Garcia C:
Experimental evidence for real-time song frequency shift in response to
urban noise in a passerine bird. *Biol Lett* 2011, **7**:36–38.

47. Catchpole CK, Slater PJB: *Bird Song: Biological Themes and Variations.* New
York: Cambridge University Press; 2008.

48. Badyaev AV, Young RL: **Complexity and integration in sexual
ornamentation: an example with carotenoid and melanin plumage
pigmentation.** *J Evol Biol* 2004, **17**:1317–1327.

49. Halfwerk W, Bot S, Buikx J, van der Velde M, Komdeur J, Cate CT,
Slabbekoorn H: **Low-frequency songs lose their potency in noisy urban
conditions.** *Proc Natl Acad Sci U S A* 2011, **108**:14549–14554.

50. Bowman R, Woolfenden GE: **Nest site selection by Florida scrub-jays in
natural and human-modified habitats.** *Wilson Bull* 2002, **114**:128–135.

51. Bokoni V, Seress G, Szabolcs N, Lendvai AZ, Liker A: **Multiple indices of
body condition reveal no negative effect of urbanization in adult house
sparrows.** *Landscape Urban Plann* 2012, **104**:75–84.

52. Stefanov WL, Ramsey MS, Christensen PR: **Monitoring urban land cover
change: an expert system approach to land cover classification of
semiarid to arid urban centers.** *Remote Sens Environ* 2001, **77**:173–185.

53. Stefanov WL, Netzband M: **Assessment of ASTER land cover and MODIS
NDVI data at multiple scales for ecological characterization of an arid
urban center.** *Remote Sens Environ* 2005, **99**:31–43.

54. Stefanov WL, Netzband M, Möller MS, Redman CL, Mack C: **Phoenix,
Arizona, USA: Applications of Remote Sensing in a Rapidly Urbanizing
Desert Region.** In *Applied Remote Sensing for Urban Planning, Governance
and Sustainability.* Edited by Netzband M, Stefanov WL, Redman CL. Berlin:
Springer; 2007:137–164.

55. Nakagawa S: **A farewell to Bonferroni: the problems of low statistical
power and publication bias.** *Behav Ecol* 2004, **15**:1044–1045.

6

Differentiated adaptive evolution, episodic relaxation of selective constraints, and pseudogenization of umami and sweet taste genes *TAS1Rs* in catarrhine primates

Guangjian Liu[1], Lutz Walter[2,3], Suni Tang[4], Xinxin Tan[1,5], Fanglei Shi[1], Huijuan Pan[6], Christian Roos[2,3]*, Zhijin Liu[1,2]* and Ming Li[1]

Abstract

Background: Umami and sweet tastes are two important basic taste perceptions that allow animals to recognize diets with nutritious carbohydrates and proteins, respectively. Until recently, analyses of umami and sweet taste were performed on various domestic and wild animals. While most of these studies focused on the pseudogenization of taste genes, which occur mostly in carnivores and species with absolute feeding specialization, omnivores and herbivores were more or less neglected. Catarrhine primates are a group of herbivorous animals (feeding mostly on plants) with significant divergence in dietary preference, especially the specialized folivorous Colobinae. Here, we conducted the most comprehensive investigation to date of selection pressure on sweet and umami taste genes (*TAS1Rs*) in catarrhine primates to test whether specific adaptive evolution occurred during their diversification, in association with particular plant diets.

Results: We documented significant relaxation of selective constraints on sweet taste gene *TAS1R2* in the ancestral branch of Colobinae, which might correlate with their unique ingestion and digestion of leaves. Additionally, we identified positive selection acting on Cercopithecidae lineages for the umami taste gene *TAS1R1*, on the Cercopithecinae and extant Colobinae and Hylobatidae lineages for *TAS1R2*, and on *Macaca* lineages for *TAS1R3*. Our research further identified several site mutations in Cercopithecidae, Colobinae and *Pygathrix*, which were detected by previous studies altering the sensitivity of receptors. The positively selected sites were located mostly on the extra-cellular region of TAS1Rs. Among these positively selected sites, two vital sites for TAS1R1 and four vital sites for TAS1R2 in extra-cellular region were identified as being responsible for the binding of certain sweet and umami taste molecules through molecular modelling and docking.

Conclusions: Our results suggest that episodic and differentiated adaptive evolution of *TAS1Rs* pervasively occurred in catarrhine primates, most concentrated upon the extra-cellular region of TAS1Rs.

Keywords: Catarrhine primates, *TAS1Rs*, Adaptive evolution, Positive selection, Episodic relaxation of selective constraints, Pseudogenization

* Correspondence: croos@dpz.eu; liuzj@ioz.ac.cn
[2]Primate Genetics Laboratory, German Primate Center, Leibniz Institute for Primate Research, Kellnerweg 4, 37077 Göttingen, Germany
[1]Key Laboratory of Animal Ecology and Conservation Biology, Institute of Zoology, Chinese Academy of Sciences, 1-5 Beichen West Road, Chaoyang, Beijing 100101, China
Full list of author information is available at the end of the article

Background

Umami and sweet tastes are important sensations allowing animals to recognize diets with nutritious proteins and carbohydrates, respectively. In mammals, sweet and umami taste perceptions are conferred by taste receptor cells through the use of G protein-coupled receptors (GPCRs) TAS1R, which are encoded by the *TAS1R* gene family [1,2]. Of these, TAS1R1 and TAS1R2 are expressed in separate taste receptor cells, although both are co-expressed with TAS1R3. The TAS1R1 protein forms a heterodimer with TAS1R3 to form a two-part umami taste receptor, and the TAS1R2 and TAS1R3 heterodimer functions as the sweetness receptor [3,4].

The gene structure of *TAS1R* family is conserved among species [5]. Human *TAS1R* genes span from 3 kb-20 kb with 6 exons and 5 introns, and the cDNA of *TAS1R* genes consist of ~2500 bp. TAS1R proteins are characteristic of seven domains spanning the plasma membrane, which have a large N-terminal extracellular domain composed of the Venus flytrap module (VFTM) and the small cysteine-rich domain (CRD), followed by the transmembrane domain (TMD) and the C-terminal intracellular domain (CID) [6]. These domains are required for recognizing sweeteners and umami compounds, such as aspartame, neotame, monellin, cyclamate, neohesperidin dihydrochalcone, brazzein and L-amino acids [7-12]. Furthermore, studies using mutagenesis, molecular modeling and functional expression have demonstrated multiple potential binding sites in the heterodimeric receptors [8-11,13-19]. For example, it is reported that two amino acid substitutions (A110V and R507Q) in the VFTM domain of TAS1R1 and two substitutions (F749S and R757C) in the TMD domain of TAS1R3, severely impair the response to monosodium glutamate (MSG) in humans [17]. Above all, genetic factors have been shown to play a crucial role in the variability of sensitivity to tastants.

Until recently, analyses of umami and sweet taste receptors have been performed on various domestic and wild animals. For example, *TAS1R1* is a pseudogene in the herbivorous giant panda (family Ursidae) [20,21], and the *TAS1R2* gene is inactivated in cats (family Felidae), vampire bats, chickens, zebra finches, the western clawed frog, and some carnivorous mammals [5,22,23]. However, research on the evolution of sweet and umami taste genes revealed that taste perception of sweet and umami is not be as conserved as previously thought, and the structural integrity of *TAS1R* is sometimes inconsistent with the known functions of these genes and the tastes involved [24]. For example, although *TAS1R2* is a pseudogene in some species of carnivores, some other obligate carnivores (such as ferrets and Canadian otters) still possess an intact *TAS1R2* gene [25]. Moreover, *TAS1R2* is absent in all bird genomes sequenced thus far, irrespective of their diet [21]. These puzzling cases indicate that our understanding of the physiological functions of sweet and umami tastes and/or their receptor genes is far from complete [24].

While most previous studies in mammals focused on the pseudogenization of taste genes which occur mostly in carnivores and species with absolute feeding specialization, omnivores and herbivores were more or less neglected, except for the systematic study of bats and the giant panda [26]. Additionally, it is known that plant foods usually contain much more complex and variable taste-inducing compounds than animal food [27,28]. The taste system of omnivores and herbivores may be more complex than in carnivores, and pseudogenization of taste genes is less possible to occur massively in omnivores and herbivores (except for the giant panda). To thoroughly understand the physiological functions and adaptive evolution of sweet and umami tastes and/or their receptor genes, more comprehensive studies should be performed in omnivorous and herbivorous animals with close phylogenetic relationships.

Catarrhine primates are a group of herbivorous animals (feeding mostly on plants) with a significant divergence in dietary preference. The Colobinae, also called leaf-monkeys, feed predominantly on relatively low-energy leaves and other plant parts [29], and are unique among primates in that they have a complex stomach to permit efficient digestion of leaves [30]. By contrast, the Cercopithecinae feed predominantly on relatively high-energy foods such as fruits, seeds, insects, and vertebrates. Gibbons are fruit-pulp specialists, and the foods eaten by the great apes and modern humans generally include a wide variety of items such as fruits, assorted types of vegetation, bark, seeds, insects and meat, although the great apes are predominantly frugivorous. Consequently, as a group of animals including omnivores and herbivores, catarrhine primates exhibit a certain degree of diet specialization and differentiation, which makes them ideal objects for studying the evolution of sweet and umami taste genes.

As diets have evolved and differentiated during the radiation of catarrhine primates, presumably tastes have responded adaptively in order to maximize energy intake. Species with the highest taste sensitivity for sugars and other soluble nutrients tend to improve foraging efficiency, which could be the target of natural selection. Thus, we predicted that the umami and sweet taste genes of species in catarrhine primates with different diets have experienced variation in selection pressure through evolutionary history. To test this hypothesis, we performed a comprehensive investigation of evolution in *TAS1R* genes for representative species of herbivorous catarrhine primates. Our research aimed to explore possible specific adaptive evolution of *TAS1R* in catarrhine primates with known dietary specializations, and to enrich our understanding of the physiological functions of sweet and umami tastes receptor genes.

Results

Characterization of the *TAS1R* genes in catarrhine primates

Complete coding regions of *TAS1R* were obtained from 30 catarrhine primate species. Most of them were highly conservative without premature stop codons or frame shift mutations. In total, 2523 bp, 2517 bp and 2556 bp DNA sequences were generated from six exons of *TAS1R1*, *TAS1R2* and *TAS1R3*, respectively. Additionally, we also downloaded available *TAS1R* sequences of Hominidae from GenBank: all *TAS1R* of *Pan troglodytes*, *Homo sapiens* and *Gorilla gorilla gorilla*; *TAS1R1* and *Tas1r2* of *Pan paniscus*; *TAS1R1* and *TAS1R3* of *Pongo abelii*; *Tas1r2* and *TAS1R3* of *Pongo pygmaeus* (see Additional file 1). Alignments of *TAS1R* revealed a total of 334 (13.24%), 398 (15.81%) and 469 (18.35%) variable sites for *TAS1R1*, *TAS1R2* and *TAS1R3*, respectively. Among these sites, 243, 318 and 414 synonymous mutations were found in *TAS1R1*, *TA1SR2*

and *TAS1R3*, respectively. Correspondingly, 185, 263 and 269 non-synonymous mutations were detected in *TAS1R1*, *TAS1R2* and *TAS1R3*, which revealed the most variable sites but the lowest non-synonymous ratio of *TAS1R3* (see Additional file 2). The alignments of amino acid sequences are provided as Additional files 3, 4 and 5.

ORF-disrupting mutations in *TAS1R* genes

The open reading frame (ORF) of *TAS1R* was disrupted in some species of Cercopithecidae, including insertions, deletions and transitions (Figure 1). Both of two individuals of *Pygathrix nemaeus* had an allele of *TAS1R1* with an insertion of a G in exon 3 (nucleotide position 695, codon number 232), leading to a frame shift mutation and a premature stop codon. Thus, *TAS1R1* is likely a pseudogene in *P. nemaeus*, resulting in the truncation of the protein in the extracellular N-terminus (VFTM).

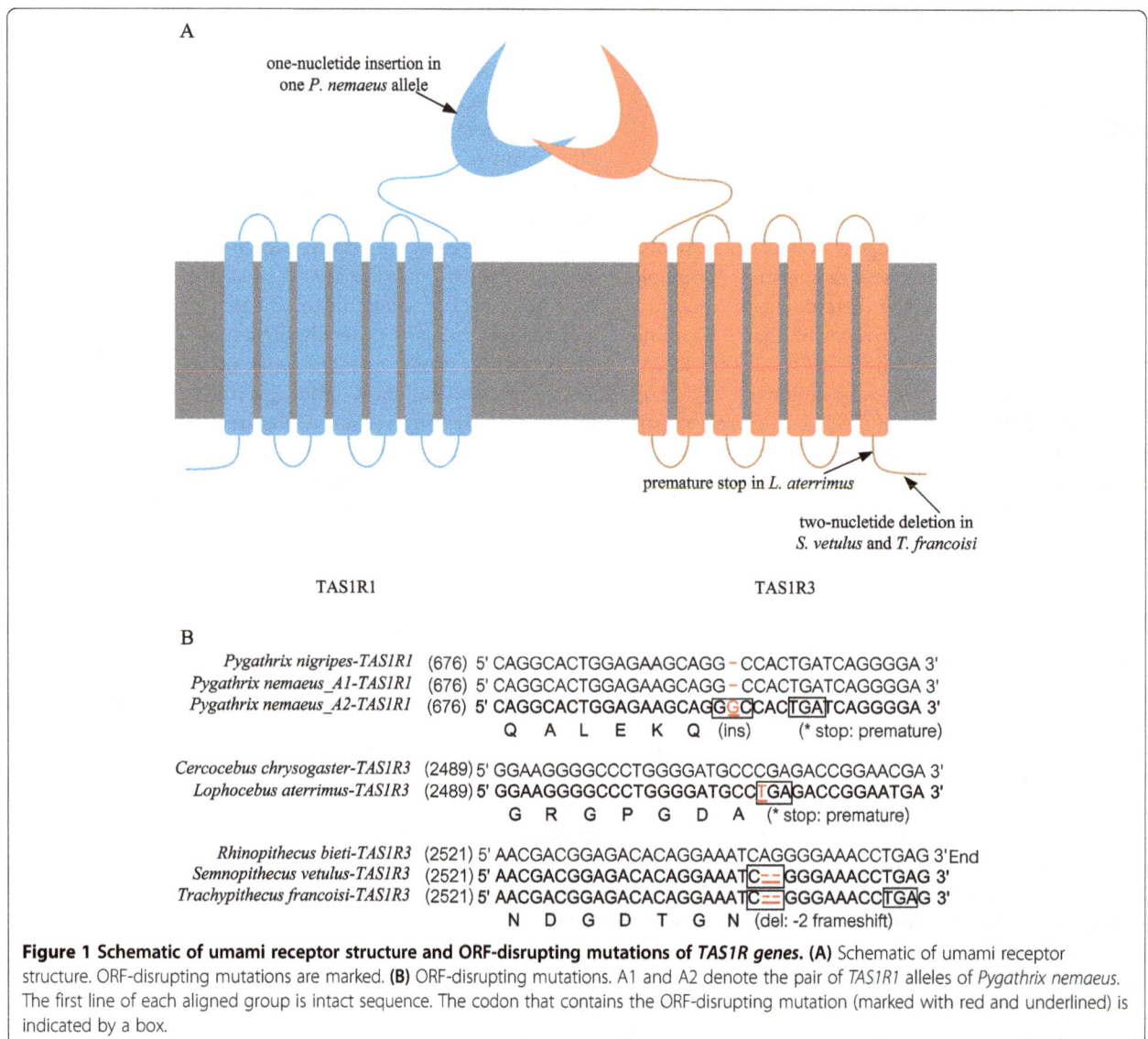

Figure 1 Schematic of umami receptor structure and ORF-disrupting mutations of *TAS1R* genes. (A) Schematic of umami receptor structure. ORF-disrupting mutations are marked. **(B)** ORF-disrupting mutations. A1 and A2 denote the pair of *TAS1R1* alleles of *Pygathrix nemaeus*. The first line of each aligned group is intact sequence. The codon that contains the ORF-disrupting mutation (marked with red and underlined) is indicated by a box.

However, this gene is intact in all other *Pygathrix* species *P. nigripes* and *P. cinerea*, indicating a recent origin of this pseudogene. *Trachypithecus francoisi* and *Semnopithecus vetulus* share a 2-nucleotide deletion at the very end of exon 6 (nucleotide position 2,543-2,544, codon number 848) of *TAS1R3*, leading to three amino acid substitutions and a two amino acid residue shorter C-terminal intracellular domain (CID) in comparison with other species. Thus, this mutation of *TAS1R3* was suggested to occur in the common ancestor of *Semnopithecus* and *Trachypithecus*, which are the closest genera in Colobinae. Interestingly, a premature stop codon was also found in *TAS1R3* of *Lophocebus*

aterrimus because of a transition from C to T in exon 6 (nucleotide position 2,512, codon number 838). All these mutations were confirmed by multiple PCR experiments and colonies of different cloning procedures.

Relaxation of selective constraint on *TAS1R* genes within catarrhine primates

In the branch model, the free-ratio model was detected to be significantly different from the one-ratio model only for *TAS1R2*, indicating that d_N/d_S ratios among lineages of *TAS1R2* were different (Table 1). These ratios varied from 0 ($d_N = 0$) to infinity ($d_S = 0$). Then, the two-ratio model was

Table 1 CODEML analyses of selective pattern on the *TAS1R2* in catarrhine primates

Models	lnL	Compared	2ΔLnL	P-value	Parameter estimates
Branch model					
M0:one-ratio	−7258.8072				$\omega = 0.1608$
M1:free-ratio	−7211.2267	M0 vs. M1	95.1610	$p < 0.05$	
Foreground branch: ancestral Colobinae					
H1: ω_0, ω_1	−7256.0276	M0 vs. H1	5.5592	$p < 0.05$	$\omega_0 = 0.1548$, $\omega_1 = 0.5583$
H0: ω_0, $\omega_1 = 1$	−7256.4033	H0 vs. H1	0.7514	$p > 0.05$	$\omega_0 = 0.1543$, $\omega_1 = 1.0000$
Foreground branch: ancestral *Hylobates*					
H1: ω_0, ω_1	−7256.8606	M0 vs. H1	3.8932	$p < 0.05$	$\omega_0 = 0.1570$, $\omega_1 = 0.5554$
H0: ω_0, $\omega_1 = 1$	−7257.1816	H0 vs. H1	0.6420	$p > 0.05$	$\omega_0 = 0.1570$, $\omega_1 = 1.0000$
Site model					
M1a	−7149.5623				$p_0 = 0.8609$, $p_1 = 0.1391$, $\omega_0 = 0.0451$, $\omega_1 = 1.0000$
M2a	−7145.1097	M1a vs. M2a	8.9052	$p < 0.05$	$p_0 = 0.8669$, $p_1 = 0.1198$, $p_2 = 0.0133$, $\omega_0 = 0.0499$, $\omega_1 = 1.0000$, $\omega_2 = 3.6223$
M8a	−7149.3577				$p_0 = 0.8725$($p_1 = 0.1275$), $p = 0.4874$, $q = 7.5964$, $\omega = 1.0000$
M8	−7143.4448	M8a vs. M8	11.8258	$p < 0.01$	$p_0 = 0.9744$($p_1 = 0.0257$), $p = 0.1750$, $q = 1.1154$, $\omega = 2.8718$
Branch-site model					
Lineages of Cercopithecidae					
Null	−7146.1093				$p_0 = 0.8463$, $p_1 = 0.0953$, $p_{2a} = 0.0525$ $p_{2b} = 0.0059$, $\omega_0 = 0.0407$, $\omega_1 = 1.0000$ $\omega_2 = 1.0000$
Alternative	−7144.2553		3.7080	$p < 0.05$	$p_0 = 0.8683$, $p_1 = 0.0924$, $p_{2a} = 0.0354$ $p_{2b} = 0.0038$, $\omega_0 = 0.0481$, $\omega_1 = 1.000$ $\omega_2 = 2.0345$
Lineages of Colobinae					
Null	−7145.5063				$p_0 = 0.8299$, $p_1 = 0.1176$, $p_{2a} = 0.0460$ $p_{2b} = 0.0065$, $\omega_0 = 0.0396$, $\omega_1 = 1.0000$, $\omega_2 = 1.0000$
Alternative	−7143.7154		3.5818	$p < 0.05$	$p_0 = 0.8542$, $p_1 = 0.1162$, $p_{2a} = 0.0261$ $p_{2b} = 0.0036$, $\omega_0 = 0.0428$, $\omega_1 = 1.0000$ $\omega_2 = 2.4997$,
Lineages of Cercopithecinae					
Null	−7148.7133				$p_0 = 0.8512$, $p_1 = 0.1292$, $p_{2a} = 0.0170$ $p_{2b} = 0.0026$, $\omega_0 = 0.0440$, $\omega_1 = 1.0000$, $\omega_2 = 1.0000$
Alternative	−7143.4285		10.5696	$p < 0.01$	$p_0 = 0.8639$, $p_1 = 0.1228$, $p_{2a} = 0.0116$ $p_{2b} = 0.0017$, $\omega_0 = 0.0466$, $\omega_1 = 1.0000$ $\omega_2 = 5.8766$
Lineages of Hylobatidae					
Null	−7149.1766				$p_0 = 0.8489$, $p_1 = 0.1326$, $p_{2a} = 0.0161$ $p_{2b} = 0.0025$, $\omega_0 = 0.0442$, $\omega_1 = 1.000$ $\omega_2 = 1.0000$
Alternative	−7143.6646		11.0240	$p0.01$	$p_0 = 0.8612$, $p_1 = 0.1297$, $p_{2a} = 0.0079$ $p_{2b} = 0.0012$, $\omega_0 = 0.0462$, $\omega_1 = 1.000$ $\omega_2 = 11.8832$

used to identify whether lineages with $0.50 < \omega < 1$ (detected by free-ratio) had underwent different selection pressure from other lineages, which was addressed as hypothesis 1 (H1) in Table 1. After comparing with the one-ratio model, we found that selection pressure acting on the ancestral lineage of Colobinae for *TAS1R2* (branch ACo in Figure 2) was significantly different from other lineages ($\omega_1 = 0.5583$, $p < 0.05$, Table 1). We also hypothesized a fixing $\omega_1 = 1$ for the ancestral lineage of *TAS1R2* in Colobinae (H0, Table 1), and H1 was not significantly supported compared with H0. The analysis revealed that selection constraints were relaxed on *TAS1R2* of ancestral Colobinae, but it was not totally removed. The relaxation of selective constraints was also found in *TAS1R1* of the common ancestor of human, chimpanzee and gorilla ($\omega_1 = 0.7772$, $p < 0.05$, Table 2, Figure 3), and in *TAS1R2* of the common ancestor of the genus *Hylobates* ($\omega_1 = 0.5554$, $p < 0.05$, Table 1, Figure 2).

Positively selected sites on *TAS1R* genes within catarrhine primates

In site models, LRTs showed that the incorporate selections (i.e. M2a and M8) fitted significantly better than neutral models (i.e. M1a and M8a) for *TAS1R1* and *TAS1R2*, whereas no significant evidence of positive selection was found for *TAS1R3* (Tables 1, 2 and 3). In the M2a model, six and one sites for *TAS1R1* and *TAS1R2*, respectively, were under positive selection. Model M8 revealed six and 10 positively selected sites for *TAS1R1* and *TAS1R2*, respectively, identified by the BEB approach with posterior probabilities larger than 0.85 (see Additional file 6).

The branch-site model was then used to test for positive selection in potential codons in lineages of separate groups of catarrhine primates, i.e., in Cercopithecidae, Cercopithecinae, Colobinae, Hylobatidae, and Hominidae (Tables 1, 2 and 3). The LRT results showed that Hylobatidae-specific lineages for *TAS1R2*, and Cercopithecidae-specific lineages for both *TAS1R1* and *TAS1R2* were subjected to strong positive selection. Further analysis indicated that lineages of Cercopithecinae and Colobinae for *TAS1R2* (Figure 2) and *Macaca*-specific lineages for *TAS1R3* were also under strong positive selection (Figure 4). In addition, 11 codons (*TAS1R1*: 391; *TAS1R2*: 8, 21, 175, 404, 411, 413, 510 and 733; *TAS1R3*: 195 and 225) were also

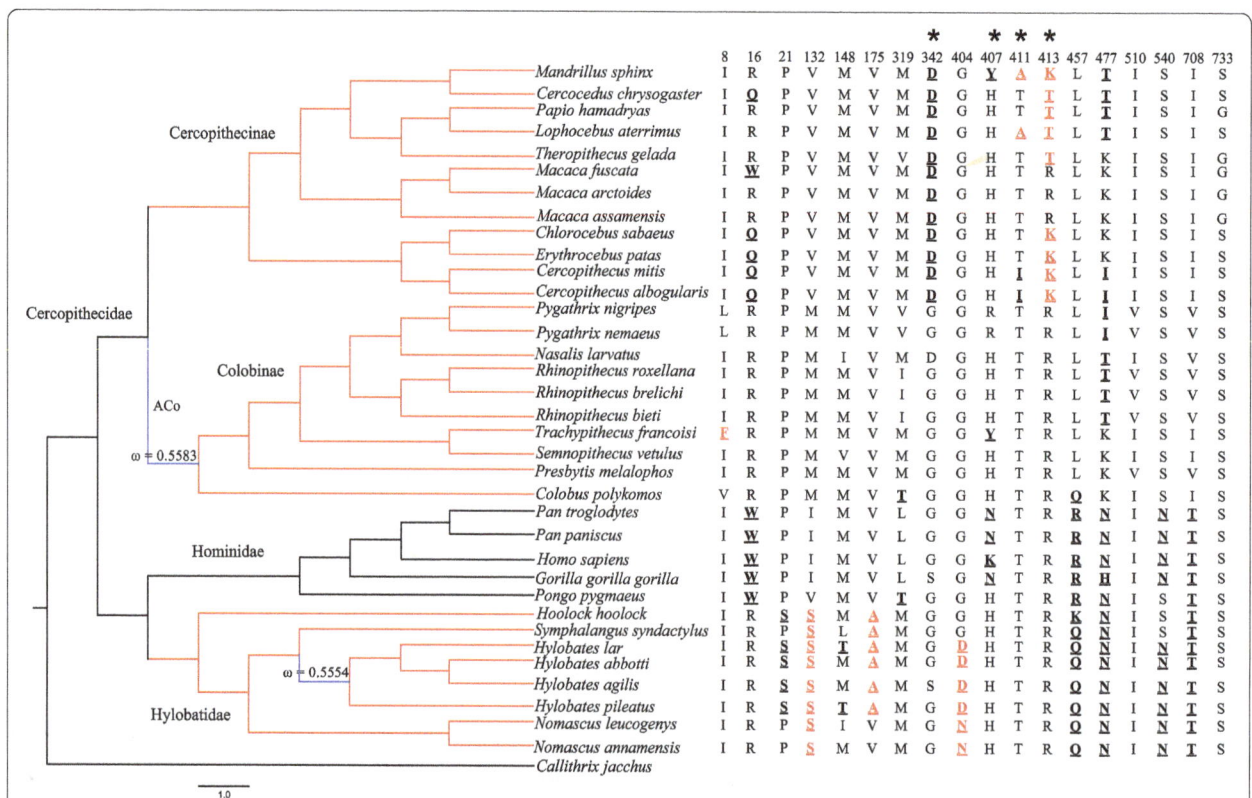

Species	8	16	21	132	148	175	319	342	404	407	411	413	457	477	510	540	708	733
Mandrillus sphinx	I	R	P	V	M	V	M	D	G	Y	A	K	L	T	I	S	I	S
Cercocebus chrysogaster	I	Q	P	V	M	V	M	D	G	H	T	T	L	T	I	S	I	S
Papio hamadryas	I	R	P	V	M	V	M	D	G	H	T	T	L	T	I	S	I	G
Lophocebus aterrimus	I	R	P	V	M	V	M	D	G	H	A	T	L	T	I	S	I	S
Theropithecus gelada	I	R	P	V	M	V	M	D	G	H	T	T	L	K	I	S	I	G
Macaca fuscata	I	W	P	V	M	V	M	D	G	H	T	R	L	K	I	S	I	G
Macaca arctoides	I	R	P	V	M	V	M	D	G	H	T	R	L	K	I	S	I	G
Macaca assamensis	I	R	P	V	M	V	M	D	G	H	T	R	L	K	I	S	I	G
Chlorocebus sabaeus	I	Q	P	V	M	V	M	D	G	H	T	K	L	K	I	S	I	S
Erythrocebus patas	I	Q	P	V	M	V	M	D	G	H	T	K	L	K	I	S	I	S
Cercopithecus mitis	I	Q	P	V	M	V	M	D	G	H	I	K	L	I	I	S	I	S
Cercopithecus albogularis	I	Q	P	V	M	V	M	D	G	H	I	K	L	I	I	S	I	S
Pygathrix nigripes	L	R	P	M	M	V	V	G	G	R	T	R	L	I	V	S	V	S
Pygathrix nemaeus	L	R	P	M	M	V	V	G	G	R	T	R	L	I	V	S	V	S
Nasalis larvatus	I	R	P	M	I	V	M	D	G	H	T	R	L	T	I	S	V	S
Rhinopithecus roxellana	I	R	P	M	M	V	I	G	G	H	T	R	L	T	V	S	V	S
Rhinopithecus brelichi	I	R	P	M	M	V	I	G	G	H	T	R	L	T	V	S	V	S
Rhinopithecus bieti	I	R	P	M	M	V	I	G	G	H	T	R	L	T	V	S	V	S
Trachypithecus francoisi	F	R	P	M	M	V	M	G	G	Y	T	R	L	K	I	S	I	S
Semnopithecus vetulus	I	R	P	M	V	V	M	G	G	H	T	R	L	K	I	S	I	S
Presbytis melalophos	I	R	P	M	M	V	M	G	G	H	T	R	L	K	V	S	V	S
Colobus polykomos	V	R	P	M	M	V	T	G	G	H	T	R	Q	K	I	S	I	S
Pan troglodytes	I	W	P	I	M	V	L	G	G	N	T	R	R	N	I	N	T	S
Pan paniscus	I	W	P	I	M	V	L	G	G	N	T	R	R	N	I	N	T	S
Homo sapiens	I	W	P	I	M	V	L	G	G	K	T	R	R	N	I	N	T	S
Gorilla gorilla gorilla	I	W	P	I	M	V	L	S	G	N	T	R	R	H	I	N	T	S
Pongo pygmaeus	I	W	P	V	M	V	T	G	G	H	T	R	R	N	I	S	T	S
Hoolock hoolock	I	R	S	M	A	M	G	G	H	T	R	K	N	I	S	T	S	
Symphalangus syndactylus	I	R	P	S	L	A	M	G	G	H	T	R	Q	N	I	S	T	S
Hylobates lar	I	R	S	S	T	A	M	G	G	D	H	T	R	Q	N	I	N	T	S
Hylobates abbotti	I	R	S	S	M	A	M	G	G	D	H	T	R	Q	N	I	N	T	S
Hylobates agilis	I	R	S	S	M	A	M	S	G	D	H	T	R	Q	N	I	N	T	S
Hylobates pileatus	I	R	S	S	T	A	M	G	G	D	H	T	R	Q	N	I	N	T	S
Nomascus leucogenys	I	R	P	S	I	V	M	G	G	N	H	T	R	Q	N	I	N	T	S
Nomascus annamensis	I	R	P	S	M	V	M	G	G	N	H	T	R	Q	N	I	N	T	S
Callithrix jacchus																		

Figure 2 Radical amino acid changes in selected sites across the catarrhine primates phylogeny of *TAS1R2*. The input tree used for pressure analysis of *TAS1R2* is shown on the left. Positively selected sites detected by site models and branch-site models are shown on the right. Radical changes are shown in bold and underlined, and changes detected by branch-site models are marked in red additionally. *Callithrix jacchus* was used as outgroup, but not included in the CODEML analysis. Lineages under positive selection are marked in red, while lineages experienced relaxation of selective constraints are marked in blue.

Table 2 CODEML analyses of selective pattern on the *TAS1R1* in catarrhine primates

Models	lnL	Compared	2ΔLnL	P-value	Parameter estimates
Branch model					
M0:one-ratio	−6347.5127				$\omega = 0.2486$
M1:free-ratio	−6316.3339	M0 vs. M1	62.3576	$p > 0.05$	
Foreground branch: all lineages of Cercopithecidae except for the ancestral branch					
H1: ω_0, ω_1	−6344.8700	M0 vs. H1	5.2854	$p < 0.05$	$\omega_0 = 0.3112$, $\omega_1 = 0.1981$
H0: $\omega_0, \omega_1 = 1$	−6412.3663	H0 vs. H1	134.9926	$p < 0.01$	$\omega_0 = 0.3149$, $\omega_1 = 1.0000$
Foreground branch: ancestral of *Mandrillus, Cercocebus, Papio, Lophocebus, Theropithecus* and *Macaca*					
H1: ω_0, ω_1	−6343.4866	M0 vs. H1	8.0522	$p < 0.01$	$\omega_0 = 0.2419$, $\omega_1 = +\infty$
Foreground branch: ancestral of *Pan, Homo* and *Gorilla*					
H1: ω_0, ω_1	−6345.0477	M0 vs. H1	4.9300	$p < 0.05$	$\omega_0 = 0.2379$, $\omega_1 = 0.7772$
H0: $\omega_0, \omega_1 = 1$	−6345.1667	H0 vs. H1	0.2380	$p > 0.05$	$\omega_0 = 0.2379$, $\omega_1 = 1.0000$
Site model					
M1a	−6316.1817				$p_0 = 0.8312$, $p_1 = 0.1688$, $\omega_0 = 0.0859$, $\omega_1 = 1.0000$
M2a	−6310.5399	M1a vs. M2a	11.2836	$p < 0.01$	$p_0 = 0.8708$, $p_1 = 0.1159$, $p_2 = 0.0133$, $\omega_0 = 0.1137$, $\omega_1 = 1.0000$, $\omega_2 = 4.7335$
M8a	−6316.2094				$p_0 = 0.8318(p_1 = 0.1682)$, $p = 9.4660$, $q = 99.0000$, $\omega = 1.0000$
M8	−6310.7199	M8a vs. M8	10.9790	$p < 0.01$	$p_0 = 0.9794(p_1 = 0.0206)$, $p = 0.4478$, $q = 1.7622$, $\omega = 4.0147$
Branch-site model					
Lineages of Cercopithecidae					
Null	−6316.1817				$p_0 = 0.8312$, $p_1 = 1688$ $p_{2a} = 0.0000$, $p_{2b} = 0.0000$ $\omega_0 = 0.0859$, $\omega_1 = 1.0000$ $\omega_2 = 1.0000$
Alternative	−6308.6453		15.0728	$p < 0.01$	$p_0 = 0.8384$, $p_1 = 1583$ $p_{2a} = 0.0027$, $p_{2b} = 0.0005$ $\omega_0 = 0.0936$, $\omega_1 = 1.0000$ $\omega_2 = 13.4610$

$+\infty$ means infinite, and in this case $d_S=0$ thus ω $(d_N/d_S)= +\infty$.

Table 3 CODEML analyses of selective pattern on the *TAS1R3* in catarrhine primates

Models	lnL	Compared	2ΔLnL	P-value	Parameter estimates
Branch model					
M0:one-ratio	−7676.8782				$\omega = 0.1342$
M1:free-ratio	−7638.3982	M0 vs. M1	76.9600	$p > 0.05$	
Site model					
M1a	−7623.3051				$p_0 = 0.8913$, $p_1 = 0.1087$, $\omega_0 = 0.0632$, $\omega_1 = 1.0000$
M2a	−7623.3051	M1a vs. M2a	0.0000	$p > 0.05$	$p_0 = 0.8913$, $p_1 = 0.0609$, $p_2 = 0.0478$, $\omega_0 = 0.0632$, $\omega_1 = 1.000$, $\omega_2 = 1.0000$
M8a	−7619.6648				$p_0 = 0.9461(p_1 = 0.0539)$, $p = 0.2391$, $q = 1.8178$, $\omega = 1.0000$
M8	−7619.7175	M8a vs. M8	0.1054	$p > 0.05$	$p_0 = 0.9678(p_1 = 0.0322)$, $p = 0.2634$, $q = 1.9053$, $\omega = 1.1121$
Branch-site model					
Lineages of *Macaca*					
Null	−7622.2362				$p_0 = 0.8389$, $p_1 = 0.1004$ $p_{2a} = 0.0543$, $p_{2b} = 0.0065$ $\omega_0 = 0.0619$, $\omega_1 = 1.000$, $\omega_2 = 1.0000$
Alternative	−7620.4248		3.6228	$p < 0.05$	$p_0 = 0.8877$, $p_1 = 0.1058$ $p_{2a} = 0.0059$, $p_{2b} = 0.0007$ $\omega_0 = 0.0621$, $\omega_1 = 1.000$, $\omega_2 = 21.4208$

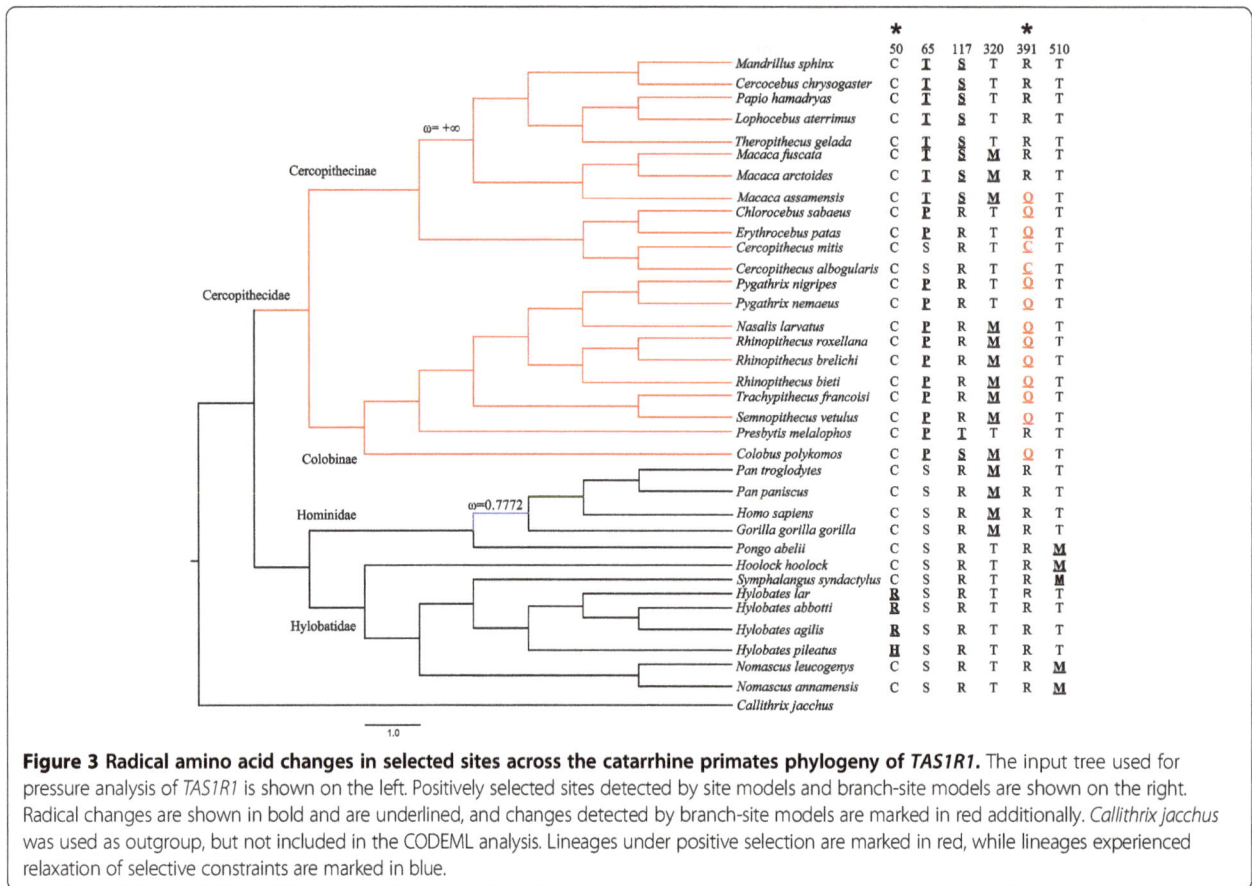

Figure 3 Radical amino acid changes in selected sites across the catarrhine primates phylogeny of *TAS1R1*. The input tree used for pressure analysis of *TAS1R1* is shown on the left. Positively selected sites detected by site models and branch-site models are shown on the right. Radical changes are shown in bold and are underlined, and changes detected by branch-site models are marked in red additionally. *Callithrix jacchus* was used as outgroup, but not included in the CODEML analysis. Lineages under positive selection are marked in red, while lineages experienced relaxation of selective constraints are marked in blue.

identified by the BEB approach with posterior probabilities larger than 0.85 in branch-site analysis (see Additional file 6). Finally, most of these changes detected by both site models and branch-site models were identified to be critical (see Additional file 6).

Prediction of amino acid sites of TAS1R1 and TAS1R2 responsible for umami and sweet taste molecules binding

In order to examine whether positively selected sites detected in this study were responsible for binding of some umami and sweet taste molecules, we built homology models of TAS1R1 and TAS1R2 VFTM domain and docked with some sweet and umami taste molecules (Figure 5). Then, we got a series of docking results and found that most of these molecules bound with taste receptors at specific regions of TAS1R1 and TAS1R2. Finally, we mapped positive selection sites detected in our study onto models of TAS1R1 and TAS1R2 and found that sites 50 and 391 of TAS1R1 may interact with inosine monophosphate (IMP) and L-glutamate (see Additional file 7). Similarly, sites 342, 407, 411 and 413 of TAS1R2 were also found to be within binding domains of D-tryptophan, D-galactose, D-glucose, fructose, galactose and sucrose (see Additional file 8).

Critical sites influencing the response of *TAS1R* genes to tastants

Previous functional expression data showed that mutations in two amino acid residues (S40T and I67S) of TAS1R2 were predicted to affect receptor response to aspartame and neotame, i.e., S40T abolished aspartame sensitivity and I67S reduced response to neotame slightly [31] (Table 4). Investigation of the amino acid residue 40 of TAS1R2 revealed a serine residue (S) in Hylobatidae and Hominidae, but a threonine residue (T) in Cercopithecidae. Furthermore, residue 67 is isoleucine (I) in most species, but methionine (M) in Colobinae (Table 4). Substitution at residue 733 (733A in Hylobatidae and Hominidae, 733 V in Cercopithecidae) of TAS1R3 was previously suggested to reduce its sensitivity to lactisole [32]. Furthermore, substitution R757C in human TAS1R3 seems to reduce responses to monosodium glutamate (MSG) and to increase binding with monopotassium glutamate (MPG) [13,16] (Table 4). This site shows histidine (H) in the genus *Pygathrix* and arginine (R) in other species (Table 4).

Discussion

The present study conducted the most comprehensive investigation to date of selection pressure on sweet and

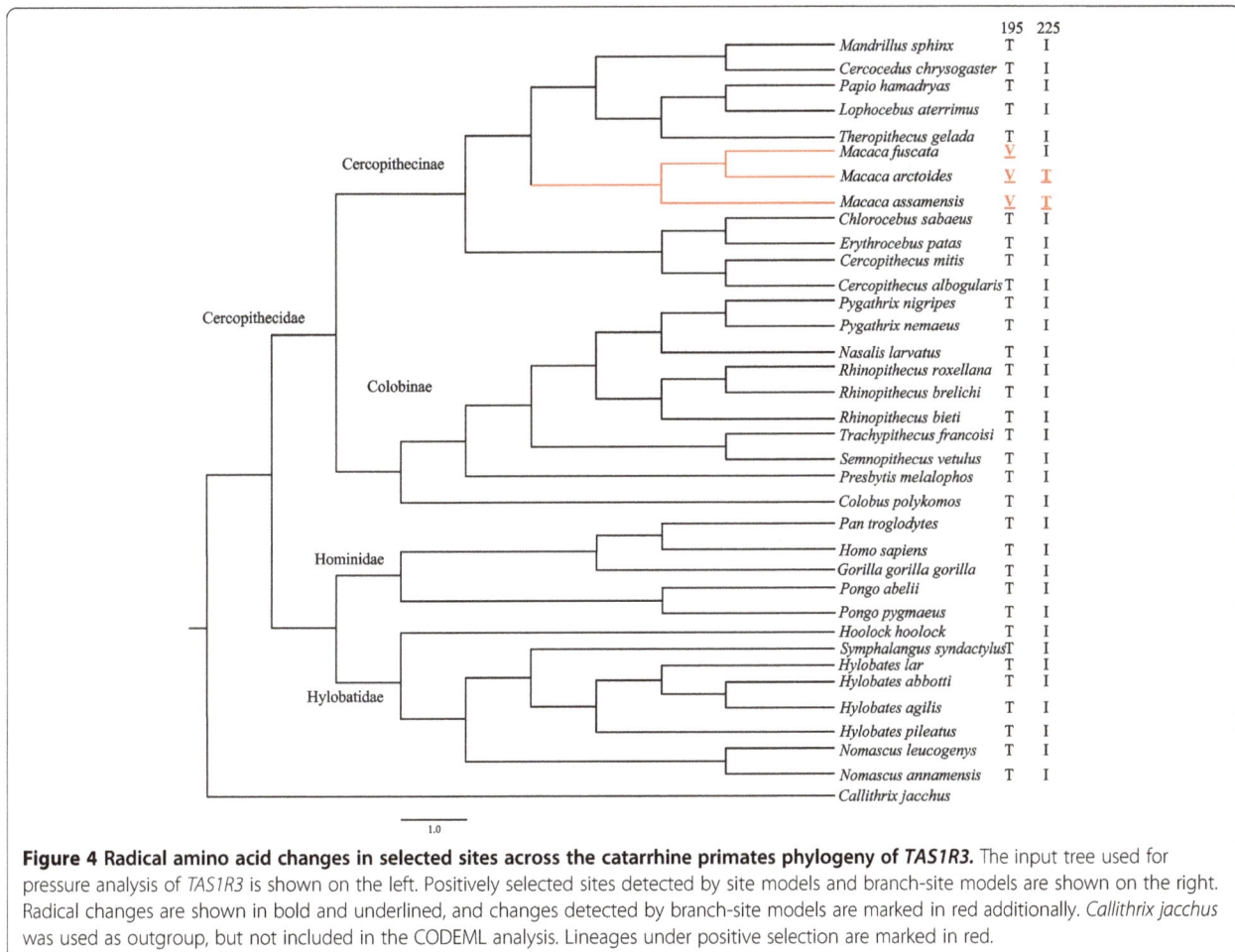

Figure 4 Radical amino acid changes in selected sites across the catarrhine primates phylogeny of *TAS1R3*. The input tree used for pressure analysis of *TAS1R3* is shown on the left. Positively selected sites detected by site models and branch-site models are shown on the right. Radical changes are shown in bold and underlined, and changes detected by branch-site models are marked in red additionally. *Callithrix jacchus* was used as outgroup, but not included in the CODEML analysis. Lineages under positive selection are marked in red.

umami taste genes (*TAS1Rs*) in catarrhine primates to test whether specific adaptive evolution occurred during their diversification in association with particular plant diets. Our results support differentiated evolution and episodic relaxation of selective constraints on *TAS1R* in herbivorous/omnivorous catarrhine primates.

Non-massive pseudogenization of *TAS1Rs* in catarrhine primates

As predicted above, no massive pseudogenization of *TAS1Rs* in catarrhine primates were founded in our data. However, disrupting mutations of *TAS1Rs* were revealed in *P. nemaeus*, *S. vetulus*, *T. francoisi* and *L. aterrimus*, which were supposed to influence species-specific functional differences in sweet and/or umami tastes (Figure 1). In the two species of *Pygathrix* (*P. nigripes* and *P. nemaeus*) tested in this study, one pseudogenized allele resulting in the truncation of the protein in VFTM was found in *TAS1R1* of *P. nemaeus*, which was otherwise intact in *P. nigripes*. Subsequent analysis of the *TAS1R1* gene in *P. cinerea* (data not shown) confirmed that this pseudogenized allele occurred only in *P. nemaeus*. Thus,

considering their close phylogenetic relationships, the cause and biological influence of this one-allele pseudogenization of *TAS1R1* seems to be informative for a comparative analysis of *Pygathrix*. This pseudogenization of one *TAS1R1* allele in *P. nemaeus* likely weakens the biological function of this gene and decreases categories of discernible umami compounds. Suggestively, *P. nemaeus* shows the lowest diversity of recorded consumed plants compared to *P. cinerea* and *P. nigripes* [33].

Unlike the pseudogenized allele of *TAS1R1* in *P. nemaeus* resulting in truncation of the protein in VFTM, the two-nucleotide deletion of *TAS1R3* in *S. vetulus* and *T. francoisi* and the transition from C to T of *TAS1R3* in *L. aterrimus* occurred in the C-terminus of TAS1R3 and leaded to a premature stop codon in C-terminal intracellular domain (CID). Until now, mutations in VRTM, CRD, TMD of TAS1Rs have been reported to influence the sensitivity of TAS1Rs to a variety of substances [13,15,17,31]. However, the d_N/d_S ratio is significantly lower than 1 for CID in rodent TAS2R genes, demonstrating the operation of purifying selection and a vital role of CID [34]. The experimental verification of the function of CID for TAS1Rs is lacking,

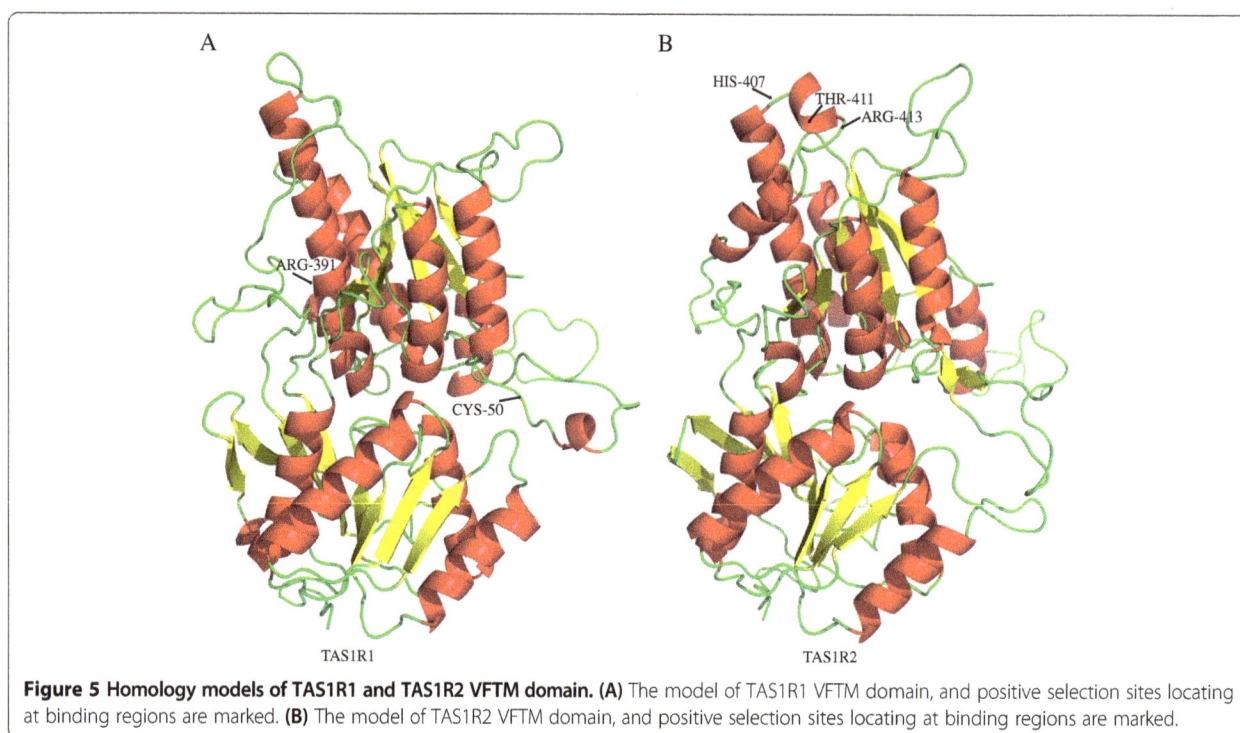

Figure 5 Homology models of TAS1R1 and TAS1R2 VFTM domain. **(A)** The model of TAS1R1 VFTM domain, and positive selection sites locating at binding regions are marked. **(B)** The model of TAS1R2 VFTM domain, and positive selection sites locating at binding regions are marked.

and the influences of these mutations on the CID of TAS1R3 await further research.

Pseudogenization of *TAS1R* genes has been reported in various species with dietary specialization, such as the giant panda (*Ailuropoda melanoleuca*), vampire bats (*Desmodus* spp.) and whales (Cetacea) [21,23,24]. Although the Colobinae display a certain degree of dietary specialization and feeds mainly on leaves, no massive loss of *TAS1R* genes was found. Furthermore, the pseudogene of *TAS1R2* found in *P. nemaeus* was heterozygous. For Cercopithecidae, the premature stop codons occurring in CID of *TAS1R3* in *S. vetulus*, *T. francoisi* and *L. aterrimus* seemed not to disrupt the main structure of TAS1R3, as they only shortened the C-terminus. While, as endangered species, it is difficult to get living organisms for experimental analysis to examine whether these stop codons would terminate *TAS1R3* function in these species. But as the only species of catarrhine primates with amino acids length mutations in such sequence conservative *TAS1Rs*, formation mechanisms and biology function of these mutations are interesting and need further research. Moreover, neither pseudogene nor premature stop codon was found in Hylobatidae, which are typical frugivores. These results imply a functional role of *TAS1R* genes in most (or even all) species of catarrhine primates.

Episodic relaxation of selective constraints and subsequent positive selection on *TAS1R2* in leaf-eating Colobinae

As specific leaf-eating primates, Colobinae attracted our primary attention in this study. The lineage-specific

analysis showed that the ancestral *TAS1R2* of Colobinae (branch ACo in Figure 2) has undergone significantly different selection pressure and relaxation of selective constrains ($\omega_1 = 0.5583$, $p < 0.05$) compared with other lineages (Table 1). The folivorous colobines mostly prefer leaves, and some unripe fruits with low simple and water-soluble sugars; microbes in the foregut of colobines can also degrade crude fibers of leaves and produce sugar, proteins and vitamins [35,36]. Additionally, colobines can recover nutrients by breaking down and digesting bacteria in a true stomach using various enzymes [37,38]. Lysozyme, one of those bacteriolytic enzymes, was documented to be under positive natural selection in a common ancestor of the foregut-fermenting colobine monkeys, which coincided with the establishment of leaf-eating and foregut-fermentation [39]. Therefore, sugar content in food was apparently not very important for ancestral colobines. Moreover, colobines are vulnerable to dietary change because the foregut microbes are not buffered by acidic defenses of the stomach [40]; the high simple sugar content of ripe and pulpy fruits will lower fore-stomach pH value, which can disrupt fermentation and induce bloating or acidosis, resulting in death [41-44]. As a consequence, the function or sensitivity of sweet taste may have been less necessary for the common ancestor of Colobinae, and selection pressure acting on functional sweet taste gene underwent a certain degree of relaxation. Consistent with this assumption, we indeed detected such a relaxation in the ancestral Colobinae *TAS1R2* gene. In addition, we also

Table 4 Functional TAS1R binding sites for sweet and umami tastants from previous studies and the mutation in catarrhine primates

Compounds	Interact receptors	Amino acid mutations	Results	Mutations in catarrhine monkeys	References
Aspartame	hTAS1R2[1]	S40T	abolish aspartame sensitivity	S40T in Cercopithecidae	Liu et al. [31]
Neotame	hTAS1R2	I67S	reduce response to neotame sightly	I67M only in Colobinae	Liu et al. [31]
Lactisole	hTAS1R3	A733V	reduce sensitivity to lactisole	A733V in Cercopithecidae	Jiang et al. [32]
Monosodium glutamate (MSG)	hTAS1R3	R757C	reduce sensitivity to MSG	R757H in *Pygathrix*	Raliou et al. [16]
Monopotassium glutamate (MPG)	hTAS1R3	R757C	increase sensitivity to MPG	R757H in *Pygathrix*	Chen et al. [13]

[1]h denotes human.

found fourteen specific amino acid substitutions in the ancestral *TAS1R2* gene of Colobinae (I67M, E118D, D119N, D225N, P348S, E423K, I436L, S547R, M616V, A635V, A649T, T686M, M697I and V741L) through alignment with the ancestral *TAS1R2* of Cercopithecidae and Cercopithecinae. Interestingly, except for D225N and A649T, all other amino acid substitutions were unique and fixed in colobine species, suggesting that these amino acid changes reflect a necessary adaption to their specific diet.

Interestingly, branch-site analysis indicated that *TAS1R2* was subjected to strong positive natural selection (ω_2 = 3.5818, p <0.05) in lineages of extant colobines, and also suggested substitutions I8F and I510V (posterior probability $\geq$0.850) as candidate sites for selection (Table 1). It seems to be a paradox that *TAS1R2* would have undergone positive selection in extant leaf-eating colobines, which prefer low simple and water-soluble sugars. However, they actually prefer young leaves and only few fruits (especially ripe fruits), which contain certain proportion of proteins and sugars and are also preferred by other catarrhine primates [43-46]. It has been reported that *P. cinerea* eat 49.5% young leaves, 21.9% ripe fruits, 19.1% unripe fruits and only 9.3% mature leaves; the nutritional component in selected food of *P. cinerea* consists of protein at 11.4%, dry matter, lipids at 2.6%, minerals at 5.0%, sugars at 4.9%, starch at 12.8%, and neutral detergent fibers at 40.8% [33]. The sense of sweetness is associated with sugars, some amino acids, a few proteins, and glucosides [47-49]. Perhaps the sensitivity of TAS1R2 for sweeteners in foods is essential for extant colobines. After the establishment of leaf-eating and foregut-fermenting, extant colobines would have undergone further selection to detect new nutrient substance through TAS1R2.

Positive selection of *TAS1R1* in Cercopithecidae

The branch-site model revealed that the *TAS1R1* lineage of all Cercopithecidae underwent positive selection, and that substitutions R391C and R391Q are candidate sites for selection. Primates can sense the presence of protein through the taste of umami, which is elicited by glutamate and glutamic acid [50,51]. Commonly incorporated in leaf protein, glutamic acid is also reported from nectar

and figs [52,53]. Besides dietary differences between Cercopithecinae and Colobinae, both groups also require proteins in plant foods (and animal foods). According to Yeager & Kool [54], colobines select foods of high nutritional value, and young leaves are preferred over mature leaves. Young leaves contain more protein and less fiber than mature leaves [44]. Furthermore, it was reported that *Rhinopithecus bieti* tended to choose leaves with high content of crude protein, as well as insects and even birds [55]. Thus, it is assumed that the taste of umami is important not only for Cercopithecidae but also for Colobinae.

Positive selection of *TAS1R2* in Cercopithecinae

Branch-site analysis showed that *TAS1R2* in Cercopithecinae is subject to strong positive natural selection (ω_2 = 5.8766, p <0.01), and the substitutions T411A, R413I and R413K (posterior probability $\geq$0.850) were indicated as candidate sites for selection (Table 1). However, compared to colobines, for which *TAS1R2* is also under positive natural selection, the candidate sites of positive natural selection in Cercopithecinae are different. The functions of the different positively selected sites of *TAS1R2* in Cercopithecinae and Colobinae await further research, and perhaps these mutations have changed sensitivity for certain substances. It might reflect the common importance of sweetness for Cercopithecinae and extant colobines, along with variation in types and proportions of different kind of sweet substances.

Positive selection of *TAS1R3* in Macaques

TAS1R3 showed positive natural selection in the genus *Macaca*. As described above, *TAS1R3* is the most conserved member among the *TAS1R* gene family due to its basic function in both sweet and umami taste receptors. Therefore, as the only group of lineages in which *TAS1R3* was detected to be under strong positive natural selection, the genus *Macaca* was suggested to possess special sweet and umami tastes and food perceptions different to that of other taxa. The genus *Macaca* represents one of the most successful primate radiations. While most cercopithecines are confined in Africa, except for the Barbary macaque of northern Africa [56], macaques are widely distributed in Asia, north to Japan and west to Afghanistan, which represented the only Asian linage in Cercopithecinae [57-59]. According to fossil data, the earliest macaque originated in northeast Africa around 7 million years ago [60,61], and spread through most of Eurasia [62]. Now, they are found in a wide range of habitats, from evergreen forests to grassland and even areas modified by humans, from tropical forests to temperate ecosystems, and from continents to deep-water islands [57]. This range is likely reflected in dietary diversity as well.

Positive selection of *TAS1R2* in Hylobatidae

Branch-site analysis showed that lineages of Hylobatidae were under significant positive natural selection for *TAS1R2* ($\omega_2 = 11.8832$, $p < 0.01$), indicating adaptive evolution of sweet taste in gibbons. As well-documented frugivores, fruits (especially ripe fruits), account for more than 50% of gibbons' diet, except for the more folivorous siamang (*Symphalangus syndactylus*) [63-66]. It is commonly agreed that food choices of primates are correlated with the nutritional and toxic contents [65,67,68]. Compared with other parts of plant, fruits typically have the highest content of soluble carbohydrates, i.e., sweet tasting compounds [69,70]. As typical frugivorous primates, gibbons seem to intake food with more soluble sweet taste contents than other primates. For example, white-handed gibbons were reported to include a high proportion of carbohydrate-rich fruits in their natural diet [71] and showed clear preferences toward ripe fruits, which typically have the highest content of soluble carbohydrates [69,70]. Furthermore, a food preference test of captive white-handed gibbons with ten types of fruits revealed a highly significant positive correlation between the food preference ranking and total carbohydrates [67]. Accordingly, total carbohydrate content in foods might be an important determinant of food choice for gibbons [72-77], which may contribute to specific adaptive evolution of *TAS1R2*. Four amino acid substitutions (M132S, V175A, G404D and G404N) were detected as candidates for natural selection, which may have influenced the sensitivity of TAS1R2 in gibbons and adapted to their special frugivorous diet.

Why is there no positive selection of *TAS1R* genes in Hominidae?

Interestingly, no signs of positive selection were detected in Hominidae. This could be explained by external and internal reasons. Hominid species are able to acquire, process and consume a wide variety of foods such as fruits, assorted types of vegetation, bark, seeds, insects and meat. They are able to change the composition of diet due to local ecological conditions as they have the ability of tool usage to acquire and consume foods [78].

Thus, sensitivity to sweet and umami taste might not play an overwhelming role in their feeding ecology. From another aspect, sensitivity of taste is also correlated with the body size of primates [47]. A large area of the lingual mucosa in large animals may increase taste performance. This means that the larger the species, the better their taste acuity [47]. As the largest species among the primates, the sensitivity of umami and sweet taste in Hominidae might be sufficiently sensitive for foraging.

Additionally, an episodic relaxation of selective constraints ($\omega_1 = 0.7772$, $p < 0.05$) was also found in *TAS1R1* for the common ancestor of human, chimpanzee, bonobo and gorilla (Table 2, Figure 3). In contrast, diet and food availability for orang-utans exert a much more restrictive influence. These large, arboreal great apes rely predominantly on fruit, such as figs (*Ficus* spp.) and durian (*Durio* spp.) This difference in diet might explain the variation in the evolutionary history of *TAS1R1* between orang-utans and other great apes (humans included).

Functional sites concentrated on extra-cellular region of taste receptors and phylogenetically scattered

Nearly all of the positive selected sites (92.3%, 24 out of 26 positive selected sites, see Additional file 2) were localized in the VFTM and CRD domains (extra-cellular region) of TAS1R, which are responsible for binding of small molecules [79]. In order to examine potential influence of these positively selected mutations, we conducted the commonly used molecular modeling and docking methods to detect binding of some sweet and umami taste molecular with TAS1Rs receptors. Interestingly, we located some positive selection sites at binding regions of several sweet and umami taste molecules based on the molecular modelling and docking results. Sites 50 and 391 of TAS1R1, and sites 342, 407, 411 and 413 of TAS1R2 were identified to interact with several taste molecules, such as D-tryptophan, IMP, L-glutamate, fructose and sucrose (see Additional files 7, 8). It is remarkable that these amino acid sites were coincidently revealed to be positively selected sites. All of these results offer us potential active regions and important amino acid sites which are likely to influence the binding of several tastants, which also implies that the extra-cellular regions play a vital role in adaptive evolution. Certainly, comparing functional expression experiments, molecular modeling and docking have certain limitations. However, as the wildly used molecular structure analysis methods, molecular modeling and docking can predicate some potential binding domains and binding sites between receptors and ligands, theoretically. In our study, as the complementary analysis to predicate potential influence on binding with ligands of positively selected sites detected in our studies, the modeling and docking studies were suggested to be necessary and critical.

Remarkably, most of these radical changes in amino acids occurred in a few sub-terminal and terminal lineages across the phylogeny of catarrhine primates, without obvious phylogeny-related occurrence (Figures 2, 3 and 4). In other words, relatives belonging to the same phylogenetic group (family or genus) usually did not exhibit uniform amino acids in a single radical change site, and even several of these radical changes occurred irregularly in a single or a few distantly related species. This suggests specific adaptive evolution of *TAS1R* genes in catarrhine primates, resulting from habitat change and complexity of plant–based diets.

Potential variation in tasting of aspartame, neotame, lactisole, monosodium glutamate and monopotassium glutamate among catarrhine primates

Although a potential functional change induced by these positively selected sites was not substantiated in this study, we referred to multiple mutation/substitution sites of *TAS1R* identified by previous functional expression studies in other species, and some of them also occurred in the sequences described here (Table 4). For example, the site substitution S40T of TAS1R2 was shown to abolish the receptor response to aspartame [31], suggesting that Cercopithecidae species showing this substitution might have lost the taste of aspartame (Table 4). However, other studies suggested that aspartame is perceived as sweet by humans, apes and some species of Cercopithecidae [20,80]. Variable perception strength to aspartame between Cercopithecidae and other species of catarrhine primates was suggested to be the best explanation for this 'contradictory' situation. We hypothesized that this amino acid substitution did not really abolish response of Cercopithecidae to aspartame, but rather led to differences in sensitivity of aspartame perception between Cercopithecidae and other species. Further functional expression studies about the influence of S40T on sensitivity of TAS1R2 in catarrhine primates should be launched in future. Additionally, the A733V substitution in TAS1R3 was one of the important changes detected for diminished receptor sensitivity to lactiosole, according to research conducted in humans and rodents [32]. The same substitution (A733V) occurred in Cercopithecidae, where A was found in Hylobatidae and Hominidae. It is reasonable to hypothesize that the A733V substitution in TAS1R3 of Cercopithecidae potentially led to more diminished sensitivity to lactisole than in Hylobatidae and Hominidae.

The substitution I67S of TAS1R2 in humans was confirmed to slightly reduce the response to neotame. Furthermore, the substitution R757C of TAS1R3 in humans reduces sensitivity to monosodium glutamate (MSG) and increases sensitivity to monopotassium glutamate. Substitutions at the same sites of TAS1R2 and TAS1R3 but not the same amino acid changes were found in Colobinae (Table 4) and their potential influence on TAS1R2 and

TAS1R3 in Colobinae is expected to be evaluated during future studies.

Though some of these tastants were artificial compounds and have less of relationship with adaption of the sweet and umami tastes, these work supplied examples which reflect the mutation-induced sensitivity variation for certain compounds no matter natural or artificial. It highlights the importance of a number of sites and domains for ligands binding. Similar work can be implemented in research on catarrhine primates incorporating more natural compounds to examine the potential influence of positive selection sites detected in our study.

Conclusions

Unlike animal food, the number and diversity of nutritional compounds of leaves and fruit varies with plant species, location, position on the tree, stage of development, and even time of day [27,81-84]. Therefore, it is reasonable to assume that catarrhine primates require more sensitive and specific sweet and umami tastes when compared to carnivorous animals. Our results demonstrate family-, subfamily-, genus- and even species-specific adaptive evolution of *TAS1R* genes, and suggest pervasively differentiated evolution of sweet and umami tastes during the divergence of catarrhine primates. Episodic relaxation of selective constraints and pseudogenization were also found, both of which might be accompanied by and even contribute to dietary transition. Remarkably, positive selected sites were concentrated on extra-cellular region of taste receptors, which indicated that extra-cellular region of TAS1Rs play a vital role in adaption for variations in sweet and umami taste molecules in different diets.

The relationships among gustatory sense, food intake and foraging efficiency are complicated [85]. Both physical and chemical properties of food play important roles in food choice in primates [86]. Thus, the influence of both external and internal factors on taste receptor gene evolution is complex. Given such multiple factors, the impact of any particular factor is likely to be quantitative rather than qualitative, and a small number of counterexamples do not automatically refute any potential. Given these complications, to avoid spurious results, it is imperative to examine a large and diverse group of species when testing the potential impact of any given ecological factor on taste receptor gene evolution [87]. Our study provides new insights into the evolutionary history of taste genes in primates, as well as a database of mutations/substitutions in taste genes which will facilitate future work assessing the ecological correlates between taste sensitivity and food choice in primates.

Methods
Ethical approval
Blood samples were taken during routine health checks by experienced veterinarians and not specifically for this

study. Fresh tissue samples were taken from deceased animals. All research complied with protocols approved by the Animal Welfare Body of the DPZ in Germany and the Wildlife Conservation Association in China, and adhered to the legal requirements of the countries, in which research was conducted. The study was carried out in compliance with the principles of the American Society of Primatologists for the ethical treatment of non-human primates (https://www.asp.org/society/resolutions/EthicalTreatmentOfNonHumanPrimates.cfm). No animals were sacrificed for this study.

Polymerase chain reaction and DNA sequencing

Based on an alignment of currently available *TAS1R* sequences of human, chimpanzee, and macaque, we designed a set of primer pair to amplify *TAS1R* sequences from DNA samples of 30 species in catarrhine monkeys (see Additional file 1). The polymerase chain reaction (PCR) mixtures (50 ul) contained 5 ul (50 ng/ul) genomic DNA, 25 ul of $2 \times$ buffer, 7.5 ul (50 mM) $MgCl_2$, 5ul (10 mM) of each primer, and 1 U Taq DNA polymerase (Takara). PCRs were performed in a DNA Engine Dyad Cycler (BioRad) under the following condition: 5 min of initial denaturation, 30 cycles of denaturation at 94°C for 30 s, annealing at 55°C for 30 s, extension at 72°C for 60 s; and a final extension at 72°C for 5 min. PCR products were examined on agarose gels and subsequently cloned in the PMD18-T (Promega) cloning vector. Positive clones were sequenced on an ABI 3130 xl DNA Sequencer. Three to five clones of each PCR product were sequenced from both directions to validate the results. All intact *TAS1R* sequences of Hominidae were downloaded from GeneBank (see Additional file 1). Additionally, we also aligned our sequences with available primate genomes from GeneBank and the genome assembly of snub-nosed monkeys (unpublished), confirming that *TAS1R* are single-copy genes in genomes of catarrhine primates. The nucleotide and deduced amino acid sequences of each gene were aligned with CLUSTALX 1.81 [88] and modified with Bio-Edit 7.0.4 [89].

Selective pressure detection

A powerful approach to detecting molecular evolution by positive selection derives from comparison of the relative rates of synonymous (d_S) and non-synonymous substitutions (d_N/) [90-92]. The rate ratio ω (d_N/d_S) is a measure of selective pressure, where $\omega = 1$, $\omega < 1$ and $\omega > 1$ correspond to neutral evolution, purifying and positive selection [93]. The ω ratio was estimated using a codon-based maximum-likelihood method implemented in CODEML program of the PAML v. 4.4 package [94]. A well-accepted phylogeny of primates was used as the input tree in all analyses [95]. Because only one or two intact *TAS1R* were available in some hominoids, input trees used in

the analysis of *TAS1R1*, *TAS1R2*, and *TAS1R3* were slightly different within lineages of Hominidae (Figures 2, 3 and 4).

A combination of branch, site and branch-site models was used to analyze the selection of *TAS1R* in catarrhine primates. Firstly, to test whether the ω ratio of each gene was different among lineages, the 'free-ratios' model (M1), which assumes an independent ω ratio for each branch, was compared with the 'one-ratio' model (M0), which assumes the same ω ratio for all branches [93]. Subsequently, site models in which ω can vary among sites were implemented to identify candidates of positively selected sites in the *TAS1R* genes [96,97]. Therefore, two pairs of site models were tested: M1a (nearly neutral) versus M2a (positive selection), and M8a (nearly neutral; beta distribution) versus M8 (positive selection; beta distribution).

Additionally, positive selection often operates episodically on a few amino acid sites in a small number of lineages in a phylogenetic tree [98]. Therefore, in the case of branch-site model, modified branch-site model A was performed for each gene in the lineages of Cercopithecidae, Cercopithecinae, Colobinae, Hylobatidae and Hominidae, separately. Finally, two-ratio model assuming the branches of interest having a d_N/d_S (ω_1) different from the background ratio (ω_0) was used for lineages with $0.50 < \omega < 1$ (detecting by free ratio model) to identify potential functional relaxation [99,100]. Ancestral *TAS1R* sequences were inferred based on empirical Bayesian methods implemented in the CODEML program of the PAML package.

The likelihood ratio test (LRT), which calculates twice the log-likelihood ($2\Delta LnL$) of the difference following a chi-square distribution, was used to evaluate the significance of differences between each pair of models. All of the pseudogenes were removed in the PAML-based analysis. To evaluate the probabilities of positively selected sites on *TAS1R* sequenced in this research, the Bayes empirical Bayes (BEB) analysis was used to calculate posterior probabilities of positively selected sites implemented in the CODEML program of PAML. Based on BEB analysis, sites with a posterior probability >0.85 were considered as candidates for selection. Finally, according to charge, polarity and volume of amino acids, positive selection sites detected by site models and branch-site models were used to estimate amino acids change patterns (conservative or radical substitutions) along the evolution lineages of primates [98].

Molecular modelling and docking

Homology models of TAS1R1 and TAS1R2 VFTM (amino acids 29–496 and 24–494, respectively) were built with the Modeller9.11 [101,102] and EasyModeller4.0 [103] using the mGluR1-VFTM crystal structure (PDB entry: 2U4E) [104] as the template. Then, automatic molecular docking programs, Autodock Vina1.1.2 [105] and MGLTools1.5.6 [106,107] were used for the docking studies with some sweet and umami taste molecules (umami molecules: glycine, IMP, L-alanine, L-glutamate; sweet molecules: D-tryptophan, fructose, D-fructose, galactose, D-galactose, D-glucose and, sucrose, neotame and aspartame), and these results were viewed modified by PyMol (The PyMOL Molecular Graphics System, Version 1.7 Schrödinger, LLC).

Additional files

Additional file 1: Information about the species and genes present in this study.

Additional file 2: Summary of mutation sites and positive selected sites in TAS1Rs proteins.

Additional file 3: Alignment of TAS1R1 amino acid sequences of 35 catarrhine primates. Variant residues are marked with red.

Additional file 4: Alignment of TAS1R2 amino acid sequences of 35 catarrhine primates. Variant residues are marked with red.

Additional file 5: Alignment of TAS1R3 amino acid sequences of 35 catarrhine primates. Variant residues are marked with red.

Additional file 6: Candidate amino acid sites under positive selection identified in Tas1rs using the site models and branch-site models.

Additional file 7: Molecular docking results of TAS1R1. Positive selection sites detected in our research are marked with red.

Additional file 8: Molecular docking results of TAS1R2. Positive selection sites detected in our research are marked with red.

Competing interests
The authors declare that they have no competing interests.

Authors' contributions
ML, ZL, CR, LW and GL participated in the study design. GL and ZL conducted the experiments, analysed the data and wrote the manuscript. ML, LW and CR offered samples, helped to draft the manuscript and revised it critically. ST, XT and FS helped to analyse data and provided valuable suggestions. All authors read and approved the final manuscript.

Acknowledgements
This project was supported by the Natural Science Foundation of China (No. 31272301), Foundation of Chinese Academy of Sciences (KSCX3-IOZ-1001 and KSCX2-EW-Q-7-2) and the German Primatology Center (Göttingen, Germany). Thanks to Robert Dudley (University of California, USA) and Jens Gruber (German Primatology Center, Germany) for technical guidance and data analysis and to Guangfeng Liu, Yanhua Li, Mechthild Pohl, Christiane Schwarz and Rasmus Liedigk for their laboratory assistance and suggestions.

Author details
[1]Key Laboratory of Animal Ecology and Conservation Biology, Institute of Zoology, Chinese Academy of Sciences, 1-5 Beichen West Road, Chaoyang, Beijing 100101, China. [2]Primate Genetics Laboratory, German Primate Center, Leibniz Institute for Primate Research, Kellnerweg 4, 37077 Göttingen, Germany. [3]Gene Bank of Primates, German Primate Center, Leibniz Institute for Primate Research, Kellnerweg 4, 37077 Göttingen, Germany. [4]Department of Biomedical Sciences, School of Pharmacy, Texas Tech University Health Sciences Center, 1300 S. Coulter St, Amarillo, TX 79106, USA. [5]Institute of Health Sciences, Anhui University, Hefei, Anhui Province 230601, China. [6]College of Nature Conservation, Beijing Forestry University, Haidian, Beijing 100083, China.

References

1. Bachmanov AA, Beauchamp GK: Taste receptor genes. *Annu Rev Nutr* 2007, **27**:389–414.
2. Lindemann B: Taste reception. *Physiol Rev* 1996, **76**:719–766.
3. Li XD, Staszewski, Xu H, Durick K, Zoller M, Adler E, Affiliations A: Human receptors for sweet and umami taste. *Proc Natl Acad Sci U S A* 2002, **99**:4692–4696.
4. Nelson G, Hoon MA, Chandrashekar J, Zhang YF, Nicholas JP, Ryba, Zuker CS: Mammalian sweet taste receptors. *Cell* 2001, **106**:381–390.
5. Shi P, Zhang J: Contrasting modes of evolution between vertebrate sweet/umami receptor genes and bitter receptor genes. *Mol Biol Evol* 2006, **23**:292–300.
6. Pin JP, Galvez T, Prezeau L: Evolution, structure, and activation mechanism of family 3/C G-protein-coupled receptors. *Pharmacol Ther* 2003, **98**:325–354.
7. Jingami H, Nakanishi S, Morikawa K: Structure of the metabotropic glutamate receptor. *Curr Opin Neurobiol* 2003, **13**:271–278.
8. Jiang P, Ji Q, Liu Z, Snyder LA, Benard LM, Margolskee RF, Max M: The cysteine-rich region of T1R3 determines responses to intensely sweet proteins. *J Biol Chem* 2004, **279**:45068–45075.
9. Jiang P, Cui M, Zhao B, Snyder LA, Benard LM, Osman R, Max M, Margolskee RF: Identification of the cyclamate interaction site within the transmembrane domain of the human sweet taste receptor subunit T1R3. *J Biol Chem* 2005, **280**:34296–34305.
10. Winnig M, Bufe B, Kratochwil NA, Slack JP, Meyerhof W: The binding site for neohesperidin dihydrochalcone at the human sweet taste receptor. *BMC Struct Biol* 2007, **7**:66.
11. Xu H, Staszewski L, Tang H, Adler E, Zoller M, Li X: Different functional roles of T1R subunits in the heteromeric taste receptors. *Proc Natl Acad Sci U S A* 2004, **101**:14258–14263.
12. Zhang F, Klebansky B, Fine RM, Liu H, Xu H, Servant G, Zoller M, Tachdjian C, Li X: Molecular mechanism of the sweet taste enhancers. *Proc Natl Acad Sci U S A* 2010, **107**:4752–4757.
13. Chen QY, Alarcon S, Tharp A, Ahmed OM, Estrella NL, Greene TA, Rucker J, Breslin PAS: Perceptual variation in umami taste and polymorphisms in *TAS1R* taste receptor genes. *Am J Clin Nutr* 2009, **90**:770S–779S.
14. Cui M, Jiang P, Maillet E, Max M, Margolskee RF, Osman R: The heterodimeric sweet taste receptor has multiple potential ligand binding sites. *Curr Pharm Des* 2006, **12**:4591–4600.
15. Jiang P, Cui M, Ji Q, Snyder L, Liu Z, Benard L, Margolskee RF, Osman R, Max M: Molecular mechanisms of sweet receptor function. *Chem Senses* 2005, **30**:I17–I18.
16. Raliou M, Grauso M, Hoffmann B, Schlegel-Le-Poupon C, Débat H, Belloir C, Wiencis A, Sigoillot M, Bano SP, Trotier D, Pernollet JC, Montmayeur JP, Faurion A, Briand L: Human genetic polymorphisms in T1R1 and T1R3 taste receptor subunits affect their function. *Chem Senses* 2011, **36**:527–537.
17. Assadi-Porter FM, Tonelli M, Maillet EL, Markley JL, Max M: Interactions between the human sweet-sensing T1R2-T1R3 receptor and sweeteners detected by saturation transfer difference NMR spectroscopy. *Biochim Biophys Acta* 2010, **1798**:82–86.
18. Nie Y, Vigues S, Hobbs JR, Conn GL, Munger SD: Distinct contributions of T1R2 and T1R3 taste receptor subunits to the detection of sweet stimuli. *Curr Biol* 2005, **15**:1948–1952.
19. Winnig M, Bufe B, Meyerhof W: Valine 738 and lysine 735 in the fifth transmembrane domain of rTas1r3 mediate insensitivity towards lactisole of the rat sweet taste receptor. *BMC Neurosci* 2005, **6**:22.
20. Li X, Bachmanov AA, Maehashi K, Li W, Lim R, Brand JG, Beauchamp GK, Reed DR, Thai C, Floriano BW: Sweet taste receptor gene variation and aspartame taste in primates and other species. *Chem Senses* 2011, **36**:453–475.
21. Zhao H, Zhou Y, Pinto M, Dominique PC, Gonzalez JG, Zhang S, Zhang J: Evolution of the sweet taste receptor gene Tas1r2 in bats. *Mol Biol Evol* 2010, **27**:2642–2650.
22. Li X, Li WH, Cao J, Maehashi K, Huang LQ, Bachmanov AA, Reed DR, Beauchamp GK, Brand JG: Pseudogenization of a sweet-receptor gene accounts for cats' in difference toward sugar. *PLoS Genet* 2005, **1**:27–35.
23. Zhao H, Yang JR, Xu H, Zhang J: Pseudogenization of the umami taste receptor gene Tas1r1 in the giant panda coincided with its dietary switch to bamboo. *Mol Biol Evol* 2010, **27**:2669–2673.
24. Zhao H, Zhang J: Mismatches between feeding ecology and taste receptor evolution: An inconvenient truth. *Proc Natl Acad Sci U S A* 2012, **109**:E1465.

25. Jiang P, Josue J, Li X, Glaser D, Li W, Brand JG, Margolskee RF, Reed DR, Beauchamp GK: **Major taste loss in carnivorous mammals.** *Proc Natl Acad Sci U S A* 2012, **109**:4956–4961.

26. Bachmanov AA, Bosak NP, Lin C, Matsumoto I, Ohmoto M, Reed DR, Nelson TM: **Genetics of taste receptors.** *Curr Pharm Des* 2014, **20**:2669–2683.

27. Chapman CA, Chapman LJ, Rode K, Hauck EM, Mcdowell LR: **Variation in the nutritional value of primate foods: among trees, time periods, and areas.** *Int J Primatol* 2003, **24**:317–333.

28. Pereira PM, Vicente AF: **Meat nutritional composition and nutritive role in the human diet.** *Meat Sci* 2013, **93**:586–592.

29. Struhsaker TT: **Colobine monkeys: their ecology, behaviour and evolution.** *Int J Primatol* 1995, **16**:1035–1037.

30. Bauchop T: **Stomach microbiology of primates.** *Annu Rev Microbiol* 1971, **25**:429–436.

31. Liu B, Ha M, Meng XY, Kaur T, Khaleduzzaman M, Zhang Z, Jiang P, Li X, Cui M: **Molecular mechanism of species-dependent sweet taste toward artificial sweeteners.** *J Neurosci* 2011, **31**:11070–11076.

32. Jiang P, Cui M, Zhao B, Liu Z, Snyder LA, Benard LMJ, Osman R, Margolskee RF, Max M: **Mechanisms of signal transduction: lactisole interacts with the transmembrane domain of human T1R3 to inhibit sweet taste.** *J Biol Chem* 2005, **280**:15238–15246.

33. Tinh NT, Long HT, Tuan BV, Vy TH, Tam NA: **The feeding behavior and phytochemical food content of grey-shanked douc langurs (*Pygathrix cinerea*) at Kon Ka Kinh National Park, Vietnam.** *Vietnamese J Primatol* 2012, **2**:25–35.

34. Wang XX, Thomas SD, Zhang JZ: **Relaxation of selective constraint and loss of function in the evolution of human bitter taste receptor genes.** *Hum Mol Genet* 2004, **13**:2671–2678.

35. Blackburn TH: **Nitrogen metabolism in the rumen.** In *Physiology of Digestion in the Ruminant*. Edited by Dougherty RW. Washington, D.C: Butterworths; 1965:322–334.

36. Hungate RE: **Ruminal fermentation.** In *Handbook of Physiology*. Edited by Cole CF. Baltimore: Waverly; 1967:2725–2745.

37. Barnard EA: **Biological function of pancreatic ribonuclease.** *Nature* 1969, **221**:340–344.

38. Beintema JJ: **The primary structure of langur (*Presbytis entellus*) pancreatic ribonuclease: adaptive features in digestive enzymes in mammals.** *Mol Biol Evol* 1990, **7**:470–477.

39. Messier W, Stewart CB: **Episodic adaptive evolution of primate lysozymes.** *Nature* 1997, **385**:151–154.

40. Kool KM: **Food selection by the silver leaf monkey, *Trachypithecus auratus sondaicus*, in relation to plant chemistry.** *Oecologia* 1992, **90**:527–533.

41. Collins L, Roberts M: **Arboreal folivores in captivity-maintenance of a delicate minority.** In *The Ecology of Arboreal Folivores*. Edited by Montgomery GG. Washington, D.C: Smithsonian Institution; 1978:5–12.

42. Danish L, Chapman CA, Hall MB, Rode KD, Worman CO: **The role of sugar in diet selection in red tail and red colobus monkeys.** In *Feeding Ecology in Apes and Other Primates*. Edited by Hohmann G, Robbins MM, Boesch C. Cambridge: Cambridge University Press; 2006:471–485.

43. Davies G, Oates JF: *Colobine Monkeys: Their Ecology, Behaviour and Evolution*. Cambridge: Cambridge University Press; 1994.

44. Milton K: **Physiological ecology of howlers (*Alouatta*): energetic and digestive considerations and comparison with the Colobinae.** *Int J Primatol* 1998, **19**:513–548.

45. Dominy NJ, Lucas PW, Osorio D, Yamshita N: **The sensory ecology of primate food perception.** *Evol Anthropol* 2001, **10**:171–186.

46. Workman C: **The foraging ecology of the Delacour's langur (*Trachypithecus delacouri*) in Van Long Nature Reserve, Vietnam.** In *PhD Dissertation*. Durham, North Carolina: Duke University; 2010.

47. Hladik CM, Simmen B: **Taste perception and feeding behavior in nonhuman primates and human populations.** *Evol Anthropol* 1997, **5**:58–71.

48. Rui-Lin N, Tanaka T, Zhou J, Tanaka O: **Phlorizin and trilobatin, sweet dihydrochalcone-glucosides from leaves of *Lithocarpus litseifolius* (Hance) Rehd. (Fagaceae).** *Agric Biol Chem* 1982, **46**:1933–1934.

49. Schiffman S: **Magnitude estimation of amino acids for young and elderly subjects.** In *Olfaction and Taste VII*. Edited by Van der Starre H. London: IRL Press; 1980:379–383.

50. Chaudhari N, Landin AM, Roper SD: **A metabotropic glutamate receptor variant functions as a taste receptor.** *Nat Neurosci* 2000, **3**:113–119.

51. Rolls ET: **The representation of umami taste in the taste cortex.** *J Nutr* 2000, **130**:960S–965S.

52. Simmen B, Sabatier D: **Diets of some French Guianan primates: food composition and food choices.** *Int J Primatol* 1996, **17**:661–693.

53. Wendeln MC, Runkle JR, Kalko EKV: **Nutritional values of 14 fig species and bat feeding preferences in Panama.** *Biotropica* 2000, **32**:489–501.

54. Yeager CP, Kool K: **The behavioural ecology of Asian colobines.** In *Catarrhine Monkeys*. Edited by Whitehead J. Cambridge: Cambridge University Press; 2000:496–528.

55. Ren BP, Li DY, Liu ZJ, Li BG, Wei FW, Li M: **First evidence of prey capture and meat eating by wild Yunnan snub-nosed monkeys *Rhinopithecus bieti* in Yunnan, China.** *Curr Zool* 2010, **56**:227–231.

56. Fooden J: **Ecogeographic segregation of macaque species.** *Primates* 1982, **23**:574–579.

57. Abegg C, Thierry B: **Macaque evolution and dispersal in insular south-east Asia.** *Biol J Linn Soc* 2002, **75**:555–576.

58. Tosi AJ, Morales JM, Melnick DJ: **Paternal, maternal, and biparental molecular markers provide unique windows onto the evolutionary history of macaque monkeys.** *Evolution* 2003, **57**:1419–1435.

59. Hassel-Finnegan H, Borries C, Zhao Q, Phiapalath P, Koenig A: **Southeast Asian primate communities: the effects of ecology and Pleistocene refuges on species richness.** *Integr Zool* 2013, **8**:417–426.

60. Delson E: **Evolutionary history of the Cercopithecidae.** In *Approaches to Primate Paleobiology. Contributions to Primatology*, Volume 5. Edited by Szalay FS. Basel: Karger; 1975:167–217.

61. Delson E: **Fossil macaques, phyletic relationships and a scenario of deployment.** In *The Macaques: Studies in Ecology, Behavior, and Evolution*. Edited by Lindburg DG. New York: Van Nos-trand Reinhold; 1980:10–30.

62. Delson E: **The oldest monkeys in Asia.** In *International Symposium: Evolution of Asian Primates*. Inuyama, Aichi, Japan: Freude and Kyoto University Primate Research Institute; 1996:40.

63. Fan PF, Fei HL, Scott MB, Zhang W, Ma CY: **Habitat and food choice of the critically endangered cao vit gibbon (*Nomascus nasutus*) in China: implications for conservation.** *Biol Conserv* 2011, **144**:2247–2254.

64. Fan PF, Ai HS, Fei HL, Zhang D, Yuan SD: **Seasonal variation of diet and time budget of Eastern hoolock gibbons (*Hoolock leuconedys*) living in a northern montane forest.** *Primates* 2013, **54**:137–146.

65. Jildmalm R, Amundin M, Laska M: **Food preferences and nutrient composition in captive white-handed gibbons, *Hylobates lar*.** *Int J Primatol* 2008, **29**:1535–1547.

66. Ni QY, Huang B, Liang ZL, Wang XW, Jiang XL: **Dietary variability in the Western Black Crested Gibbon (*Nomascus concolor*) inhabiting an isolated and disturbed forest fragment in southern Yunnan, China.** *Am J Primatol* 2014, **76**:217–229.

67. Leighton M: **Modeling dietary selectivity by Bornean orangutans: evidence for integration of multiple criteria in fruit selection.** *Int J Primatol* 1993, **14**:257–313.

68. Bollard EG: **The physiology and nutrition of developing fruits.** In *The Biochemistry of Fruits and Their Products*, Volume 1. Edited by Hulme AC. London: Academic Press, London; 1970:387–425.

69. Stevenson PR: **Fruit choice by woolly monkeys in Tinigua National Park, Colombia.** *Int J Primatol* 2003, **35**:367–381.

70. Simmen B, Hladik A, Ramasiarisoa PL, Iaconelli S, Hladik CM: **Taste discrimination in lemurs and other primates, and the relationships to distribution of the plant allelochemicals in different habitats of Madagascar.** In *New Directions in Lemur Studies*. Edited by Rakotosamimanana H. New York: Kluwer. Souci SW, Fachmann W; 1999:201–219.

71. Chivers DJ: **Feeding and ranging in gibbons: a summary.** In *The Lesser Apes*. Edited by Preuschoft H, Chivers DJ, Brockelman WY, Creel N. Edinburgh: Edinburgh University Press; 1984:267–281.

72. Bartlett TQ: **The hylobatidae, small apes of Asia.** In *Primates in Perspective*. Edited by Campbell CJ, Fuentes A, Mackinnon KC, Panger M, Bearder SK. New York: Oxford University Press; 2007:274–289.

73. Carpenter CR: **A field study in Siam of the behavior and social relations of the gibbon (*Hylobates lar*).** In *Comparative Psychology Monographs*. Edited by Dorcus RM. Baltimore: The Johns Hopkins Press; 1940:84–206.

74. Jolly A: **Food and feeding.** In *The Evolution of Primate Behavior*. Edited by Jolly A. New York: Macmillan; 1985:45–71.

75. Raemaekers J: **Changes through the day in the food choice of wild gibbons.** *Folia Primatol* 1978, **30**:194–205.

76. Richard AF: **Primate diets: patterns and principles.** In *Primates in Nature*. Edited by Richard AF. New York: W. H. Freeman; 1985:163–205.
77. Ungar PS: **Fruit preferences of four sympatric primate species at Ketambe, Northern Sumatra, Indonesia.** *Int J Primatol* 1995, **16**:221–245.
78. Kleiman DG, Geist V, McDade MC: *Grzimek's Animal Life Encyclopedia, Mammals (III)*, Volume 14. 2nd edition. Michigan: The Gale Group, Inc; 2004.
79. Temussi PA: **Sweet, bitter and umami receptors: a complex relationship.** *Cell* 2009, **34**:296–302.
80. Hellekant G, Danilova V, Ninomiya Y: **Primate sense of taste: behavioral and single chorda tympani and glossopharyngeal nerve fiber recordings in the rhesus monkey, *Macaca mulatta*.** *J Neurophysiol* 1997, **77**:978–993.
81. Fernandez-Escobar R, Moreno R, Garcia-Creus M: **Seasonal changes of mineral nutrients in olive leaves during the alternate-bearing cycle.** *Sci Hort* 1999, **82**:25–45.
82. Klages K, Donnison H, Wunsche J, Boldingh H: **Diurnal changes in non-structural carbohydrates in leaves, phloem exudate and fruit in 'Braeburn' apple.** *Aust J Plant Physiol* 2001, **28**:131–139.
83. Marquis RJ, Newell EA, Villegas AC: **Non-structural carbohydrate accumulation and use in an understorey rain-forest shrub and relevance for the impact of leaf herbivory.** *Funct Ecol* 1997, **11**:636–643.
84. Woodwell GM: **Variation in the nutrient content of leaves of *Quercus alba*, *Quercus coccinea*, and *Pinus rigida* in the Brookhaven forest from bud-break to abscission.** *Am J of Bot* 1974, **61**:749–753.
85. Xiang Z, Liang W, Nie S, Li M: **Short notes on extractive foraging behavior in gray snub-nosed monkeys.** *Integr Zool* 2013, **8**:389–394.
86. Blaine KP, Lambert JE: **Digestive retention times for Allen's swamp monkey and L'Hoest's monkey: data with implications for the evolution of cercopithecine digestive strategy.** *Integr Zool* 2012, **7**:183–191.
87. Li DY, Zhang JZ: **Diet shapes the evolution of the vertebrate bitter taste receptor gene repertoire.** *Mol Biol Evol* 2014, **31**:303–309.
88. Thompson JD, Gibson TJ, Plewniak F, Jeanmougin F, Higgins DG: **The CLUSTAL_X windows interface: flexible strategies for multiple sequence alignment aided by quality analysis tools.** *Nucleic Acids Res* 1997, **25**:4876–4882.
89. Hall TA: **BioEdit: a user-friendly biological sequence alignment editor and analysis program for windows 95/98/NT.** *Nucleic Acids Symp Ser* 1999, **41**:95–98.
90. Kimura M: *The Neutral Theory of Molecular Evolution*. Cambridge, UK: Cambridge University Press; 1983.
91. Gillespie JH: *The Causes of Molecular Evolution*. Oxford: Oxford University Press; 1991.
92. Ohta T: **The nearly neutral theory of molecular evolution.** *Annu Rev Ecol Syst* 1993, **23**:263–286.
93. Yang Z: **Likelihood ratio tests for detecting positive selection and application to primate lysozyme evolution.** *Mol Biol Evol* 1998, **15**:568–573.
94. Yang Z: **PAML 4: phylogenetic analysis by maximum likelihood.** *Mol Biol Evol* 2007, **24**:1586–1591.
95. Perelman P, Johnson WE, Roos C, Seuánez HN, Horvath JE, Moreira MA, Kessing B, Pontius J, Roelke M, Rumpler Y, Schneider MP, Silva A, O'Brien SJ, Pecon-Slattery J: **A molecular phylogeny of living primates.** *PLoS Genet* 2011, **7**:e1001342.
96. Nielsen R, Yang Z: **Likelihood models for detecting positively selected amino acid sites and applications to the HIV-1 envelope gene.** *Genetics* 1998, **148**:929–936.
97. Yang Z, Nielsen R, Goldman N, Pedersen AMK: **Codon-substitution models for heterogeneous selection pressure at amino acid sites.** *Genetics* 2000, **155**:431–449.
98. Zhang J: **Rates of conservative and radical nonsynonymous nucleotide substitution in mammalian nuclear genes.** *J Mol Evol* 2000, **50**:56–68.
99. Bielawski JP, Yang Z: **Maximum likelihood methods for detecting adaptive protein evolution.** In *Statistical Methods in Molecular Evolution*. Edited by Nielsen R. New York, NY: Springer; 2005:103–124.
100. Feng P, Zheng JS, Rossiter SJ, Wang D, Zhao H: **Massive losses of taste receptor genes in toothed and baleen whales.** *Genome Biol Evol* 2014, **6**:1254–1265.
101. Sali A, Blundell TL: **Comparative protein modelling by satisfaction of spatial restraints.** *J Mol Biol* 1993, **234**:779–815.
102. Marti-Renom MA, Stuart AC, Fiser A, Sánchez R, Melo F, Sali A: **Comparative protein structure modeling of genes and genomes.** *Annu Rev Biophys Biomol Struct* 2000, **29**:291–325.
103. Kuntal BK, Polamarasetty A, Reddanna P: **EasyModeller: a graphical interface to MODELLER.** *BMC Res Notes* 2010, **3**:226.
104. Muto T, Tsuchiya D, Morikawa K, Jingami H: **Structures of the extracellular regions of the group II/III metabotropic glutamate receptors.** *Proc Natl Acad Sci U S A* 2007, **104**:3759–3764.
105. Trott O, Olson AJ: **AutoDock Vina: improving the speed and accuracy of docking with a new scoring function, efficient optimization and multithreading.** *J Comput Chem* 2010, **31**:455–461.
106. Sanner MF: **Python: a programming language for software integration and development.** *J Mol Graphics Mod* 1999, **17**:57–61.
107. Morris GM, Huey R, Lindstrom W, Sanner MF, Belew RK, Goodsell DS, Olson AJ: **Autodock4 and AutoDockTools4: automated docking with selective receptor flexiblity.** *J Comput Chem* 2009, **30**:2785–2791.

7

Rapid evolution of *Anguillicola crassus* in Europe: species diagnostic traits are plastic and evolutionarily labile

Urszula Weclawski[1*], Emanuel G Heitlinger[1], Tobias Baust[2], Bernhard Klar[2], Trevor Petney[1], Yu-San Han[3] and Horst Taraschewski[1*]

Abstract

Background: Since its introduction from Taiwan to Europe around 1980, *Anguillicola crassus*, a natural parasite of the Japanese eel (*Anguilla japonica*), has acquired the European eel (*Anguilla anguilla*) as a novel definitive host. In this host the nematode differs noticeably in its body mass and reproductive capacity from its Asian conspecifics. We conducted a common garden experiment under a reciprocal transplant design to investigate whether differences in species-diagnostic morphological traits exist between two European and one Asian population of *A. crassus* and if yes whether these have a genetically fixed component.

Results: We found that worms from Germany, Poland and Taiwan differ in the size and shape of their body, oesophagus and buccal capsule. These changes are induced by both phenotypic plasticity and genetic divergence: in the European eel, nematodes from Europe as well as from Taiwan responded plastically with larger body and oesophagus dimensions compared to infections in the Japanese eel. Interestingly, the oesophagus simultaneously shows a high degree of genetically based changes being largest in the Polish strain kept in *A. anguilla*. In addition, the size and shape of the buccal capsule has undergone a rapid evolutionary change. Polish nematodes evolved a genetically fixed larger buccal capsule than the German and Taiwanese populations. The German strain had the smallest buccal capsule.

Conclusions: This study provides evidence for the genetic divergence of morphological traits in *A. crassus* which evolved over a timescale of about 30 years. Within Europe and in the European eel host these alternations affect characters used as diagnostic markers for species differentiation. Thus we provide an explanation of the discrepancy between morphological and molecular features reported for the parasitic nematode featured here, demanding general caution in morphological diagnosis of parasites discovered in new hosts.

Keywords: Anguillicola crassus, Anguillicoloides, Introduction, Biological invasion, Host switch, Rapid evolution, Genetic divergence, Phenotypic plasticity, Evolutionary labile traits

Background

Anguillicola crassus, a natural swim bladder parasite of the Japanese eel (*Anguilla japonica*) [1], is one of the most successful aquatic parasitic aliens in the history of globalization. Within about 30 years of its introduction from Taiwan to Germany around 1980 this nematode has spread throughout the European continent and North Africa colonizing almost all populations of the European eel (*Anguilla anguilla*) [2,3].

Invasive pathogens and most prominently *A. crassus* have thus been proposed as an additional cause for the decline of European eel stocks, along with overfishing and habitat destruction [4].

The development of *A. crassus* requires copepod or ostracod intermediate hosts and a final host belonging to the genus *Anguilla* in the natural as well as the novel range [5]. After the ingestion of an infected intermediate or a paratenic host by an eel [6], the L3 larvae migrate

* Correspondence: weclawski.urszula@googlemail.com; horst.taraschewski@kit.edu
[1]Department of Ecology and Parasitology, Zoological Institute, Karlsruhe Institute of Technology, Kornblumenstrasse 13, Karlsruhe, Germany
Full list of author information is available at the end of the article

through the intestinal wall and body cavity towards the swim bladder. In the swim bladder wall the L3 develop via the L4 stage to pre-adults that enter the swim bladder lumen. Sexually dimorphic adults eventually mate and L2 are released by females. These L2 reach the water via the eel's faeces and are ingested by an intermediate host in which they moult to the L3 stage [5].

Field studies in Europe and Asia have revealed that the worms infecting the European eel are much larger than their Taiwanese conspecifics in the Japanese eel [7]. In experiments with European parasites in *A. anguilla* and *A. japonica* infectivity, body mass, weight gain and reproductive output were significantly higher in the European eel than in the natural host, while larval mortality was lower in the colonized host [8].

Similarly, the American eel is more heavily affected by *A. crassus* than the Japanese eel [9].

Recently, we demonstrated rapid genetic divergence of the introduced European parasite population compared to conspecifics from Taiwan in terms of infectivity and developmental dynamics, using a reciprocal transplant experiment under common garden conditions [10]. In this paper, we report on the morphological traits of parasites from these studies.

The oesophagus (i.e. pharynx) and the buccal capsule are involved in the ingestion of food, in adult *A. crassus*, namely blood from host capillaries [11]. The oesophagus is a muscular structure that likely creates a vacuum while sucking blood into the worm's gut. The buccal capsule is a sclerotinized structure that bears the teeth and acts as an abutment to the applied pressure [12,13].

The different feeding behaviour of larger nematode groups (esp. in the Spriurina and the whole of the Rhabditida) is reflected by structural variability of the oesophagus and the buccal capsule. The taxonomy of nematodes thus traditionally relies on these characters [14].

Based on these features *A. crassus*, *Anguillicola novaezelandiae* and *Anguillicola papernai* can be differentiated (in addition to differences in natural host usage and distributional range). In contrast, *Anguillicola australiensis* showed an additional specific character. Its anterior end is inflated, forming a bulb around the mouth opening [15]. Later, the subgenus *Anguillicoloides*, to which all 4 species belonged, was transferred to the status of a genus, while the genus *Anguillicola* retained only one species, *Anguillicola globiceps* [16]. According to molecular evidence, however, the new genus and the previous subgenus are paraphyletic and the original generic grouping is more appropriate [17]. For this reason the generic name *Anguillicola* is used in this paper. In the present study we investigated whether taxonomically important morphological characters are affected by rapid evolution following the colonization of a novel host in a recipient area.

Results

In the European eel 1,136 adult worms were chosen for morphological investigations: 424 (45.4% female, 54.6% male) worms belonging to the German, 387 (41.4% female, 58.6% male) to the Polish and 325 (30% female, 70% male) to the Taiwanese worm population. In the Japanese eels from 430 (37.1% female, 62.9% male) adult worms that were chosen for morphological studies 179 (44.4% female, 55.6% male) worms belonged to the German population, 142 (47.4% female, 52.6% male) to Polish and 111 (43.5% female, 56.5% male) to the Taiwanese population.

Body

In the Japanese eel, the worms were generally smaller for both body dimensions (width and length) and also grew less quickly than in the European eel (Figure 1, Additional file 1: Table S1, Additional file 2: Table S2, Additional file 3: Figure S1). They also had a higher length/width ratio (Table 1). Further, since the (estimated) coefficient for dpi is almost precisely the negative of the coefficient for the interaction between Japanese eel and dpi, the coefficient for dpi for the Japanese eel (which is the sum of both values) is roughly zero (Table 1). This means that the shape of the worms in the Japanese eel did not change significantly over time.

Under the same host conditions no differences between the 3 nematode strains were found: in both eel species the parasite populations did not differ from one another with respect to length or width of their body (Additional file 1: Table S1, Additional file 2: Table S2; Figure 1, Additional file 3: Figure S1).

The only significant difference found was a smaller length/width ratio for the Taiwanese population infecting the Japanese eel, indicating a shorter body of Taiwanese worms in this host species compared to both European populations (Table 1).

Oesophagus

In both eel species the Polish *A. crassus* population had a wider but shorter oesophagus than did the German and Taiwanese strains (Table 2, Additional file 4: Table S3, Additional file 5: Table S4, Figure 2, Additional file 6: Figure S2). There were no differences in length, width and aspect ratio of the oesophagus between the German and Taiwanese strains. The larger dimensions and shorter shape of the oesophagus of the Polish parasite population are apparently genetically fixed.

The oesophagus of the worms in the Japanese eel was thinner (Figure 2, Additional file 5: Table S4, Additional file 6: Figure S2) and grew less quickly (Additional file 4: Table S3, Additional file 5: Table S4) than that of worms in European eels. In addition, there was a non-significant trend (p = 0.097) towards a shorter oesophagus in worms

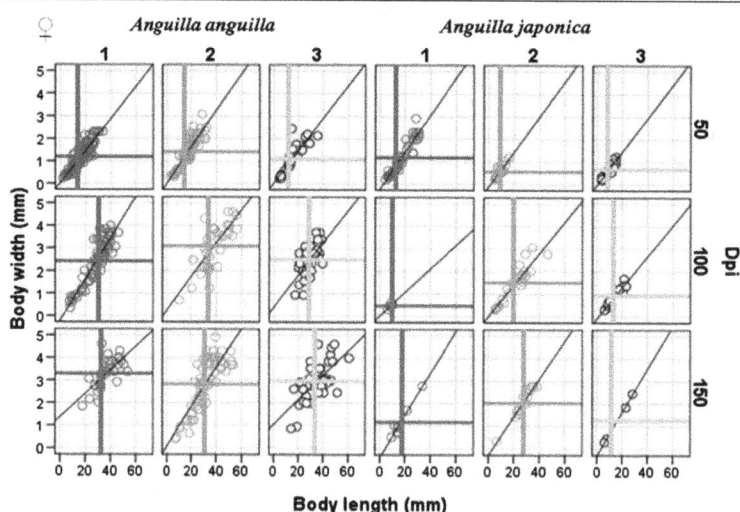

Figure 1 Body dimensions of female adults of *Anguillicola crassus* in *Anguilla anguilla* and *A. japonica.* Body length and width of female adults of *Anguillicola crassus* in *Anguilla anguilla* and *A. japonica* with the arithmetic mean values (horizontal and vertical lines) and the linear regression lines. 1 – German parasite population, 2 - Polish parasite population, 3 – Taiwanese parasite population. Dpi – days post infection. For same visualization for male worms see: Additional file 3: Figure S1.

from this host species. In the Japanese eel the oesophagus had a higher length/width ratio (Table 2).

Buccal capsule
Neither of the dimensions (width and length) of the worms' buccal capsule differed between specimens harboured in the different host species (Figure 3, Additional file 7: Table S5, Additional file 8: Table S6, Additional file 9: Figure S3). Only male worms had a slightly but significantly shorter buccal capsule in the Japanese eel (Table 3).

In both eel species the Polish nematode strain had the biggest buccal capsule (longer and wider) than that of the German and Taiwanese strains. The buccal capsule of the Taiwanese worms showed a trend (p = 0.08) towards wider shape than that of the German conspecifics

but it was significantly longer (Figure 3, Additional file 7: Table S5, Additional file 8: Table S6, Additional file 9: Figure S3).

The shape of the buccal capsule did not differ significantly between the Polish and Taiwanese worms, but both populations had a longer buccal capsule than the German specimens (Table 3). Thus, again, the two European populations differed from one another.

Discussion
A. crassus has a historically well documented invasion time line. The first published reports of its presence in Europe date from the early 1980s [18,19]. It had therefore been in Europe for about 30 years at the time our experiments were carried out.

Table 1 Minimal adequate mixed-effects linear model for body ratio of *Anguillicola crassus*: reference group: German parasite population in the European eel

Explanatory variables and interactions	Estimate	SD	t-value	p-value
(Intercept)	13.993017	0.344908	40.570	**0.0000**
Japanese eel	2.864821	0.597873	4.792	**0.0000**
Polish parasite population	−0.351104	0.384348	−0.914	0.3618
Taiwanese parasite population	0.415774	0.407112	1.021	0.3081
Dpi	−0.017952	0.004705	−3.816	**0.0002**
Male	2.515979	0.294228	8.551	**0.0000**
Japanese eel*Polish parasite population	−0.565012	0.721843	−0.783	0.4345
Japanese eel*Taiwanese parasite population	−2.825598	0.738580	−3.826	**0.0002**
Japanese eel*Dpi	0.017957	0.007626	2.355	**0.0193**
Dpi*Male	0.011858	0.004458	2.660	**0.0079**
Japanese eel*Male	−1.012030	0.422709	−2.394	**0.0168**

Significant effects are in bold.

Table 2 Minimal adequate mixed-effects linear model for oesophagus ratio of *Anguillicola crassus*: reference group: German parasite population in the European eel

Explanatory variables and interactions	Estimate	SD	t-value	p-value
(Intercept)	3.317222	0.032135	103.229	**0.0000**
Japanese eel	0.354854	0.043646	8.130	**0.0000**
Polish parasite population	−0.152373	0.033900	−4.495	**0.0000**
Taiwanese parasite population	−0.030428	0.035232	−0.864	0.3886
Dpi	−0.004029	0.000415	−9.701	**0.0000**
Male	−0.048802	0.022015	−2.217	**0.0268**
Japanese eel*Dpi	0.002353	0.000802	2.933	**0.0037**

Significant effects are in bold.

Using a cross-infection experiment with both Asian and European hosts and parasites we postulate divergence linked to the origin of the parasite and not to the species of the experimental host to be genetically fixed. We show that the morphology of the body, oesophagus and buccal capsule of *A. crassus* show a mixture of both plastic and genetically fixed differences.

We report a larger body mass and size of worms raised in the European eel compared to the Japanese eel. A similar finding with *A. anguilla* and *A. japonica* infected with *A. crassus* from Germany has previously been documented [8]. We conclude that this feature of the worm is based only on phenotypic plasticity since we show that it is linked only to the species of experimental host and not to the origin of the parasite. For *A. japonica* we assume that the worms are subjected to a strong concomitant immunity limiting the weight gain of the parasite. This hypothesis is supported by the lower prevalence and intensities of *A. crassus* in its natural host [7], by a stronger and more

rapid production of antibodies directed against the parasite [20,21] and by the ability of the Japanese eel to encapsulate and eliminate visceral larvae of *A. crassus* [22].

Concerning the size and shape of the oesophagus, phenotypic modification also seems to be present. It is not surprising that the development and growth of this muscular structure is influenced by the same factors as the overall body dimensions as shown, for example, in *Halicephalobus cf. gingivalis* [23].

However, we could not detect any influence of the host on size or shape of the buccal capsule, and we thus conclude a lack of environmental influence on that structure. The size and shape of this sclerotinized structure seems to be genetically determined and fully independent of the worms overall size. Growth of nematodes and other organisms keeping sclerotinized organs in constant dimensions is a known phenomenon, and these features are thus traditionally regarded as suitable for morphological taxonomy [24,25].

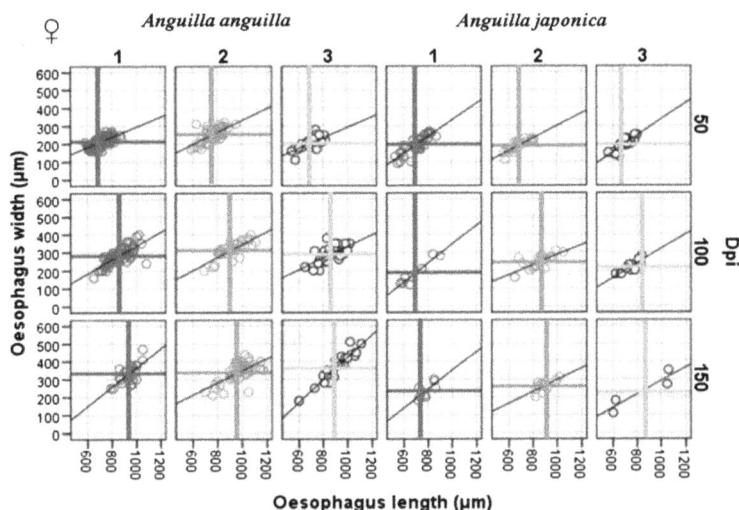

Figure 2 Oesophagus dimensions of female adults of *Anguillicola crassus* in *Anguilla anguilla* and *A. japonica*. Oesophagus length and body width of female adults of *Anguillicola crassus* in *Anguilla anguilla* and *A. japonica* with the arithmetic mean values (horizontal and vertical lines) and the linear regression lines. 1 – German parasite population, 2 - Polish parasite population, 3 – Taiwanese parasite population. Dpi – days post infection. For the same visualization for male worms see: Additional file 5: Figure S2.

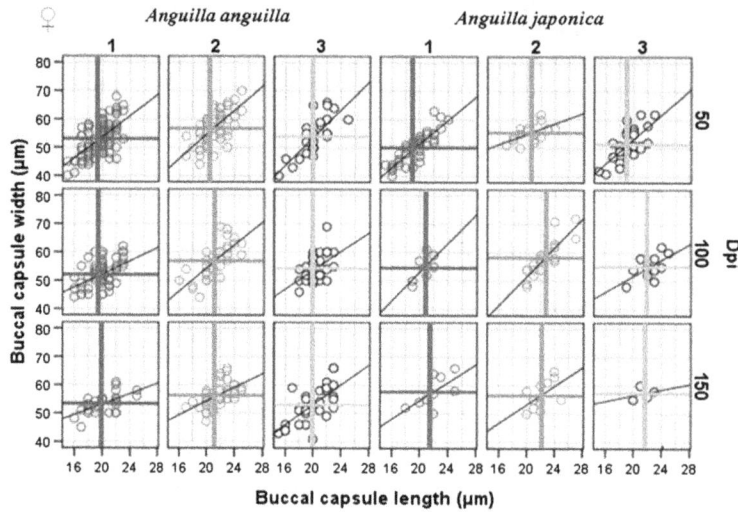

Figure 3 Buccal capsule dimensions of female adults of *Anguillicola crassus* in *Anguilla anguilla* and *A. japonica*. Buccal capsule length and width of female adults of *Anguillicola crassus* in *Anguilla anguilla* and *A. japonica* with the arithmetic mean values (horizontal and vertical lines) and the linear regression lines. 1 – German parasite population, 2 - Polish parasite population, 3 – Taiwanese parasite population. Dpi – days post infection. For the same visualization for male worms see: Additional file 9: Figure S3.

Coming to the question of the most likely mode of the observed rapid evolution, the shape and size of the oesophagus and the buccal capsule of the worms from a German river directly linked to the place of original introduction [18] resemble the traits of the conspecifics from the donor area in Taiwan. Only a Polish strain of the parasite displayed rapid divergence of morphological features, as reflected by an enlargement of the buccal cavity and the oesophagus. Furthermore, transplantation experiments indicate that morphological changes in the Polish strain genetically fixed and intrinsic to the worms than a result of the host environment.

The host-induced sources of selective pressure are the same between Poland and Germany, but differ between European and Asian locations. This is especially true in the eel, a (nearly) panmictic species due to its spawning behaviour [26,27], excluding the possibility of local host-parasite adaptation. However, *A. crassus* seems to have colonized Poland via the brackish water of the Baltic

[28,29]. Our observations could be explained either via a bottleneck and genetic drift within the colonized eastern areas or by selective pressure exerted by the abiotic environment [30]. It appears that the rapid evolution of morphological traits in the parasite does not have to be directional as an adaptation to the host.

Indeed a molecular genetic survey using microsatellite markers found that allelic diversity of European populations reflected a moderate bottleneck associated with the introduction of the parasite into a novel area followed successive moderate loss of diversity during the southward expansions of its populations [30]. This argues for drift as an explanation of our findings.

It was recently shown that genetically fixed changes in infectivity and developmental traits separate European *A. crassus* from their Taiwanese conspecifics [10]. The selective pressure of the novel host (European eel), not the genetic drif, was thus considered to have modified the developmental features in the European strains of *A. crassus*.

Table 3 Minimal adequate mixed-effects linear model for buccal capsule ratio of *Anguillicola crassus*: reference group: German parasite population in the European eel

Explanatory variables and interactions	Estimate	SD	t-value	p-value
(Intercept)	0.375781	0.002190	171.579	**0.0000**
Japanese eel	0.009065	0.003123	2.902	**0.0040**
Polish parasite population	−0.004939	0.002317	−2.131	**0.0340**
Taiwanese parasite population	−0.005019	0.002402	−2.089	**0.0377**
Dpi	0.000078	0.000024	3.214	**0.0015**
Male	0.015202	0.002018	7.534	**0.0000**
Japanese eel*Male	−0.012043	0.003956	−3.044	**0.0024**

Significant effects are in bold.

We show in the present publication that the mode (neutral or directional) of rapid evolution may differ for different traits in the same species. The differentiation between *A. crassus*, *A. novaezelandiae* and *A. papernai* for example is based mainly on features of the buccal capsule and oesophagus [5]. Our findings thus provide a possible explanation for the incongruence between morphological and molecular phylogeny [17] in the special case of the genus *Anguillicola*. If a diagnostic trait can repeatedly undergo rapid evolutionary change in one lineage but not in the other, species may be grouped wrongly in higher order taxa, such as (sub-) genera [31]. Our findings might point to an explanation for the common incongruence between phylogeny and morphological characteristics: Rapid evolution of a phenotypically stable character.

In addition, our results indicate that it is not well advised to use morphology alone for species discrimination and higher order systematics in nematodes and other taxonomic groups with limited morphological differentiation. Even if the morphological characters used do not vary in different environments (are phenotypically stable), the same characters might be evolutionarily labile potentially creating uninformative patterns of similarity. Furthermore, parasite evolution does not necessarily have to be directional, leading to host adaptation, but selection exerted by the abiotic environment or by neutral evolution and genetic drift can lead to divergence in ecological time.

Conclusions

A reciprocal transplant experiment under common-garden conditions allowed us to disentangle phenotypic modification from genetically fixed divergence of morphological traits in *A. crassus*. The nematode is larger in the recently acquired European host *A. anguilla* compared to the original Asian host *A. japonica*. Overall body dimensions seem only plastically modified and size differences showed no genetically fixed component.

The oesophagus was also affected by this environmentally induced plasticity. This trait, however, also showed genetically fixed differences between European populations of *A. crassus*. The sclerotinized structure of the buccal capsule, on the other hand is not modified by the host environment, but has genetically diverged between parasite populations within Europe.

We conclude that evolutionary change in characters important for morphological taxonomy has occurred in *A. crassus* in its novel European range. Moreover, we generalize our findings to suggest that traits which are normally phenotypically stable in helminths can be evolutionarily labile on an ecological time scale and are thus likely to introduce artefacts of taxonomic relevance. Species with limited morphological differentiation thus require the additional use of molecular methods for species

differentiation, taxonomy and systematics. The molecular genetic architecture characterized by the traits discussed here should enable a broad characterisation of genome wide differences between the different *A. crassus* populations [32]. Our findings show also that this invasive species was able to change at the evolutionary level in ecological time, which is probably an important component of the dispersal success of introduced taxa.

Methods
Experimental design
We conducted a series of experiments infecting *A. japonica* and *A. anguilla* with *A. crassus* originating from Germany, Poland and Taiwan, as described in [10].

Briefly, L2 larvae were collected from wild eels from the Rhine River near Karlsruhe in the southwest of Germany, the Kao-ping River in the southwest of Taiwan and the Lake Śniardwy in northern Poland. Copepod intermediate hosts were infected with these L2 and infective L3 larvae were harvested. Eels were then infected via a stomach tube with 50 L3 larvae each.

Infected eels were kept in 160-liter tanks in groups of 20 individuals at a constant temperature of 22°C and a 12:12 photoperiod. The eels chosen for the experiment were 37.7 cm ± 0.2 (± SE) and 49.4 cm ± 0.3 long for the European and the Japanese eel, respectively. At 25, 50, 100 and 150 dpi 20 eels were dissected, resulting in total of 239 European and 216 Japanese eels. Eels that died during the trial were not considered in the statistical analysis. The tanks were continuously provided with oxygenated tap water and fed *ad libitum* with commercial fish pellets (Dan-Ex 2848, Dana Feed A/S Ltd, Horsens, Denmark). The experiment had been approved by the responsible authorities (Regierungspräsidium Karlsruhe).

At dissection, the swim bladder was opened, adult parasites were sexed and preserved in 70% ethanol. Body dimensions were measured at 0.65×, 2.0× or 5.0× magnification using a dissecting microscope (STEMI 2000, Carl Zeiss) equipped with a measuring ocular. Afterwards the anterior end of the worms was removed and covered with Berlese-mixture (Waldeck GmbH & Co. KG, Division Chroma, 3D 101) on a glass slide. Dimensions of the oesophagus and buccal capsule were measured at 10× and 100× magnification under a compound microscope (Axiolab, Carl Zeiss) using a measuring ocular. In order to describe differences in the shape of the morphological features, the ratio of each measurement pair (length/width) was calculated.

Statistics
The models were performed separately for each term acting as response variable (the length, width and ratio of body, oesophagus and buccal capsule); the German parasite population in the European eel served as a

Table 4 Set up of the mixed-effects linear models morphological dimensions of *Anguillicola crassus*: response and explanatory variables in the minimal adequate models after simplification *(Continued)*

Response variable	Referenced in text as	Explanatory variables in the minimal adequate models
Body length	Additional file 1: Table S1	Eel species
		Dpi
		Number of adults recovered alive
		Number of L3 recovered alive
		Worm sex
		Eel species*dpi
		Dpi*worm sex
		Eel species*worm sex
Body width	Additional file 2: Table S2	Eel species
		Dpi
		Number of adults recovered alive
		Number of L3 recovered alive
		Worm sex
		Eel species*dpi
		Dpi*worm sex
		Eel species*worm sex
Ratio body length/width	Table 1	Eel species
		Population
		Dpi
		Worm sex
		Eel species*population
		Eel species*dpi
		Dpi*worm sex
		Eel species*worm sex
Oesophagus length	Additional file 4: Table S3	Eel species
		Population
		Dpi
		Number of adults recovered alive
		Number of L3 recovered alive
		Worm sex
		Eel species*dpi
		Dpi*worm sex
Oesophagus width	Additiolan file 5: Table S4	Eel species
		Population
		Dpi
		Number of adults recovered alive
		Worm sex
		Eel species*dpi
		Dpi*worm sex
Ratio oesophagus length/width	Table 2	Eel species
		Population
		Dpi
		Worm sex
		Eel species*dpi
Buccal capsule length	Additional file 7: Table S5	Eel species
		Population
		Dpi
		Number of adults recovered alive
		Number of L3 recovered alive
		Worm sex
		Eel species*dpi
Buccal capsule width	Additional file 8: Table S6	Eel species
		Population
		Dpi
		Number of L3 recovered alive
		Worm sex
		Eel species*dpi
		Population*dpi
Ratio buccal capsule length/width	Table 3	Eel species
		Population
		Dpi
		Worm sex
		Eel species*worm sex

*Refers to the interactions between explanatory variables given in the table. For more information please see Methods.

reference group. As measurements of different worms from one eel are not statistically independent, mixed-effects linear models were chosen for statistical modeling. The models were fitted for response variables presented in Table 4. After exclusion of extreme outliers, models were fitted by stepwise simplification starting from maximal models including the following explanatory variables: eel species, length of eel, parasite population, dpi (as a continuous variable), eel specimen, worm sex, number of L3, L4 and adult worms. Additionally, all three-way interactions between eel species, parasite population and time in dpi and sex of the worms and dpi and sex of the worms and eel species were allowed.

As the number of observations at 25 dpi was statistically insufficient, the starting time for all statistical models were set to 50 dpi. All count data were modeled as numerical variables. Eel specimen was modeled as a random effects explanatory variable, all other factors were modeled as fixed effects explanatory variables. Significant explanatory variables left after simplification are presented in Table 4. All statistics were executed in R (R Development Core Team, 2009) using the packages MASS [33] and lme4 [34]. Significance was assumed if $p < 0.05$.

Additional files

Additional file 1: Table S1. Minimal adequate mixed-effects linear model for body length of *Anguillicola crassus*; reference group: German parasite population in the European eel. Significant effects are in bold.

Additional file 2: Table S2. Minimal adequate mixed-effects linear model for body width of *Anguillicola crassus*; reference group: German parasite population in the European eel. Significant effects are in bold.

Additional file 3: Figure S1. Body length and body width of male adults of *Anguillicola crassus* in *Anguilla anguilla* and *Anguilla japonica* with the arithmetic mean values (horizontal and vertical lines) and the linear regression lines. 1 – German parasite population, 2 - Polish parasite population, 3 – Taiwanese parasite population. Dpi – days post infection.

Additional file 4: Table S3. Minimal adequate mixed-effects linear model for oesophagus width of *Anguillicola crassus*; reference group: German parasite population in the European eel. Significant effects are in bold).

Additional file 5: Table S4. Minimal adequate mixed-effects linear model for oesophagus length of *Anguillicola crassus*; reference group: German parasite population in the European eel. Significant effects are in bold.

Additional file 6: Figure S2. Oesophagus length and body width of male adults of *Anguillicola crassus* in *Anguilla anguilla* and *Anguilla japonica* with the arithmetic mean values (horizontal and vertical lines) and the linear regression lines. 1 – German parasite population, 2 - Polish parasite population, 3 – Taiwanese parasite population. Dpi – days post infection.

Additional file 7: Table S5. Minimal adequate mixed-effects linear model for buccal capsule length of *Anguillicola crassus*; reference group: German parasite population in the European eel. Significant effects are in bold.

Additional file 8: Table S6. Minimal adequate mixed-effects linear model for buccal capsule width of *Anguillicola crassus*; reference group: German parasite population in the European eel. Significant effects are in bold.

Additional file 9: Figure S3. Buccal capsule length and width of male adults of *Anguillicola crassus* in *Anguilla anguilla* and *Anguilla japonica* with the arithmetic mean values (horizontal and vertical lines) and the linear regression lines. 1 – German parasite population, 2 - Polish parasite population, 3 – Taiwanese parasite population. Dpi – days post infection.

Abbreviation
Dpi: Days post infection.

Competing interests
The authors declare that they have no competing interests.

Authors' contributions
UW planned and conducted the experiments, collected results, made the preliminary statistical analyses and wrote the manuscript. HT designed and supervised the experiments, and participated in interpretation of the results. EGH made the figures and participated in the statistical analysis. TB and BK carried out the statistical evaluation. TP helped with the preliminary statistical analyses, participated in the interpretation of the results and helped writing the manuscript. YSH organized the acquisition of The Japanese eels and supervised their dissection. All authors read and approved the final manuscript.

Acknowledgements
We would like to thank Dr. Albert Keim for the taxonomic classification of the copepods used for harvesting of the infective L3 stage of the parasite.

Author details
[1]Department of Ecology and Parasitology, Zoological Institute, Karlsruhe Institute of Technology, Kornblumenstrasse 13, Karlsruhe, Germany. [2]Department of Stochastics, Karlsruhe Institute of Technology, Kaiserstrasse 89, Karlsruhe, Germany. [3]Institute of Fisheries Science, College of Life Science, National Taiwan University, Taipei, Taiwan.

References
1. Kuwahara A, Niimi A, Itagaki H: **Studies of a nematode parasitic in the air bladder of the eel: 1. Description of *Anguillicola crassa* n. sp. (Philometridae, Anguillicolidae).** *Jpn J Parasitol* 1974, **23**:275–279.
2. Kirk RS: **The impact of *Anguillicola crassus* on European eels.** *Fish Manag Ecol* 2003, **10**:385–394.
3. Taraschewski H: **Hosts and parasites as aliens.** *J Helminthol* 2006, **80**:99–128.
4. Sures B, Knopf K: **Parasites as a threat to freshwater eels?** *Science* 2004, **304**:209–211.
5. Moravec F: *Parasitic nematodes of freshwater fishes of Europe.* Prague: Academia; 2013.
6. Sures B, Knopf K, Taraschewski H: **Development of *Anguillicola crassus* (Dracunculoidea, Anguillicolidae) in experimentally infected Balearic congers *Ariosoma balearicum* (Anguilloidea, Congridae).** *Dis Aquat Org* 1999, **39**:75–78.
7. Münderle M, Taraschewski H, Klar B, Chang CW, Shiao JC, Shen KN, He JT, Lin SH, Tzeng WN: **Occurrence of *Anguillicola crassus* (Nematoda: Dracunculoidea) in Japanese eels *Anguilla japonica* from a river and an aquaculture unit in SW Taiwan.** *Dis Aquat Org* 2006, **71**:101–108.
8. Knopf K, Mahnke M: **Differences in susceptibility of the European eel (*Anguilla anguilla*) and the Japanese eel (*Anguilla japonica*) to the swim-bladder nematode *Anguillicola crassus*.** *Parasitology* 2004, **129**:491–496.
9. Han Y-S, Chang Y-T, Taraschewski H, Chang S-L, Chen C-C, Tzeng W-N: **The swimbladder parasite *Anguillicola crassus* in native Japanese eels and exotic American eels in Taiwan.** *Zool Stud* 2008, **47**:667–675.
10. Weclawski U, Heitlinger EG, Baust T, Klar B, Petney T, San Han Y, Taraschewski H: **Evolutionary divergence of the swim bladder nematode *Anguillicola crassus* after colonization of a novel host, *Anguilla anguilla*.** *BMC Evol Biol* 2013, **13**:78.
11. Polzer M, Taraschewski H: **Identification and characterization of the proteolytic enzymes in the developmental stages of the eel pathogenic nematode *Anguillicola crassus*.** *Parasitol Res* 1993, **79**:24–27.
12. Bruňanská M, Fagerholm H-P, Moravec F: **Structure of the pharynx in the adult nematode *Anguillicoloides crassus* (Nematoda: Rhabditida).** *J Parasitol* 2007, **93**:1017–1028.
13. Bruňanská M, Fagerholm H-P, Moravec F, Vasilková Z: **Ultrastructure of the buccal capsule in the adult female *Anguillicoloides crassus* (Nematoda: Anguillicolidae).** *Helminthologia* 2010, **47**:170–178.
14. Andrássy I: *Klasse Nematoda:(Ordnungen Monhysterida, Desmoscolecida, Araeolaimida, Chromadorida, Rhabditida).* Stuttgart: Fischer Verlag; 1984.
15. Moravec F, Taraschewski H: **Revision of the genus *Anguillicola* Yamaguti, 1935 (Nematoda: Anguillicolidae) of the swimbladder of eels, including descriptions of two new species, *A. novaezelandiae* sp. n. and *A. papernai* sp. n.** *Folia Parasitol* 1988, **35**:125–146.
16. Moravec F: *Dracunculoid and Anguillicoloid nematodes parasitic in vertebrates.* Prague: Academia; 2006.
17. Laetsch DR, Heitlinger EG, Taraschewski H, Nadler SA, Blaxter ML: **The phylogenetics of Anguillicolidae (Nematoda: Anguillicoloidea), swimbladder parasites of eels.** *BMC Evol Biol* 2012, **12**:60.
18. Neumann W: **Schwimmblasenparasit *Anguillicola* bei Aalen.** *Fischer Teichwirt* 1985, **11**:322.
19. Taraschewski H, Moravec F, Lamah T, Anders K: **Distribution and morphology of two helminths recently introduced into European eel populations: *Anguillicola crassus* (Nematoda, Dracunculoidea) and *Paratenuisentis ambiguus* (Acanthocephala, Tenuisentidae).** *Dis Aquat Org* 1987, **3**:167–176.

20. Nielsen ME: Infection status of the swimbladder worm, *Anguillicola crassus* in silver stage European eel, *Anguilla anguilla*, from three different habitats in Danish waters. *J Appl Ichthyol* 1997, **13**:195–196.

21. Knopf K, Lucius R: Vaccination of eels (*Anguilla japonica* and *Anguilla anguilla*) against *Anguillicola crassus* with irradiated L3. *Parasitology* 2008, **135**:633–640.

22. Heitlinger EG, Laetsch DR, Weclawski U, Han Y-S, Taraschewski H: Massive encapsulation of larval *Anguillicoloides crassus* in the intestinal wall of Japanese eels. *Parasit Vectors* 2009, **2**:48.

23. Fonderie P, Steel H, Moens T, Bert W: Experimental induction of intraspecific morphometric variability in a single population of *Halicephalobus cf. gingivalis* may surpass total interspecifc variability. *Nematology* 2013, **15**:529–544.

24. Fagerholm HP: Intra-specific variability of the morphology in a single population of the seal parasite *Contracaecum osculatum* (Rudolphi) (Nematoda, Ascaridoidea), with a redescription of the species. *Zool Scripta* 1989, **18**:33–41.

25. Inglis WG: Allometric growth in the Nematoda. *Nature* 1954, **173**:957–957.

26. Maes GE, Pujolar JM, Hellemans B, Volckaert FA: Evidence for isolation by time in the European eel (*Anguilla anguilla* L.). *Mol Ecol* 2006, **15**:2095–2107.

27. Wirth T, Bernatchez L: Genetic evidence against panmixia in the European eel. *Nature* 2001, **409**:1037–1040.

28. Reimer LW, Hildebrand A, Scharberth D, Walter U: *Anguillicola crassus* in the Baltic sea: field data supporting transmission in brackish waters. *Dis Aquat Org* 1994, **18**:77–79.

29. Rolbiecki L, Rokicki J: *Anguillicola crassus* - an alien nematode species from the swim bladders of eel *Anguilla anguilla* in the Polish zone of the southern Baltic and in the waters of northern Poland. *Oceanol Hydrobiol Stud* 2005, **34**:121–136.

30. Wielgoss S, Taraschewski H, Meyer A, Wirth T: Population structure of the parasitic nematode *Anguillicola crassus*, an invader of declining north Atlantic eel stocks. *Mol Ecol* 2008, **17**:3478–3495.

31. Larson A: The comparison of morphological and molecular data in phylogenetic systematics. In *Molecular Approaches to Ecology and Evolution*. Edited by DeSalle DR, Schierwater DB. Basel: Birkhäuser Verlag; 1998:275–296.

32. Heitlinger E, Bridgett S, Montazam A, Taraschewski H, Blaxter M: The transcriptome of the invasive eel swimbladder nematode parasite *Anguillicola crassus*. *BMC Genomics* 2013, **14**:87.

33. Venables WN, Ripley BD: *Modern Applied Statistics with S*. New York: Springer; 2002.

34. Bates D, Maechler M, Bolker B, Walker S: lme4: Linear Mixed-Effects Models Using Eigen and S4. *J Stat Software* 2014, http://CRAN.R-project.org/package=lme4.

8

A male pheromone-mediated trade-off between female preferences for genetic compatibility and sexual attractiveness in rats

Yao-Hua Zhang and Jian-Xu Zhang[*]

Abstract

Introduction: Chemosensory signals play a vital role in socio-sexual interactions of rodents. Females rely heavily on chemosensory signals to evaluate genetic similarity and quality of potential mates, but their olfactory preferences for these criteria often conflict in mate choice.

Results: Using two inbred strains of rats, Brown Norway (BB) and Lewis (LL) and their F1 reciprocal hybrids (BL, BB♀ breed with LL♂; LB, LL♀ breed with BB♂) as genetic models, we found that the chemosensory preferences of BB and LL females between these 4 strains of rats could be predicted on the basis of genetic compatibility benefits, except that LL females exhibited incestuous preferences for male urine odor of LL rats over that of the BB strain and the F1 hybrids. Seven ketone components of major urine volatiles proved to be potential male pheromones and were enriched in LL males compared to BB males or the F1 hybrid males. We hypothesize that these ketones produced an extravagant male trait that attracts LL females, overriding compatibility traits. This conclusion was corroborated by adding three synthetic pheromone analogues, 4-heptanone, 2-heptanone and 9-hydroxy-2-nonanone of these 7 components, which resulted in equalization of the sexual attractiveness of BB male urine and LL male urine. Additionally, in the genetically diverse F2 hybrids (BL♀ breed with BL♂), the pheromones-enriched males could consistently attract the F2 females.

Conclusions: We suggest that the exaggerated male pheromones serve as a "sexual chemical ornament" to attract females, independent of genetic compatibility, whereas genetic dissimilarity could influence the preferences only when male pheromones varied on a small scale.

Keywords: Compatibility, Male pheromones, Olfactory preference, Quality

Introduction

In rodents, male urine odor can signal genetic relatedness, based on genotype-specific odortype and genetic quality indicated by sexually selected signals for female mate assessment and choice [1-5]. However, mating preferences for compatibility are distinguished from mating preferences for sexual attractiveness: compatibility depends on the genotypic matching between males and females, whereas sexually attractive traits are uniformly favored by all the potential mates [4,6,7]. Females prefer genetically dissimilar males and/or males whose sexually selected signals indicate high genetic quality; however, a sexually attractive partner is not necessarily the compatible one and vice versa, so mating preferences for compatibility and sexual attractiveness criteria are often in conflict [4]. However, the way in which male pheromone molecules act as sexually selected signals to interact with genotype-correlated odorant signals in determining female mate choice decision has seldom been tested empirically [1,4,5,7-10].

Several hypotheses have been put forward to explain the evolutionary process of mate choice to obtain indirect genetic benefits, such as good genes, sexy son and heterozygous benefits [11]. The nocturnal rodents have a well-developed sense of smell; female mate choice mainly depends on male odorant signals, which contain information both about genetic compatibility and genetic quality (indicating the possibility of "good genes" and

* Correspondence: zhangjx@ioz.ac.cn
State Key Laboratory of Integrated Management of Pest Insects and Rodents in Agriculture, Institute of Zoology, Chinese Academy of Sciences, 1# Bei-Chen West Road, Beijing 100101, China

sexy son benefits) [4,6,11,12]. Avoidance of incestuous mating is widely reported across animal taxa, where the major histocompatibility complex (MHC) has provided an excellent example to illustrate the genetic compatibility-based mate choice mechanism [4,12-15]. However, theory predicts that an individual that mates with a relative will spread genes identical by descent, which also offers a positive effect on the inclusive fitness of parents [16,17]. In some animal species, for instance, the fruit fly (*Drosophila melanogaster*), the cestode (*Schistocephalus solidus*) and the cichlid fish (*Pelvicachromis taeniatus*), the females display significant preferences for mating with their brothers over unrelated males [18-21]. In natural populations of animals, mating among relatives occurs more often than expected by chance, suggesting that genetic compatibility effects are limited [6,22].

Almost all research on odor-based mate choice in mice (*Mus musculus*) and rats (*Rattus norvegicus*) have shown that females chose genetically compatible mates to obtain heterozygous offspring and high reproductive success [1,2,4,5,13,14,23]. Exceptionally, Roberts and Gosling [4] found that "compatible genes" and "good genes" interact in female mate choice in mice, and their relative influence can vary with the degree of variability of each trait among available males. In rodents, both genetic compatibility and male sexual attractiveness can be signaled by urinary volatiles derived from bladder urine (BU) and preputial gland secretion (PGS) and major urinary protein (MUP) from liver; however, the ways in which genetic relatedness- and quality-related signaling molecules interact in female mate choice has not been tested experimentally [1,4,5,8-11,24,25].

Numerous inbred strains of laboratory mice and rats provide ideal animal models to study odor-based mate choice [4,13,23]. Individuals of an inbred strain are nearly genetically identical to each other (like identical twins), but genetically distinct across different inbred strains [26,27]. The reciprocal hybrids share the same complement of genes, and have intermediate relatedness between the two parental strains. Previous work in our laboratory using two inbred strains of rats, Lewis (LL) and Brown Norway (BB), showed that BB females preferred the male urine odor of LL male to that of their own strain [5]. Here, unlike BB females, LL females display an olfactory preference for the males of their own strain over BB males, violating genetic compatibility benefits. We use these inbred strains, their reciprocal hybrids (BL, BB♀ × LL♂; LB, LL♀ × BB♂) and BL F2 hybrids (BL♀ × BL♂) as genetic models to examine whether advantages of mating with close kin override the effects of inbreeding depression, and how the male sex pheromone mediates the trade-off between preferences for genetic compatibility and genetic quality in female mate choices.

Results

Urine scent-based preference test by BB and LL females
In binary choice tests, BB females showed an olfactory preference for LL male urine over that of the other three strains (BB vs. LL: $P = 0.032$, $T = 2.455$, $N = 12$; LL vs. LB: $P = 0.021$, $T = 2.688$, $N = 12$; LL vs. BL: $P = 0.015$, $Z = 2.433$, $N = 12$). BB females preferred BL male urine odor to that of BB males, but showed no preference between either BB or BL and LB male urine (BB vs. BL: $P < 0.001$, $T = 7.097$, $N = 12$) (Figure 1A).

On the other hand, LL females displayed an incestuous preference for male urine odor of their own strain to that of the other three strains (LL vs. BB: $P = 0.009$, $Z = 2.621$, $N = 13$; LL vs. LB: $P = 0.039$, $T = 2.320$, $N = 12$; LL vs. BL: $P = 0.001$, $T = 3.180$, $N = 13$); however, they preferred male urine odor of the BB strain compared to that of the genetically more similar LB and BL strains (BB vs LB: $P = 0.005$, $Z = 2.830$, $N = 13$; BB vs. BL: $P = 0.002$, $Z = 3.040$, $N = 13$). In addition, LL females showed no preference between LB and BL male urine odor (Figure 1B).

In these rats, within-strain breeding resulted in inbreeding depression such as reduced reproductive successes and offspring (Additional file 1: Table S1).

The urine-borne volatile profiles varying among strains
Thirty BU volatiles and 24 PGS volatiles were detected by GC-MS (Additional file 2: Figure S1 and Additional file 3: Figure S2). The relative abundances (percentage of summed peak area) of these volatiles were subjected to a principle component analysis (PCA). In the case of BU volatiles, BB males were distinct from LL males but the hybrids were poorly separated from LL (Additional file 4: Figure S3A). In the case of PGS volatiles, BB males were nearly separated from LL, and the two hybrids clustered together (Additional file 5: Figure S4A). ANOVA with a *post hoc* LSD test revealed that 16 BU and 10 PGS volatile components were different in relative abundances between BB and LL male rats (Additional file 1: Table S2 and Table S3). Although the 4 strains were not clearly separated in the PCA plots, BB had 14 BU and 10 PGS volatiles that differed from LB, and 15 BU and 5 PGS volatiles that differed from BL. LL had 3 BU and 11 PGS volatiles that differed from LB, and 7 BU and 10 PGS volatiles that differed from BL. LB and BL hybrid males clustered together in the PCA plots, and neither BU nor PGS compounds varied in relative abundance (Additional file 1: Table S2 and Table S3; Additional files 4 and 5: Figure S3A and Figure S4A).

Regarding the abundances (peak areas in GC graphs) of these volatiles, PCA revealed that, in the case of BU, BB and LL males were clearly separated, but they were poorly separated from hybrid males (Additional file 4: Figure S3B). In the case of PGS, LL males were nearly separated from the other three strains, whereas the

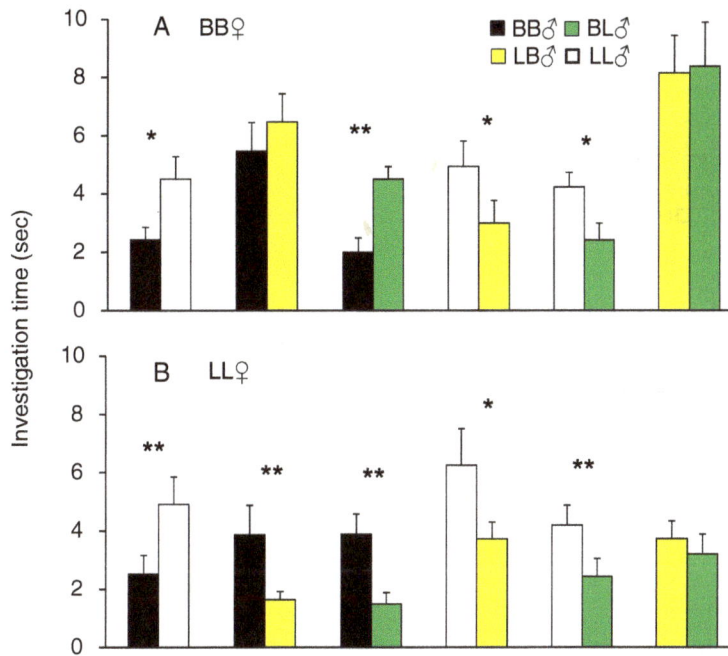

Figure 1 Duration of investigation (mean ± SE) by BB (A, $N = 12$) and LL (B, $N = 13$) female rats in binary choice tests (*$P < 0.05$, **$P < 0.01$, paired t-test or Wilcoxon signed-rank test).

hybrids and BB males were not separated (Additional file 5: Figure S4B). ANOVA with a *post hoc* LSD test revealed that 15 BU and 8 PGS volatile components were different in abundances between BB and LL male rats (Additional file 1: Table S4 and Table S5). Although the 4 strains were not clearly separated in the PCA plots, BB had 8 BU and 5 PGS volatiles that differed from LB, and 6 BU and 3 PGS that differed from BL. LL had 9 BU and 10 PGS volatiles that differed from LB, and 15 BU and 8 PGS volatiles that differed from BL. LB and BL hybrid males clustered together in the PCA plots, and neither BU nor PGS compounds varied in relative abundance (Additional file 1: Table S4 and Table S5; Additional files 4 and 5: Figure S3B and Figure S4B).

The volatiles exaggerated in LL male urine and potential male pheromones

Of the 16 BU volatile components varying between the rat strains, 7 (Peaks: 2, 3, 11, 15, 17, 19 and 22) had a relative abundance > 1% and, together accounted for 83.51% of all the 16 compounds. These 7 compounds were male-biased in BB, LL and the hybrids and had consistently higher relative abundance or absolute abundance in LL males than in BB and hybrid males (Figure 2, Additional file 6: Figure S5, Additional file 1: Table S2 and Table S4; and data not shown). However, the 14 male-biased PGS compounds showed inconsistent differences between rat strains (Additional file 1: Table S3 and Table S5). These volatiles in LB and BL males of the hybrids

from reciprocal crosses between BB and LL rats seemed to have intermediate levels (Figure 2).

Furthermore, the above-mentioned 7 BU volatiles were severely suppressed in absolute abundances (Figure 3A) or in relative abundances (Figure 3B) by orchiectomy and restored by testosterone-treatment in LL male rats (Figure 3, Additional file 1: Tables S6 and Additional file 7: Figure S6).

Meanwhile, LL females were used as odor recipients. When synthetic analogs of 4-heptanone (4H), 2-heptanone (2H) and 9-hydroxy-2-nonanone (9H2N) selected from the 7 compounds were replenished into castrated male urine so that levels were equal to those of LL males (4H = 1.5 ppm; 2H = 22 ppm; 9H2N = 7 ppm), the sexual attractiveness of castrated male urine were restored (sham vs. castrated: $P = 0.004$, $T = 3.513$, $N = 14$; castrated vs. castrated + the 3 ketones: $P = 0.041$, $Z = 2.040$, $N = 14$; sham vs. castrated + the 3 ketones: $P = 0.926$, $T = 0.095$, $N = 14$) (Figure 4, Additional file 1: Tables S6).

Male pheromones and LL female preferences

We further supplemented the the above-mentioned 3 ketones into BB male urine, which greatly boosted the attractiveness of BB male urine to LL females, especially when the 3 ketones reached 75% and 100% of the LL levels (4H = 0.06 ppm, 2H = 3.76 ppm, 9H2N = 0.86 ppm; BB vs. BB + 75% 3 ketones, $P = 0.008$, $Z = 2.669$, $N = 14$; BB vs. BB + 100% 3 ketones, $P = 0.016$, $T = 2.769$, $N = 14$; Figure 5A). At same time, the attractiveness of these

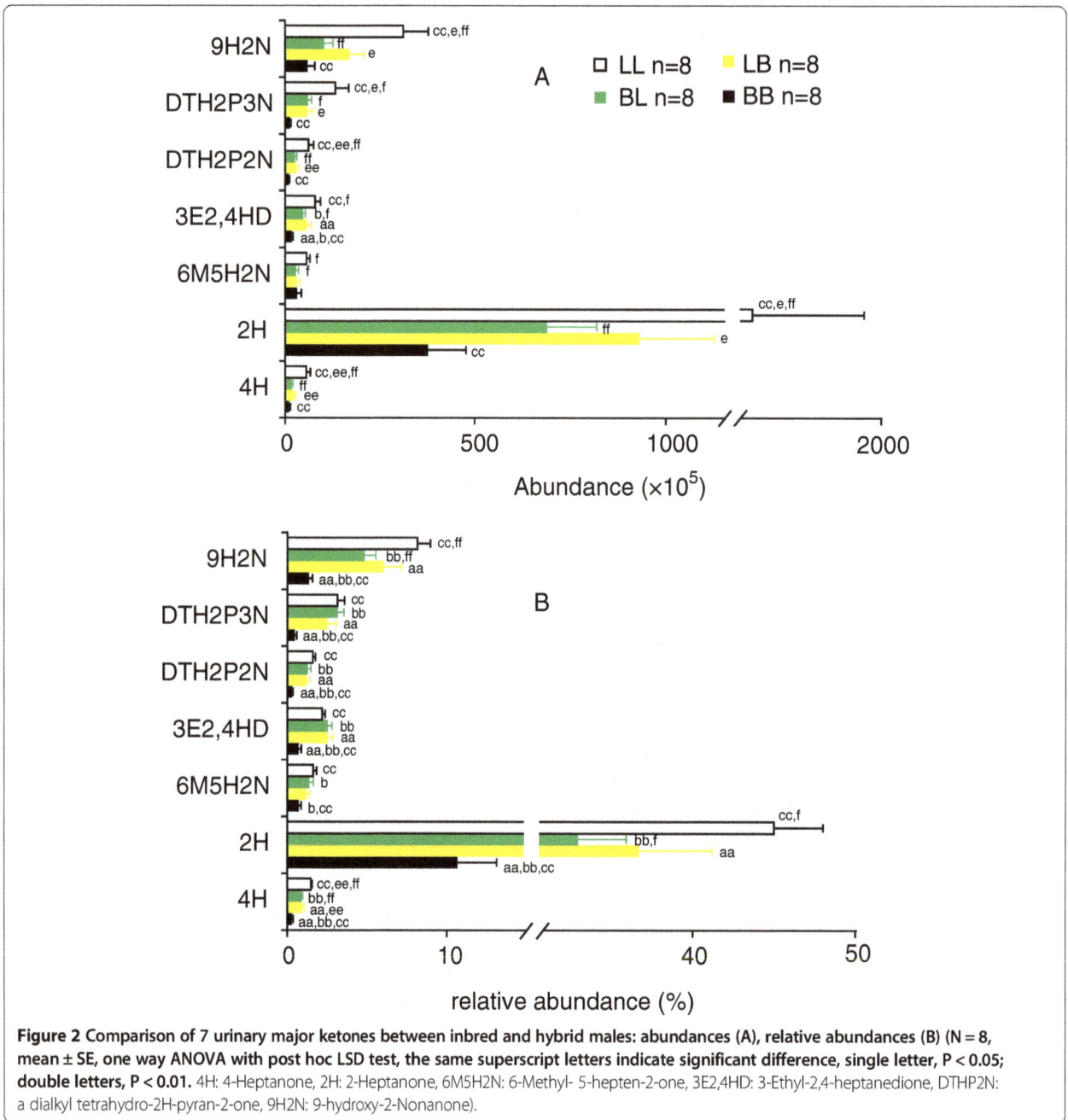

Figure 2 Comparison of 7 urinary major ketones between inbred and hybrid males: abundances (A), relative abundances (B) (N = 8, mean ± SE, one way ANOVA with post hoc LSD test, the same superscript letters indicate significant difference, single letter, P < 0.05; **double letters, P < 0.01.** 4H: 4-Heptanone, 2H: 2-Heptanone, 6M5H2N: 6-Methyl- 5-hepten-2-one, 3E2,4HD: 3-Ethyl-2,4-heptanedione, DTHP2N: a dialkyl tetrahydro-2H-pyran-2-one, 9H2N: 9-hydroxy-2-Nonanone).

treated BB urine was comparable to that of LL male urine (LL vs. BB + 75% 3 ketones, $P = 0.102$, $T = 1.761$, $N = 14$; LL vs. BB + 100% 3 ketones, $N = 14$, $P = 0.265$, $T = 1.164$, $N = 14$; Figure 5B).

Male pheromones and female preferences in F2 hybrids
In F2 hybrids crossed from F1 hybrids (BL♀ × BL♂), F2 females also preferred male urine scent of LL rats to that of BB rats ($P < 0.001$, $T = 5.403$, $N = 40$; Figure 6A). The three groups we selected from F2 males showed significant differences in their ketone levels (Figure 6B). F2 females exhibited a significant preference for male urine scent of

HIGH group to that of LOW group (HIGH vs. LOW, $P = 0.009$, $Z = 2.621$, $N = 40$), but showed no preference between other groups (MEDIUM vs. LOW, $P = 0.989$, $Z = 0.013$; MEDIUM vs. HIGH, $P = 0.129$, $Z = 1.519$; $N = 40$; Figure 6C).

Discussion
In binary choice tests, BB females exhibited an olfactory preference for more genetically dissimilar males, in most cases as expected from the genetic compatibility hypothesis [11]. Although LL females also showed preferences for compatible benefits between BB and the F1 hybrid male

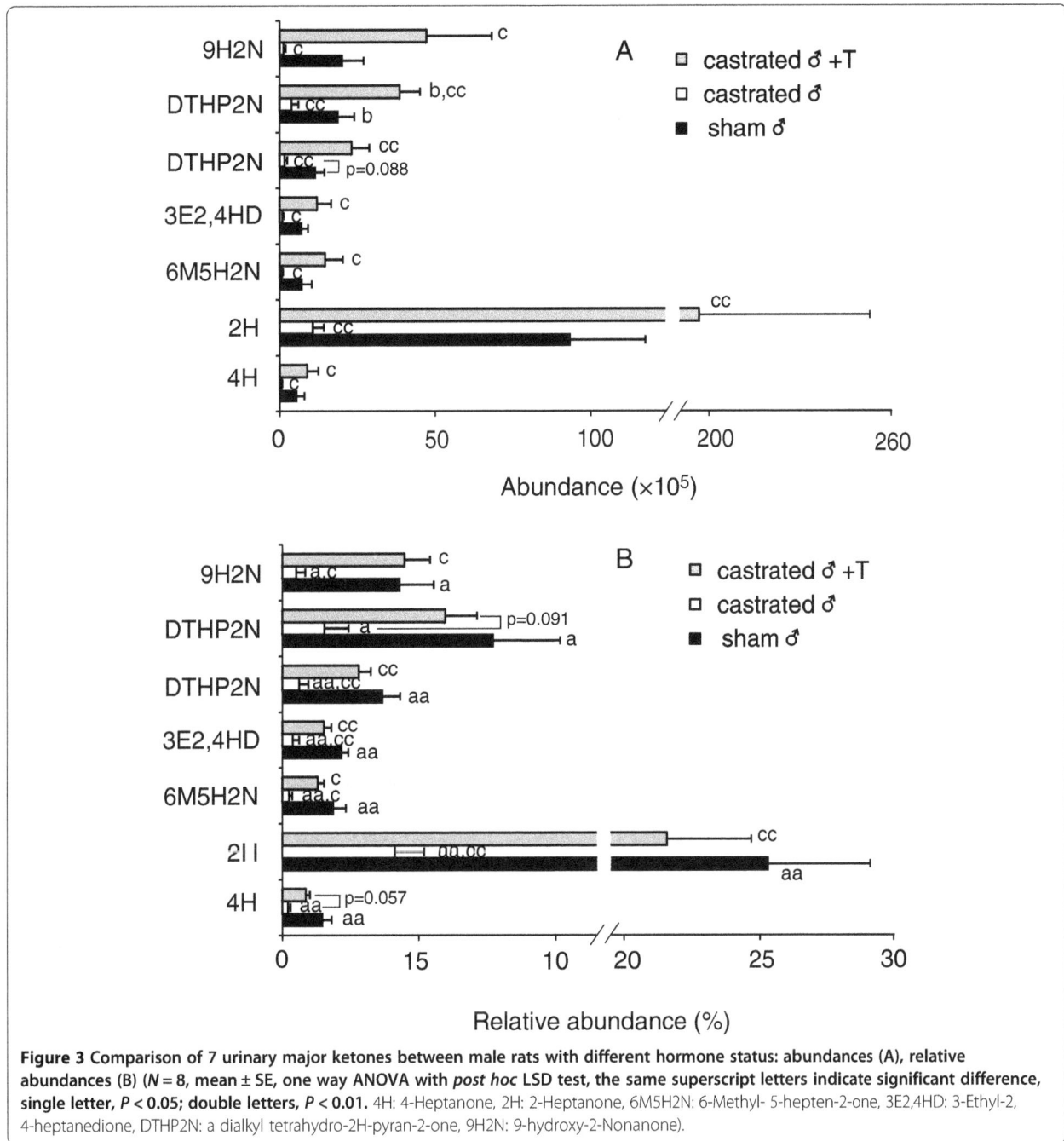

Figure 3 Comparison of 7 urinary major ketones between male rats with different hormone status: abundances (A), relative abundances (B) ($N = 8$, mean ± SE, one way ANOVA with *post hoc* LSD test, the same superscript letters indicate significant difference, single letter, $P < 0.05$; double letters, $P < 0.01$. 4H: 4-Heptanone, 2H: 2-Heptanone, 6M5H2N: 6-Methyl- 5-hepten-2-one, 3E2,4HD: 3-Ethyl-2, 4-heptanedione, DTHP2N: a dialkyl tetrahydro-2H-pyran-2-one, 9H2N: 9-hydroxy-2-Nonanone).

rats, they chose male urine of genetically identical LL rats over that of BB or the F1 hybrid rats at the cost of inbreeding. Male pheromone levels might be positively correlated with genetic quality in male animals, and thus signal the information to choosy females about genetic benefits, such as "good genes" and sexy son benefits [4,5,11,13,15,23,26,28]. Preferences for genetic compatibility strictly depend on the interplay of the genotypes of potential partners and are different from directional preferences for ornamental secondary sexual traits [4,7]. Thus, the most compatible partner may not be the one with most pronounced ornaments and vice versa [4,6,7]. Genetic compatibility criteria and quality criteria may often conflict, and are often difficult to reconcile with female mate choice [4,6,7]. In mice, genetic quality indicated by urine marking rates largely determine female mate choice, whereas genetic compatibility works only when variability in genetic dissimilarity among the males is relatively large, or conversely, when the variability in marking rates is small [4]. However, in the cichlid fish, females weighed genetic compatibility of male mates stronger than quality traits (body size) [7]. Here, the contents of male pheromone

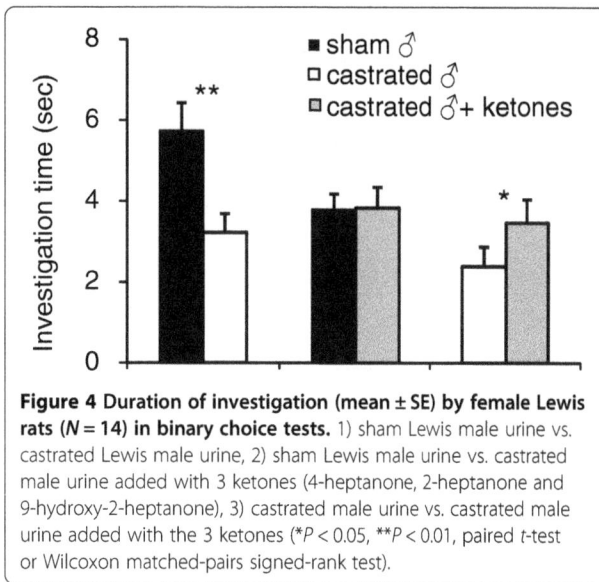

Figure 4 Duration of investigation (mean ± SE) by female Lewis rats (N = 14) in binary choice tests. 1) sham Lewis male urine vs. castrated Lewis male urine, 2) sham Lewis male urine vs. castrated male urine added with 3 ketones (4-heptanone, 2-heptanone and 9-hydroxy-2-heptanone), 3) castrated male urine vs. castrated male urine added with the 3 ketones (*P < 0.05, **P < 0.01, paired t-test or Wilcoxon matched-pairs signed-rank test).

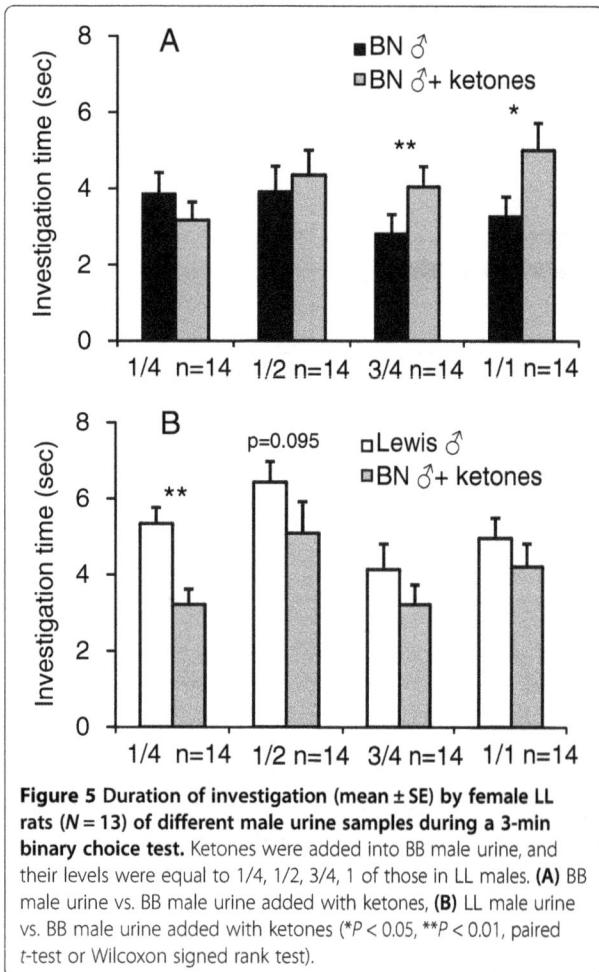

Figure 5 Duration of investigation (mean ± SE) by female LL rats (N = 13) of different male urine samples during a 3-min binary choice test. Ketones were added into BB male urine, and their levels were equal to 1/4, 1/2, 3/4, 1 of those in LL males. **(A)** BB male urine vs. BB male urine added with ketones, **(B)** LL male urine vs. BB male urine added with ketones (*P < 0.05, **P < 0.01, paired t-test or Wilcoxon signed rank test).

components quantitatively varied among the 4 strains: LL males were highest, BB males were lowest and the hybrid males were intermediate. Coincidently, LL female displayed incestuous preferences only between LL and the others. At the same time, when increasing the urinary ketones of BB male urine to 75% and 100% of those in that of LL, the treated BB males could be as attractive as LL. We therefore inferred that the genetic compatibility worked only when the sexual chemical ornaments were below the threshold. Above the threshold, the influence of a sexy trait would override the compatibility benefits.

In the binary choice test, the paired male urine scent of different strains, except the BL-LB hybrid pairs, received different responses by female LL or BB rats, and therefore must contain strain- or genotype-specific odorant signals to advertise genetic similarity and quality to females [1,2,4,5,9,13]. The chemosignals that provide olfactory information about individuality, strain and species are often composed of a large number of compounds in a mosaic manner or a few key components that vary in relative abundance [1,5,29-31]. In the current study, LL male urine might also contain some chemosignals, which might be analogous to some extravagant ornamental traits in birds and fishes and related to sexual attractiveness and high quality, and thus incur an inbreeding preference that overrides inbreeding avoidance by LL females [4,11,15,32,33]. Such conjectures are confirmed by the following chemical analysis of urine-borne volatiles.

In rats, the voided urine contains volatile compounds excreted from both bladder urine and preputial glands [5,25]. The percentage of summed GC peak areas (referred to as relative abundances) of volatiles accurately reflects genetic relatedness and thus can be used to analyze the genotype-related odortypes [1,5]. Here, the PCA and pairwise comparison of the relative abundances of BU volatiles and PG volatiles suggested that the genetic relatedness-correlated odortypes were determined by the volatile profiles with quantitative variation and/or some key urinary volatiles and thus might be used as genetic compatibility criteria in these rats.

In our previous work in rats, a few urinary volatiles, such as 2-heptanone, 4-ethyl phenol and squalene, have been identified as male pheromones, which alone or in combination can attract females and reflect genetic relatedness [5,25]. Here, since the above-mentioned 7 major BU volatiles are at higher levels in males than in females, and dependent on testis and androgen, they are male sex pheromone candidates [2,25]. On the other hand, all or most of these 7 compounds were consistently over-expressed in LL males, as compared to BB and hybrid males; therefore, they may function as an extravagant male trait to advertise high quality. Such conclusions were further exemplified by replenishing the ketones into BB and castrated male urine, which resulted

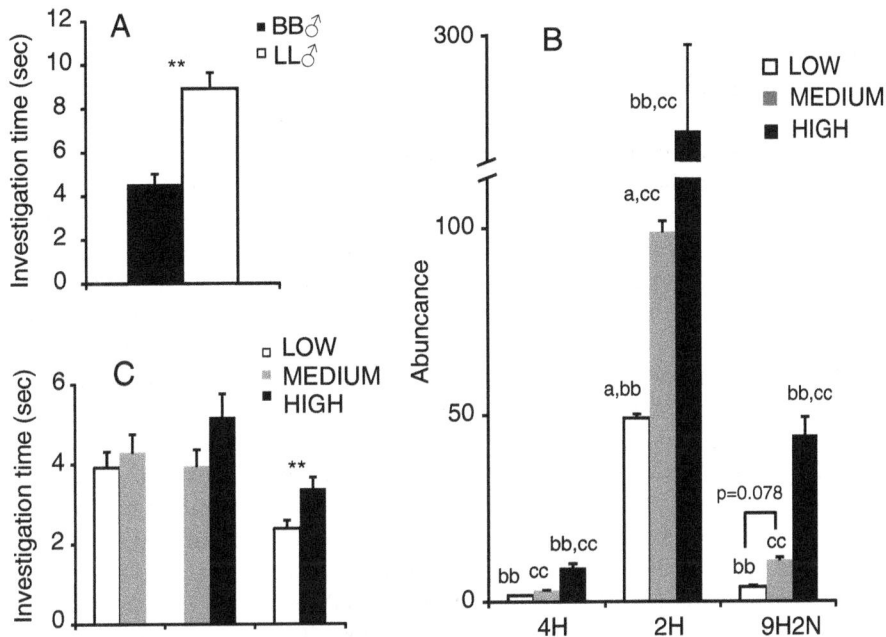

Figure 6 Behavioral and chemical data of F2 male urine. (A) Duration of investigation (mean ± SE) by F2 BL females ($N = 40$) of different male urine samples during a 3-min binary choice test (**$P < 0.01$, paired t-test). **(B)** Comparison of 3 urinary major ketones among LOW, MEDIUM and HIGH group of the BL F2 male urine ($N = 20$, mean ± SE, one way ANOVA with *post hoc* LSD test, the same superscript letters indicate significant difference, single letter, $P < 0.05$; double letters, $P < 0.01$). **(C)** Duration of investigation (mean ± SE) by F2 BL females ($N = 40$) of F2 BL male urine of LOW, MEDIUM and HIGH groups during a 3-min binary choice test (**$P < 0.01$, Wilcoxon signed rank test).

in equalization of the sexual attractiveness of the adjusted BB, adjusted castrated male urine and LL male urine. However, few of these pheromone components differed between BB and F1 hybrid male urine and thus LL female mate choice was majorly influenced by genetic compatibility. These results indicated that these compounds were of male pheromones and when above a certain concentration, could serve as an extravagant male trait. Similar results were also observed in genetically diverse F2 hybrids. Specifically, the F2 males of HIGH group were more strongly attractive to F2 females than those of LOW group, suggesting that extravagant male pheromones-based preference might be implicated in mate choice of wild populations.

Since the recipients could sense both volatile and non-volatile cues in male urine during our 2-choice tests, we also measured the major urine proteins (MUPs) of these strains and their F2 generation using sodium dodecyl sulfate-polyacrylamide gel electrophoresis (SDS-PAGE). We have not observed the between-strain differences of MUPs (data not shown). Such a 2-choice test method has been successfully used in our previous work [5]. Here, a close correlation was found between the volatiles and the female preferences. It implied that the female chemosensory preferences might be mainly shaped by the volatile ketone pheromones rather than MUPs [34].

Here, we first reported in rats that genetic compatibility and sexual attractiveness criteria signaled by urine-borne volatiles interact to affect odor-based female mate preferences [4,35]. The variation in male pheromones of potential mates allowed LL females to weigh the trade-off between genetic compatibility and quality benefits. The extravagant male pheromones in LL males served as a chemical ornament trait or high genetic quality trait to attract LL females, independent of genetic compatibility [32,33]. The sexual attractiveness indicated by male pheromones in LL males could consistently predict the preferences of LL females independent of genetic dissimilarity and heterozygosity, whereas genetic compatibility could influence the preferences only when male pheromones of potential mates varied on a small scale, for example, the preferences between BB and the F1 hybrid males [2,4,12].

Such a male pheromone-indicating quality might be associated with "good genes" and sexy son benefits, where the "good genes" benefit focuses on the increased viability of both sons and daughters and the sexy son benefit suggests an indirect effect of the attractiveness of sons at the cost of offspring fitness [11]. Thus far, theoretical studies and a recent meta-analysis of 90 studies on 55 animal species have suggested that the choosy females get "sexy sons" more often than "good genes", but few experiments have directly demonstrated the "sexy sons" mechanism [11,12,33,36,37]. When choosing attractive LL males, LL females gained the attractive traits, but took the risk of a fitness reduction due to inbreeding depression, suggesting that extravagant male pheromones reflected sexy son benefits rather than "good genes" [11,32,38].

Although male pheromones in rats are under the control of androgen, we did not detect the difference of serum testosterone levels between BB and LL males ($P = 0.482$, $Z = 0.703$, $N = 7$) [39]. Therefore, the "extravagant chemical ornament" highly expressed in homozygous LL males, might be determined by polygenes from both homozygous LL parents. Mating between relatives can spread genes identical by descent, which offers a positive effect on the inclusive fitness of parents [12,14,16,17,38,40]. The decreased fitness and the increased expression of deleterious recessive genes in the offspring would offset the benefits of sexual attractiveness for mating success [32]. The male pheromones acting as attractive traits were suppressed in BB males but highly expressed in LL males, suggesting that male pheromones and sexual attractiveness are not always reduced by inbreeding [41,42]. Most theoretical and experimental studies assume that the genes regulating emission and perception of sensory signals are separate but linked, except in the case of two studies in the medaka (*Oryzias latipes*) and the fruit fly (*Drosophila melanogaster*) showing a single-gene control of both pheromone emission and perception [43,44].

Conclusions

We suggest that the exaggerated male pheromones produce a "sexual chemical ornament" that consistently attracts females, independent of genetic compatibility. Genetic compatibility influenced the female preferences only when the male pheromones of potential mates varied on a small scale. Despite the risk of inbreeding, choosy females might gain the advantages of homozygous sexy sons by mating with relatives. The volatile signals can transmit information at a distance, and precede the first episode of mating behavior and consequently increased the potential reproductive success [34,45]. The male "chemical ornament" and "sexy genes" benefits might also give an additional explanation for inclusive fitness with inbreeding in natural population [11,22].

Methods
Subject
In experiment 1, BN rats (BB, $N = 16$, half of each sex) and Lewis rats (LL, $N = 16$, half of each sex) at 12 weeks of age, obtained from Vital River Laboratory Animal Technology Co. Ltd, were reciprocally crossed to produce F1 hybrid rats (BL = BB♀ × LL♂; LB = LL♀ × BB♂). BB ($N = 20$, half of each sex) and LL ($N = 16$, half of each sex) were used to obtain pure BB and LL F1 rats. 15 BL males were bred with 30 BL females to produce BL F2 rats. The

males were removed when the females had become noticeably pregnant. The offspring were weaned, weighed, sexed at 4 weeks, and then housed with same sexed siblings (4–6 per cage).

In experiment 2, 21 LL rats (12 weeks old; Vital River Laboratory Animal Technology Co. Ltd) were assigned to the following treatment groups: bilateral orchiectomy ($N = 7$); sham orchiectomy ($N = 7$); bilateral orchiectomy with testosterone supplement ($N = 7$). In the orchiectomy group, rats were anesthetized with sodium pentobarbital (45 mg/kg), and then the testes were excised following incision of the scrotum and ligation of the blood vessels and vas deferens. The sham orchiectomy procedure was the same, but the testes were not excised. A silastic capsule (1.57 mm inner diameter, 20 mm length, left blank or filled with testosterone, incubated in 0.9% sodium chloride solutions for 48 h) was implanted subcutaneously between the scapulae of each rat. Rats were single-housed for a 4-week postsurgical recovery period, and then served as urine and PGS donor.

In addition, 40 BB females, 60 LL females and 100 BL females (some of them brought from Vital River Laboratory Animal Technology Co. Ltd and some of them F1 and F2 offspring from the present experiment) were used as odor recipients in binary choice tests. Urine collection, binary choice test and chemical analysis were carried out at 12–16 weeks. The room had a reversed 14:10 h light: dark photoperiod (lights on at 19:00) and was maintained at $23 \pm 2°C$. Rats were kept in rat cages (37 × 26 × 17 cm). Food (standard rat chow) and water were provided *ad libitum*.

Scent collection and sample preparation
Each rat was placed in a clean cage (25 × 15 × 13.5 cm, fitted with a wire grid 1 cm above the cage floor), and urine was collected immediately and transferred to a sealed glass vial on ice. Urine contaminated by feces was discarded. Rats were decapitated, and the paired preputial glands were immediately dissected and stored in vials. Urine and preputial glands were kept at −20°C prior to analysis.

To prepare urine samples treated with the identified synthetic analogs for behavioural assays, 4-heptanone, 2-heptanone and 9-hydroxy-2-nonanone were dissolved in dichloromethane. The solution is added in to a vial and been vaporized for 3 minutes before adding urine. Meanwhile, equal amount of pure dichloromethane were added into urine controls for comparison with fortified urines. Male urine was then mixed equally from 7 urine donors, and added to the vial.

To characterize the composition of urine samples, we mixed 120 μL dichloromethane (purity >99.5%; DIMA Technology, Inc.) with 120 μL urine, stirred the mixture thoroughly, stored it at 4°C for 12 h, and then used the bottom phase (the layer with dichloromethane) for chemical analysis.

Two-choice olfactory tests

We determined the estrus cycles of female rats, and used the estrus ones as odor recipients. Double blind binary choice tests were carried out under dim red light during the dark phase of the light cycle. For each test, we left one subject in its home cage while temporarily removed its cage mates. We painted 2 μL of urine samples on one end of a glass rod (20 cm long, 4 mm diameter) and held the other end with plastic gloves. Two scented glass rods were poked through the bars of the cage, and simultaneously presented to the subject. The two ends painted with urine samples were 1.5 cm apart from each other. The investigation time for each urine sample was recorded for 3 min after the subject first sniffed or licked the rod tip. The experimenter was blind to the nature of the sample. To control for experimenter bias, we exchange the positions of odor samples when testing the next female. Recipients were randomly chosen, and each rat was used only once a day. Each female was used once in a day and given at least 4 days of rest before another test. Each subject was used in one or two tests. The tests of BB and LL females were replicated once. As many of the subjects were only tested once, we did not assess the random effects and subject variations. Recipients were allowed to freely investigate the scented glass rods in their home cage, and those whose investigating time was less than 1 s were excluded for the day.

Gas chromatography- mass spectrometry (GC–MS)/GC assay

Analysis of F1 urine and PGS were performed on an Agilent 6890 N GC System connected to the 5973 Mass Selective Detector (NIST 2002 Library; Agilent Technologies, Inc., USA). The GC was equipped with a non-polar column (HP5MS, 30 m long, 0.25 mm inner diameter, 0.25 μm film) as well as a polar column (DBWAX, 30 m long, 0.25 mm inner diameter, 0.25 μm film). The carrier gas was Helium (1.0 ml/min). The injector temperature was set at 280°C. The oven temperature was set initially at 50°C, heated by 5°C/min to 100°C, then ramped by 10°C/min to 280°C, and held for 5 min. Electron impact ionization was at 70 eV. The transfer line temperature was set at 280°C. Scanning mass ranged from 30 to 450 amu. We injected a 4 μL sample at a splitless mode for urine and 3 μL sample in a split mode (1:10) for PGS. The GC-FID and GC-MSD system are checked daily with calibration standards before running samples.

Tentative identifications were done by comparison of results on both polar and non-polar column and comparing the mass spectra of GC peaks with those in the MS library and the published literature [25,46]. The diagnostic fragments at m/z 60, m/z M^+-29, and m/z M^+-43 imply fatty acids; and m/z 58, m/z 71 imply ketones. Fourteen compounds, 4-heptanone, 2-heptanone, dimethyl sulfone, 4-methyl phenol, 4-ethyl phenol, 9-hydroxy-2-nonanone, indole, E-β-farnesene, dodecanoic acid, tetradecanoic acid, hexadecanoic acid, Z9-octadecenoic acid, octadecanoic acid, and squalene (all purity >98%; ACROS Organics) were further confirmed by matching retention times and mass spectra with the authentic analogs.

Analysis of BL F2 urine was performed on an Agilent 7890 N GC System connected to a flame ionization detector (FID) (Agilent Technologies, Inc., USA). The GC was equipped with a HP5-MS column, and the carrier gas was N_2 (1.0 ml/min). The inject volume, inject temperature and oven temperature were the same as those of 6890 N GC System. We compared the retention time of an F2 urine sample with the retention times of 4-heptanone, 2-heptanone and 9-hydroxy-2-nonanone authentic analogs to identify the 3 ketones. Acoording to the chemical data, we sorted the 140 BL F2 by according to 2-heptanone levels, and chose the LOW, MEDIUM and HIGH group ($N = 20$ for each group) for the behaviour tests (mean levels of 2-heptanone: HIGH = 20.33 ppm, MEDIUM = 7.15 ppm, LOW = 3.41 ppm).

We quantified 4-heptanone, 2-heptanone and 9-hydroxy-2-nonanone in urine by comparing their GC areas in the samples with an established standard curve (GC area vs. concentration). The abundance and relative abundances of compounds were used for statistical analysis. For a particular compound, the abundance was quantified by GC peak area, and the relative abundance was a percentage of the summed peak areas from all targeted GC peaks.

Statistical analysis

The distribution of raw data was examined using a Kolmogorov–Smirnov test and then either parametric or nonparametric tests were used in the analyses that followed. A one-way ANOVA was applied to determine the effects of strain (in experiment 1) and hormone status (in experiment 2) on each volatile compound. One way ANOVA was followed by least significant different (LSD) tests. Independent t- tests or Mann Whitney U tests were used to determine the differences in concentrations of volatiles between LL male and female. Also, a principal component analysis (PCA) was applied to the chemical data to investigate whether the whole volatile profiles were differentiated between groups. The raw data were deduced, and new variables (PCs) were generated. The first two most powerful PCs, accounting for more than 60% of the total variance were plotted in two-dimensional graphs. Paired t- tests or Wilcoxon signed-rank tests were applied to the binary choice data. All Statistical analysis was conducted using SPSS (v15.0, SPSS Inc.). Alpha was set at $P < 0.05$.

Ethical standards

The procedures of animal care and use in this study fully complied with Chinese legal requirements and were approved by the Animal Use Committee of the Institute of Zoology, Chinese Academy of Sciences (approval number IOZ12017).

Additional files

Additional file 1: Table S1. Breeding pairs and offspring; **Table S2.** The relative abundances of urinary components of 4 strains of rats; **Table S3.** The relative abundances of PGS components of 4 strains of rats; **Table S4.** The abundances of urinary components of 4 strains of rats; **Table S5.** The abundances of PGS components of 4 strains of rats; **Table S6.** The relative abundances of urinary components of male rats under different hormonal status; **Table S7.** The abundances of urinary components of male rats under different hormonal status.

Additional file 2: Figure S1. Representative GC profiles of urinary volatiles detected from BB, LL, BL, LB males (Numbered GC peaks correspond to the compounds in Table S2 and S4. 4H: 4-Heptanone, 2H: 2-Heptanone, 6M5H2N: 6-Methyl- 5-hepten-2-one, 3E2,4HD: 3-Ethyl-2, 4-heptanedione, DTHP2N: a dialkyl tetrahydro-2H-pyran-2-one, 9H2N: 9-hydroxy-2-Nonanone).

Additional file 3: Figure S2. GC chromatogram of PGS volatiles detected from BB males (upper panel) and LL males (lower panel). (Numbered GC peaks correspond to the compounds in Table S3 and S5).

Additional file 4: Figure S3. Principal component plots of BB, LL and hybrid males, based on the relative abundances (A) and absolute abundances (B) of the BU volatiles. Each symbol represents 1 individual rat, and the percentage of the total variance along each principal component is given.

Additional file 5: Figure S4. Principal component plots of BB and LL males, based on the relative abundances: (A) and absolute abundances (B) of the PGS volatiles. Each symbol represents 1 individual rat, and the percentage of the total variance along each principal component is given.

Additional file 6: Figure S5. Comparison of 7 urinary major ketones between male and female LL rats: (A) abundances, (B) relative abundances. (mean ± SE, single asterisk, $P < 0.05$; double asterisks, $P < 0.01$. 4H: 4-Heptanone, 2H: 2-Heptanone, 6M5H2N: 6-Methyl- 5-hepten-2-one, 3E2,4HD: 3-Ethyl-2,4-heptanedione, DTHP2N: a dialkyl tetrahydro-2H-pyran-2-one, 9H2N: 9-hydroxy-2-Nonanone).

Additional file 7: Figure S6. Duration of investigation (mean ± SE) by female LL rats ($N = 13$) of different male urine samples during a 3-min binary choice test: castrated male urine vs. castrated male urine added with 4-heptanone, (4H, 0.5 ppm) , 2-heptanone (2H, 4 ppm) or 9-hydro-2-nonanoe (9H2N, 9.6 ppm) at minimal effective concentrations , respectively ($*P < 0.05$, $**P < 0.01$, paired t-test).

Competing interests

The authors declare that they have no competing interests.

Authors' contributions

JXZ and YHZ designed the study. YHZ and JXZ wrote the paper. YHZ collected and analyzed the data. Both authors read and approved the final manuscript.

Acknowledgments

This work was supported by the Strategic Priority Research Program of the Chinese Academy of Sciences [XDB11010400], the grants from China National Science Foundation [91231107 and 31301887], Chinese Academy of Sciences [KSCX2-EW-N-5], and the Foundation of State Key Laboratory of IPM [ChineseIPM1303]. We are most grateful for Mr. Jin-Hua Zhang's assistance with behavioral tests and Prof. Peter Keightley' s help with improving the manuscript.

References

1. Singer AG, Beauchamp GK, Yamazaki K: **Volatile signals of the major histocompatibility complex in male mouse urine.** *Proc Natl Acad Sci USA* 1997, **94**:2210–2214.
2. Novotny M, Harvey S, Jemiolo B: **Chemistry of male dominance in the house mouse, *mus domesticus*.** *Experientia* 1990, **46**:109–113.
3. Penn D, Potts W: **MHC-disassortative mating preferences reversed by cross-fostering.** *Proc R Soc B* 1998, **265**:1299–1306.
4. Roberts SC, Gosling LM: **Genetic similarity and quality interact in mate choice decisions by female mice.** *Nat Genet* 2003, **35**:103–106.
5. Zhang YH, Zhang JX: **Urine-derived key volatiles may signal genetic relatedness in male rats.** *Chem Senses* 2011, **36**:125–135.
6. Tregenza T, Wedell N: **Genetic compatibility, mate choice and patterns of parentage: Invited review.** *Mol Ecol* 2000, **9**:1013–1027.
7. Thünken T, Meuthen D, Bakker TCM, Baldauf SA: **A sex-specific trade-off between mating preferences for genetic compatibility and body size in a cichlid fish with mutual mate choice.** *Proc R Soc B* 2012, **279**:2959–2964.
8. Cheetham SA, Thom MD, Jury F, Ollier WER, Beynon RJ, Hurst JL: **The genetic basis of individual-recognition signals in the mouse.** *Curr Biol* 2007, **17**:1771–1777.
9. Sherborne AL, Thom MD, Paterson S, Jury F, Ollier WE, Stockley P, Beynon RJ, Hurst JL: **The genetic basis of inbreeding avoidance in house mice.** *Curr Biol* 2007, **17**:2061–2066.
10. Thom MD, Stockley P, Jury F, Ollier WER, Beynon RJ, Hurst JL: **The direct assessment of genetic heterozygosity through scent in the mouse.** *Curr Biol* 2008, **18**:619–623.
11. Andersson M, Simmons LW: **Sexual selection and mate choice.** *Trends Ecol Evol* 2006, **21**:296–302.
12. Neff BD, Pitcher TE: **Genetic quality and sexual selection: an integrated framework for good genes and compatible genes.** *Mol Ecol* 2005, **14**:19–38.
13. Brown RE, Singh PB, Roser B: **The major histocompatibility complex and the chemosensory recognition of individuality in rats.** *Physiol Behav* 1987, **40**:65–73.
14. Eizaguirre C, Yeates SE, Lenz TL, Kalbe M, Milinski M: **MHC-based mate choice combines good genes and maintenance of MHC polymorphism.** *Mol Ecol* 2009, **18**:3316–3329.
15. Head ML, Hunt J, Jennions MD, Brooks R: **The indirect benefits of mating with attractive males outweigh the direct costs.** *PLoS Biol* 2005, **3**:e33.
16. Kokko H, Ots I: **When not to avoid inbreeding.** *Evolution* 2006, **60**:467–475.
17. Puurtinen M: **Mate Choice for Optimal (K) Inbreeding.** *Evolution* 2011, **65**:1501–1505.
18. Schjørring S, Jäger I: **Incestuous mate preference by a simultaneous hermaphrodite with strong inbreeding depression.** *Evolution* 2007, **61**:423–430.
19. Thünken T, Bakker TCM, Baldauf SA, Kullmann H: **Active inbreeding in a cichlid fish and its adaptive significance.** *Curr Biol* 2007, **17**:225–229.
20. Loyau A, Cornuau JH, Clobert J, Danchin E: **Incestuous Sisters: Mate Preference for Brothers over Unrelated Males in Drosophila melanogaster.** *PLoS ONE* 2012, **7**:e51293.
21. Robinson SP, Kennington WJ, Simmons LW: **Preference for related mates in the fruit fly, *Drosophila melanogaster*.** *Anim Behav* 2012, **84**:1169–1176.
22. Szulkin M, Stopher KV, Pemberton JM, Reid JM: **Inbreeding avoidance, tolerance, or preference in animals?** *Trends Ecol Evol* 2013, **28**:205–211.
23. Hepper PG: **The discrimination of different degrees of relatedness in the rat - evidence for a genetic identifier.** *Anim Behav* 1987, **35**:549–554.
24. Hurst JL, Payne CE, Nevison CM, Marie AD, Humphries RE, Robertson DH, Cavaggioni A, Beynon RJ: **Individual recognition in mice mediated by major urinary proteins.** *Nature* 2001, **414**:631–634.
25. Zhang JX, Sun L, Zhang JH, Feng ZY: **Sex- and gonad-affecting scent compounds and 3 male pheromones in the rat.** *Chem Senses* 2008, **33**:611–621.
26. Bender K, Balogh P, Bertrand MF, Den Bieman M, Von Deimling O, Eghtessadi S, Gutman GA, Hedrich HJ, Hunt SV, Kluge R, Matsumoto K, Moralejo DH, Nagel M, Portal A, Prokop C-M, Seibert RT, van Zutphen LFM: **Genetic characterization of inbred strains of the rat (*Rattus norvegicus*).** *J Exp Anim Sci* 1994, **36**:151–165.

27. Isles AR, Baum MJ, Ma D, Keverne EB, Allen ND: **Genetic imprinting - Urinary odour preferences in mice.** *Nature* 2001, **409:**783–784.

28. Hopp SL, Owren MJ, Marion JR: **Olfactory discrimination of individual littermates in rats (*Rattus norvegicus*).** *J Comp Psychol* 1985, **99:**248–251.

29. Wyatt TD: **Pheromones and signature mixtures: defining species-wide signals and variable cues for identity in both invertebrates and vertebrates.** *J Comp Physiol Sensory Neural Behav Physiol* 2010, **196:**685–700.

30. Johnston RE: **Chemical communication in rodents: from pheromones to individual recognition.** *J Mammal* 2003, **84:**1141–1162.

31. Gabirot M, Lopez P, Martin J: **Female mate choice based on pheromone content may inhibit reproductive isolation between distinct populations of Iberian wall lizards.** *Curr Zool* 2013, **59:**210–220.

32. Brooks R: **Negative genetic correlation between male sexual attractiveness and survival.** *Nature* 2000, **406:**67–70.

33. Huk T, Winkel WG: **Testing the sexy son hypothesis - a research framework for empirical approaches.** *Behav Ecol* 2008, **19:**456–461.

34. Brennan PA, Kendrick KM: **Mammalian social odours: attraction and individual recognition.** *Philos Trans R Soc Lond B Biol Sci* 2006, **361:**2061–2078.

35. Kokko H, Brooks R, McNamara JM, Houston AI: **The sexual selection continuum.** *Proc R Soc B* 2002, **269:**1331–1340.

36. Sharma MD, Griffin RM, Hollis J, Tregenza T, Hosken DJ: **Reinvestigating good genes benefits of mate choice in *drosophila simulans*.** *Biol J Linn Soc* 2012, **106:**295–306.

37. Prokop ZM, Michalczyk L, Drobniak SM, Herdegen M, Radwan J: **Meta-analysis suggests choosy females get sexy sons more than "good genes".** *Evolution* 2012, **66:**2665–2673.

38. Taylor ML, Wedell N, Hosken DJ: **The heritability of attractiveness.** *Curr Biol* 2007, **17:**R959–R960.

39. Taylor GT, Haller J, Regan D: **Female Rats Prefer an Area Vacated by a High Testosterone Male.** *Physiol Behav* 1982, **28:**953–958.

40. Wedell N, Tregenza T: **Successful fathers sire successful sons.** *Evolution* 1999, **53:**620–625.

41. Bolund E, Martin K, Kempenaers B, Forstmeier W: **Inbreeding depression of sexually selected traits and attractiveness in the zebra finch.** *Anim Behav* 2010, **79:**947–955.

42. Van Bergen E, Brakefield PM, Heuskin S, Zwaan BJ, Nieberding CM: **The scent of inbreeding: a male sex pheromone betrays inbred males.** *Proc R Soc B* 2013, **280:**20130102.

43. Fukamachi S, Kinoshita M, Aizawa K, Oda S, Meyer A, Mitani H: **Dual control by a single gene of secondary sexual characters and mating preferences in medaka.** *BMC Biol* 2009, **7:**64.

44. Bousquet F, Nojima T, Houot B, Chauvel I, Chaudy S, Dupas S, Yamamoto D, Ferveur JF: **Expression of a desaturase gene, *desat1*, in neural and nonneural tissues separately affects perception and emission of sex pheromones in *Drosophila*.** *Proc Natl Acad Sci USA* 2012, **109:**249–254.

45. Thonhauser KE, Raveh S, Hettyey A, Beissmann H, Penn DJ: **Scent marking increases male reproductive success in wild house mice.** *Anim Behav* 2013, **86:**1013–1021.

46. Zhang JX, Rao XP, Sun LX, Zhao CH, Qin XW: **Putative chemical signals about sex, individuality, and genetic background in the preputial gland and urine of the house mouse (*Mus musculus*).** *Chem Senses* 2007, **32:**293–303.

9

Behavioural response of a migratory songbird to geographic variation in song and morphology

Kim G Mortega[1,2,4*], Heiner Flinks[3] and Barbara Helm[2,4]

Abstract

Introduction: Sexually selected traits contribute substantially to evolutionary diversification, for example by promoting assortative mating. The contributing traits and their relevance for reproductive isolation differ between species. In birds, sexually selected acoustic and visual signals often undergo geographic divergence. Clines in these phenotypes may be used by both sexes in the context of sexual selection and territoriality. The ways conspecifics respond to geographic variation in phenotypes can give insights to possible behavioural barriers, but these may depend on migratory behaviour. We studied a migratory songbird, the Stonechat, and tested its responsiveness to geographic variation in male song and morphology. The traits are acquired differently, with possible implications for population divergence. Song can evolve quickly through cultural transmission, and thus may contribute more to the establishment of geographic variation than inherited morphological traits. We first quantified the diversity of song traits from different populations. We then tested the responses of free-living Stonechats of both sexes to male phenotype with playbacks and decoys, representing local and foreign stimuli derived from a range of distances from the local population.

Results: Both sexes discriminated consistently between stimuli from different populations, responding more strongly to acoustic and morphological traits of local than foreign stimuli. Time to approach increased, and time spent close to the stimuli and number of tail flips decreased consistently with geographic distance of the stimulus from the local population. Discriminatory response behaviour was more consistent for acoustic than for morphological traits. Song traits of the local population differed significantly from those of other populations.

Conclusions: Evaluating an individual's perception of geographic variation in sexually selected traits is a crucial first step for understanding reproductive isolation mechanisms. We have demonstrated that in both sexes of Stonechats the responsiveness to acoustic and visual signals decreased with increasing geographic distance of stimulus origin. These findings confirm consistent, fine discrimination for both learned song and inherited morphological traits in these migratory birds. Maintenance or further divergence in phenotypic traits could lead to assortative mating, reproductive isolation, and potentially speciation.

Keywords: Sexual selection, Population divergence, Reproductive isolation, Phenotypic traits, Geographic clines, Simulated territorial intrusion, *Saxicola torquata*, Songbird, Behavioural isolation barrier

Introduction

Phenotypic traits involved in signalling, for example aspects of song and morphology, are known to contribute to reproductive isolation between diverging populations [1,2]. Specifically, signalling in the context of mate attraction or territoriality may promote reproductive isolation through assortative mating and settlement patterns [3-5]. In birds, both sexes can be actively involved in signalling and also in discrimination of local conspecifics as potential sexual partners or sexual competitors [6].

In most songbirds, songs are a key component of signalling and are culturally transmitted across generations via vocal learning [7]. Young birds learn to produce or recognize song early in life, while still in their natal region. The geographic variation of such song traits is thought to result from the effect of imperfect song copying [8]. Accordingly, song dialects, i.e. the unique repertoire of shared songs within a population, combined with female

* Correspondence: kmortega@orn.mpg.de
[1]Department of Migration and Immuno-Ecology, Max Planck Institute for Ornithology, 78315 Radolfzell, Germany
[2]Department of Ornithology, University of Konstanz, 78457 Konstanz, Germany
Full list of author information is available at the end of the article

preference for a local dialect due to parental imprinting, may lead to reproductive divergence [9-12]. Female preference for familiar vocalizations has been shown in captive and field experiments by increased copulation-solicitation displays to standardized playback [13-16].

Often not only vocalizations but a suite of selected traits of different sensory modalities contribute to the establishment and maintenance of reproductive isolation [17]. For example, morphological traits are also proposed to facilitate pre-mating isolation barriers between related avian lineages [18]. Such traits often include plumage coloration, e.g. redness in house finches, *Carpodacus mexicanus* [19]. In golden-collared manakins, *Manacus ssp.*, the golden is preferred over the white phenotype [20]. Genetically inherited visual signals may therefore facilitate diversification [21,22]. In contrast, sexually selected traits that are inherited culturally, notably learned avian vocalizations, can change instantaneously without requiring genetic change. They may therefore be a more efficient mechanism for reproductive isolation than inherited traits [23-26].

By promoting isolation, geographically differentiated signals are thought to aid local adaptation. The local adaptation hypothesis predicts that birds which select mates from their natal regions will gain fitness advantages because their offspring will more likely express adaptations to local ecological conditions [27], for example adaptations of seasonal activities associated with local climates, or morphologies tailored to specific lifestyles [28,29]. In North American crossbills (*Loxia curvirostra* - complex) distinct song types are associated with incipient speciation [30-32]. Interestingly, the differences in song types are coupled with morphological differences relating to ecological speciation. However, the processes of local adaptation and associated signalling may be sensitive to movement behaviour [33]. Migration may counteract population divergence [34] because: a) migration is thought to correlate positively with dispersal distance, which in turn generally promotes gene exchange [33,35]; b) migrants are typically under pressure to make rapid reproductive decisions, implying that female migrants may be less choosy than female residents [36], and may therefore not pair with the best (i.e., locally adapted) mate available [37]; c) relating to acoustic signals, migratory departure after breeding limits opportunities for young males and females to learn or imprint to the local dialect. Earlier studies have reported lower song discrimination in migrant than resident species, but have also indicated mechanisms by which migrants could nonetheless learn local song dialects after dispersal [34,38].

To better understand processes of local differentiation, in particular in migratory birds, we investigated discriminatory abilities in Stonechats (*Saxicola torquata* and closely related lineages [39]). The *Saxicola* complex has a wide distribution range, comprising substantial local differentiation in seasonal and morphological traits [40]. We focused on the short-distance migrant European stonechat (*Saxicola torquata*), which is socially monogamous with seasonal pair bonds selected by females [41]. During the entire breeding season, males defend their territory with distinct behavioural responses. Females also actively respond to conspecific intruders [42]. The fact that males sometimes "punish" their mates for their response to intruders indicates a sexual context to female interest [43]. The female responsiveness allowed us to examine discriminatory abilities in both sexes. We studied song variation between Stonechat populations and tested the behavioural response of the focal European population to song recordings and stuffed decoys. Early in the breeding season we obtained and analysed song repertoires of the local population and additional populations that breed 90 and 180 km away. We experimentally tested the responsiveness of local Stonechats to song from these populations and to stimuli from African Stonechats and a control species by conducting simulated territorial intrusions with playbacks. We also conducted a decoy experiment simulating a territorial intrusion by presenting a taxidermic mount of phenotypes from populations with differing geographic distances. The experiments focused on male response, but we also report data on the latency of the female response to the stimuli. All experiments were conducted during the breeding season at defined breeding stages in the presence of both pair mates.

In view of the geographic differentiation within Stonechats, we hypothesised that despite their migratory behaviour female and male Stonechats i) can discriminate between phenotypes of geographically distinct populations during playback and decoy experiments, ii) respond most strongly to local population stimuli, and iii) may show a consistent decline in their responsiveness with geographic distance. Furthermore, we hypothesised that songs may elicit stronger responses than morphological traits in both sexes because they may have diverged more rapidly.

Results

Song traits

The Stonechat populations differed in their song traits from each other. A principal component analysis of seven traits (Table 1a) showed that several principal components explained the variation in song (PC1 = 37.26, PC2 = 29.17, PC3 = 21.62, Figure 1). Based on the first principal component, the focal population differed significantly from the neighbouring population (90 km). Differences increased further with geographic distance from the local population (Additional file 1: Table S1, Figure S1), although Stonechats from 90 km and 180 km were not significantly different from each other (Table 1b, Figure 1).

Table 1 Song traits

(a)

	PC 1	PC 2	PC 3
Song duration	0.40	0.33	0.46
No. of elements	0.46	0.38	0.29
Element rate	−0.27	−0.29	0.46
Peak frequency	−0.02	0.30	−0.63
Min. frequency	−0.23	0.61	−0.04
Max. frequency	0.53	−0.08	−0.27
Bandwidth	0.47	−0.44	−0.14
Eigenvalue	2.61	2.04	1.51
% variance	37.26	29.17	21.62

(b)

Fixed effects	estimate	s.e.m	t	p
Intercept	−1.76	0.12	**−14.89**	**<0.001**
90 km	2.28	0.29	**7.97**	**<0.001**
180 km	2.38	0.18	**13.52**	**<0.001**
African	0.64	0.19	**3.44**	**<0.001**
Control	4.79	0.19	**25.45**	**<0.001**

(a) Factor loadings of the principal component analysis for seven song traits of European Stonechats from the local population, a population from 90 km distance, a population from 180 km distance, African stonechats, and the winter wren. (b) Results of general linear model testing whether the first principal component (PC1) differed between songs from different locations, estimated by maximum likelihood methods. Estimates for the different song locations refer to differences from the intercept estimate, which represents song traits of the local population. Subjects were included as random intercepts to control for repeated measures. 'Significant' differences are shown in bold.

Playback and decoy experiments

Stonechats of the local population responded differently to stimuli from distinct populations, measured by the time they took to approach the caller or decoy (i.e., latency to approach within 5 m). In response to playback, males discriminated significantly between origins of the stimulus (z = −8.42, p <0.001, Table 2a, Figure 2). The males' latency to approach the caller was lowest when exposed to the local song and increased with distance of stimulus origin (Table 2a, Figure 2). Breeding stage, trial order (Additional file 1: Figure S4), date and time of day showed no significant effect on the males' latency to approach (Table 2a).

Likewise, females also differed significantly in their behavioural response to different playback stimuli (z = −6.28 p <0.001, Table 2b, Figure 3). The females' latency to approach the caller was lowest when presented with the local song and increased with geographic distance of the stimulus origin (Table 2b, Figure 3). Breeding stage, trial order, date and time had no significant effects on the females' latency to approach the caller (Table 2b).

During the decoy experiment, the males' discrimination was less consistent than during the playback experiment (z = −4.93, p <0,001, Table 2c, Figure 4). The latency to

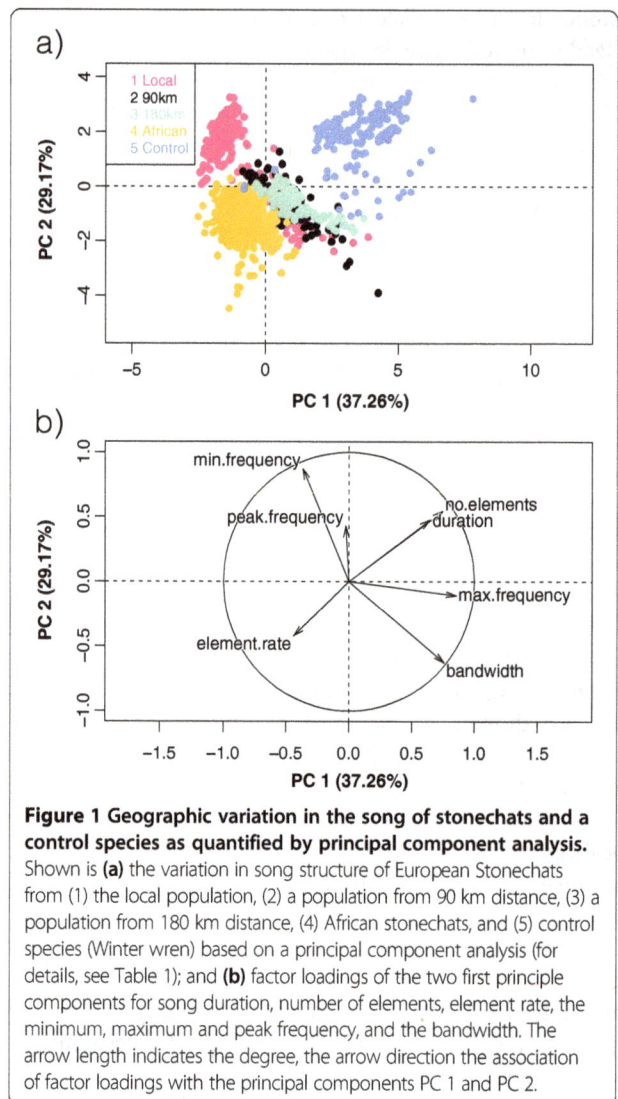

Figure 1 Geographic variation in the song of stonechats and a control species as quantified by principal component analysis. Shown is **(a)** the variation in song structure of European Stonechats from (1) the local population, (2) a population from 90 km distance, (3) a population from 180 km distance, (4) African stonechats, and (5) control species (Winter wren) based on a principal component analysis (for details, see Table 1); and **(b)** factor loadings of the two first principle components for song duration, number of elements, element rate, the minimum, maximum and peak frequency, and the bandwidth. The arrow length indicates the degree, the arrow direction the association of factor loadings with the principal components PC 1 and PC 2.

approach local decoy stimuli did not differ from other European (z = −1.03, p =0.30, Table 2c, Figure 4) and African stimuli (z = −1.52, p =0.13, Table 2c, Figure 4), but males approached the control decoy significantly later than all others (z = −4.62, p <0.001, Table 2c, Figure 4). Breeding stage, trial order, date and time had no significant effect on the latency to approach (Table 2c).

Females showed finer discrimination (z =4.84, p <0.001, Table 2d, Figure 3). They approached the local decoy with lower latency than decoys of populations from greater geographic distances (Table 2d, Figure 3). Breeding stage, trial order, date and time showed no significant effect on the females' latency to approach the decoy (Table 2d).

There were no significant differences between different pairs, neither during the playback (z = −0.54, p =0.59, Additional file 1: Table S2a, Figure 5) nor the decoy experiments (z = −0.39, p =0.70, Additional file 1: Table S2b, Figure 5). Males approached the stimuli with

Table 2 Latency to approach within 5 m

	Fixed effects	Estimate	Hazard ratio	s.e.m	z	p
Playback						
(a) males	**Origin**	−0.86	0.42	0.10	−8.42	**<0.001**
	90 km	−1.30	0.27	0.31	−4.17	**<0.001**
	180 km	−2.17	0.11	0.34	−6.36	**<0.001**
	African	−2.32	0.10	0.35	−6.66	**<0.001**
	Control	−4.18	0.02	0.53	−7.85	**<0.001**
	Breeding stage	0.60	1.82	0.59	1.02	0.31
	Trial order	0.04	1.04	0.07	0.52	0.61
	Date	0.08	1.09	0.08	1.06	0.29
	Time	0.004	1.00	0.03	0.16	0.88
(b) females	**Origin**	−1.36	0.26	0.22	−6.28	**<0.001**
	90 km	−2.11	0.12	0.49	−4.32	**<0.001**
	180 km	−2.40	0.09	0.54	−4.45	**<0.001**
	African	−4.51	0.01	0.85	−5.32	**<0.001**
	Control	−6.88	0.009	0.97	−7.84	**<0.001**
	Breeding stage	−0.75	0.47	0.41	−1.84	0.06
	Trial order	0.16	1.18	1.27	1.29	0.20
	Date	0.02	1.02	0.07	0.23	0.82
	Time	−0.09	0.91	0.06	−1.60	0.11
Decoy						
(c) males	**Origin**	−0.78	0.46	0.16	−4.93	**<0.001**
	European	−0.43	0.65	0.41	−1.03	0.30
	African	−0.65	0.52	0.43	−1.52	0.13
	Control	−2.95	0.05	0.64	−4.62	**<0.001**
	Breeding stage	0.08	1.08	0.17	0.46	0.65
	Trial order	−0.18	0.83	0.16	−1.18	0.24
	Date	−0.05	0.95	0.11	−0.49	0.62
	Time	−0.001	0.99	0.009	−0.19	0.85
(d) females	**Origin**	−0.79	0.45	0.16	−4.84	**<0.001**
	European	−1.50	0.22	0.56	−2.65	**0.007**
	African	−1.70	0.18	0.59	−2.87	**0.004**
	Control	−3.19	0.04	0.84	−3.78	**<0.001**
	Breeding stage	−0.10	0.10	0.18	−0.54	0.59
	Trial order	−0.10	0.10	0.16	−0.59	0.55
	Date	−0.15	0.86	0.10	−1.44	0.15
	Time	−0.01	0.99	0.009	−1.50	0.13

Results of cox mixed-effects model with estimates, hazard ratio, standard error, z-value, and p-value fitted by maximum likelihood for playback in (a) males and (b) females, and decoy experiment in (c) males and (d) females. Estimates refer to differences from the intercept estimate, which represents the latency to approach of the local population (not shown). 'Origin' represents the overall estimate of differences between populations. 'Significant' differences are shown in bold.

significantly lower latency than females during both, the playback (z =4.78, p <0.001, Additional file 1: Table S2a, Figure 5) and decoy experiment (z =5.88, p <0.001, Additional file 1: Figure S2b, Figure 5). Breeding stage, date and time did not influence the response patterns of pairs (Additional file 1: Table S2). Trial order had no

influence on the behavioural response of pairs during the playback, and only a slight but significant effect during the decoy experiment (Additional file 1: Table S2). Birds tended to approach the stimulus with lower latency in the first two compared to later trials (Additional file 1: Figure S4). A Spearman's correlation test was run to determine

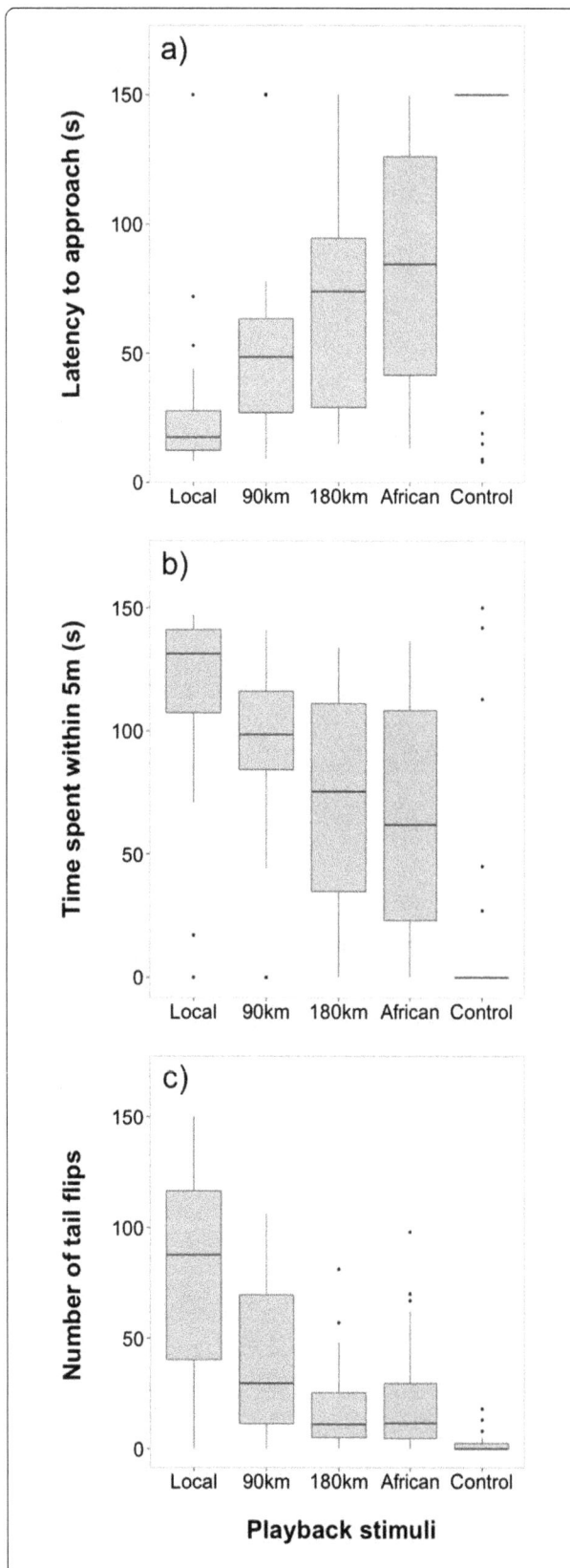

Figure 2 Playback experiment in males. Behavioural response for **(a)** latency to approach, **(b)** time spent within 5 m, and **(c)** number of tail flips in response to European Stonechats from (1) the local population, (2) a population from 90 km distance, (3) a population from 180 km distance, (4) African stonechats, and (5) control stimuli (Winter wren). Males discriminated between local and foreign stimuli by responding more strongly to song of their own population. Box plots represent, from bottom to top: minimum, lower quartile, median, upper quartile and maximum. Dots indicate observations further than one s.d. away from the mean; n =28.

the relationship between the behavioural response of female and male mates within a pair. The latency to approach was correlated between females and males during the playback experiment (r_s =0.51, p <0.001, n =15, Additional file 1: Table S3, Figure S2), but not during the decoy experiment (r_s =0.23, p <0.103, n =14, Additional file 1: Table S3, Figure S2).

Further behaviours of males also differed in response to stimuli from different populations. Males remained significantly longer within 5 m of the stimulus of the local population than of all other populations during the playback (Table 3a, Figure 2) and decoy experiment (Table 3b, Figure 4). Males of the local population also discriminated between origin of the stimuli in the number of tail flips, an indicator of agitation. In response to playback, the number of tail flips differed significantly between stimulus origins (estimate =16.65, t = −10.58, CI = −19.82, 13.45, Table 4a, Figure 2). The number of tail flips was highest when males were exposed to the local song and decreased with geographic distance of stimulus origin (Table 4a, Figure 2). Breeding stage, trial order, date and time showed no significant effect on the number of tail flips (Table 4a). Similarly, males also differed significantly in their number of tail flips during the presentation of decoy stimuli (estimate = −18.43, t = −1.30, CI = −48.35, 11.16, Table 4b, Figure 4). During trials of the local stimuli, males significantly flipped their tails more often than during all other trials (Table 4b, Figure 4). Breeding stage, trial order, date and time showed no significant effect on the number of tail flips (Table 4b).

Discussion
This study reports clear differentiation in song traits of migratory European Stonechats over relatively short distances (90 km and 180 km from the focal population). By testing the behavioural responses to acoustic and morphological stimuli, we have also demonstrated the Stonechats' ability to discriminate between geographic origins of sexually selected traits in two modalities. The responses of both sexes during playback and decoy experiments were graded and declined with increasing geographic distance from the local population. The concordance of these responses and the significant preference for the closest population suggests potential for

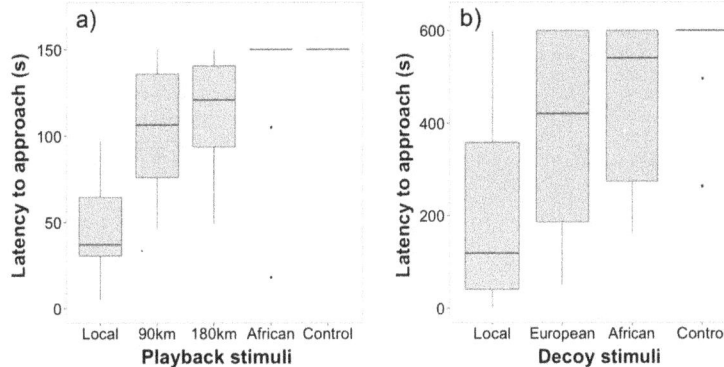

Figure 3 Playback and decoy experiment in females. Behavioural response for the latency to approach within 5 m in **(a)** playback experiments in response to song of European Stonechats from (1) the local population, (2) a population from 90 km distance, (3) a population from 180 km distance, (4) African stonechats, and (5) control stimuli (Winter wren); and **(b)** decoy experiments in response to stimuli from (1) local, (2) European, (3) African, and (4) control stimuli (European Robin). Females distinguished between stimuli by approaching the local stimuli significantly faster than all other stimuli. Detailed description of boxplots as in Figure 2; n =15.

the evolution of reproductive isolation although at present this is confirmed only for a single population.

Male and female Stonechats were similar in their behavioural discrimination, in contrast to results from other species. A recent study on Rufous-collared sparrows, *Zonotrichia capensis*, also reported discrimination between local and foreign stimuli, but the sexes differed in behaviour [16]. Females were presented with songs of the local, nearby nonlocal, and distant nonlocal dialect, and a control song from another bird species. They preferred the males' local song dialect to all other dialects tested, showing significantly more copulation solicitation displays. In contrast, males displayed only a low discrimination ability between dialects of geographically close populations [16]. Similarly, in White-crowned sparrows, *Zonotrichia leucophrys*, females were more sensitive to geographic variation in song than males [44]. A study on hybridizing Pied flycatcher, *Ficedula hypoleuca*, and Collared flycatcher, *Ficedula albicollis*, revealed that females quickly recognise male species identity by song and have a strong preference for conspecific males resulting in assortative mating, and thus preventing further hybridisation [45,46]. In contrast, males of both species courted the heterospecific female and the conspecific female with similar intensity, thereby promoting hybridisation. This lack of species recognition could be due to mating being less costly in males, which can inseminate several females over a short period, while females are constrained by the number of their eggs. Females, therefore, should not make mistakes in mate choice [37]. The fine discrimination ability of Stonechats indicates that females may mate assortatively, while males may use the fine discrimination to fight off particularly attractive sexual competitors with local dialects. We cannot disentangle male and female responses because we conducted simulated territorial intrusions in presence

of both pair members. An influence of the mate is suggested by the correlation between mates during the playback experiment (Additional file 1: Table S3, Figure 5) and has been shown previously in Stonechats [42]. Therefore, a crucial future step for a better understanding of the response to acoustic and morphological traits in Stonechats is to conduct experiments separately on females and males.

The local differentiation and consistent behavioural discrimination of song by origin of Stonechats, which migrate, was similar to that of resident species (e.g., indigobird *Vidua sp.* [47], Galapagos Sharp-beaked ground finch, *Geospiza difficilis* [6], and song sparrow, *Melospiza melodia* [24]), but differed from findings in some migratory species. Among *Zonotrichia* sparrows, long distance migrants (e.g., *Z. l. gambelii)* do not form dialects [34], whereas in sedentary *Zonotrichia* subspecies (e.g., *Z. l. nutalli)* geographic song variation occurs [13,15]. The corresponding lack of genetic diversification in *Zonotrichia* migrants, in contrast to significant genetic structuring among dialect areas in non-migrants, supports the idea that migration may counteract population divergence and isolation [48-50].

Although the fine acoustic discrimination ability of Stonechats suggests potential behavioural barriers, its implications for geographic isolation are not fully clear, partly depending on song plasticity, and ultimately on the mechanisms involved in song learning. In passerine birds, song is typically learned during a sensitive period early in life. In species like Stonechats that show geographic discrimination, males that subsequently disperse into ranges of other populations would face reduced mating prospects if an acoustic signature of the natal population remains in their repertoire [40]. However, this could be offset if the males were able to learn new songs after the sensitive phase. For example, migratory

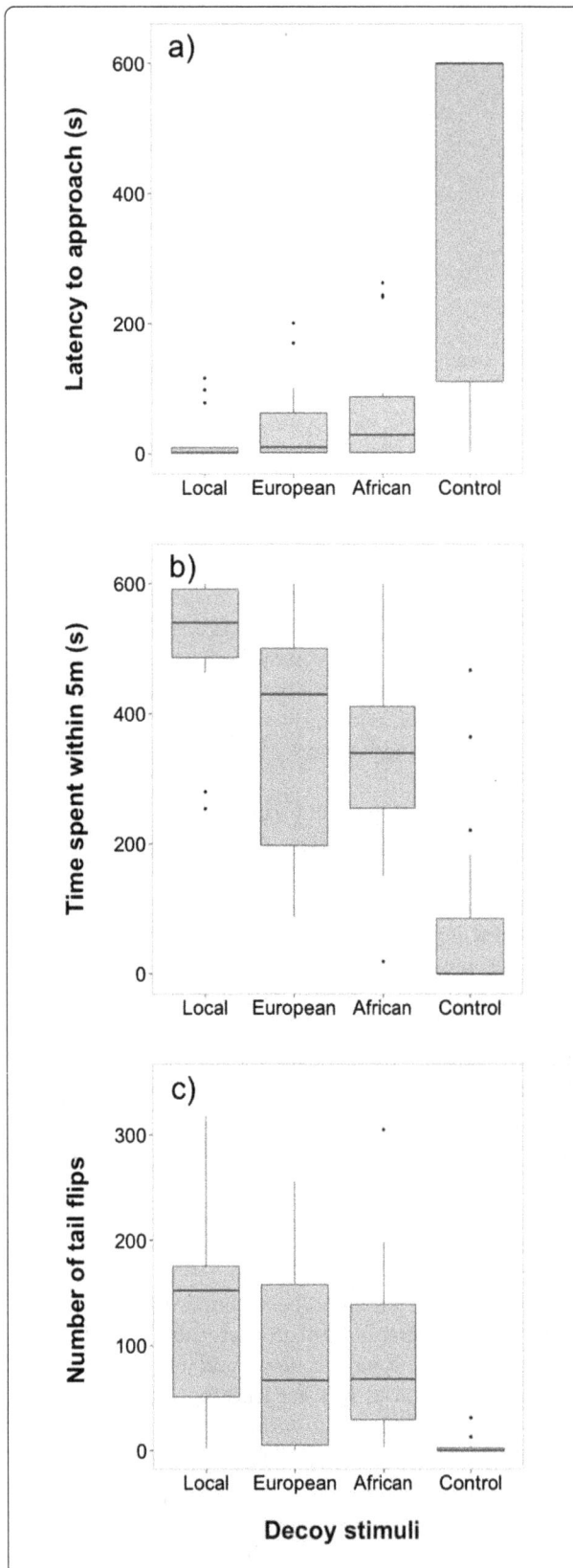

Figure 4 Decoy experiment in males. Behavioural response for **(a)** latency to approach, **(b)** time spent within 5 m, and **(c)** number of tail flips to (1) local, (2) European, (3) African, and (4) control stimuli (European robin). Males discriminated between local and foreign stimuli by responding more strongly to decoys of their own population. Detailed description of boxplots as in Figure 2; n =16.

nightingales were able to acquire new song types in their first singing season [51,52]. In some species plastic song is based on an initial overproduction of learned songs during ontogeny [53]. Such overproduction of learned songs has been suggested to be correlated with a migratory lifestyle [38,54]. If present in Stonechats, plastic song learning could therefore enable dispersing males to be sexually selected by local females, although benefits of local song in sexual selection could be partly offset by the greater aversive response of local males. Dispersing females, in turn, may have no choice but to mate with a male singing a foreign dialect, and this might reduce population divergence. A modelling study by Ellers and Slabbekoorn suggests that evolutionary implications of song dialects are not straightforward [55]. Although in the majority of scenarios genetic and vocal divergence were concordant, the type of song learning and intrasexual competition in males affected the evolutionary outcome. For Stonechats, to answer this question unambiguously would require population genetic analyses alongside analyses of song traits among populations [56].

In our study, we found that Stonechats were also able to discriminate by morphological traits. Most studies of sexual selection do not explicitly test the role of simultaneous signalling with different sensory modalities, and instead focus on a single divergent signal or a suite of signals of the same modality [57-59]. In contrast, explicitly testing for effects of multiple signals enables the detection of divergent signal use in discrimination [60,61]. In Stonechats, we expected that culturally transmitted song may evolve more quickly, and thus could play a more important role for geographic clines than do morphological traits. We found that discrimination by song was more consistent than by morphological traits. The discrimination by song was sensitive to a geographic distance of only 90 km, whereas the decoy against which the birds visually discriminated originated from a population which breeds 1,000 km away. A caveat in the interpretation of these differences are the different breeding stages during which the stimuli were tested: song stimuli were applied during egg-laying and incubation stages, when birds may be particularly responsive, whereas decoys were tested during nestling and fledgling stages. However, Stonechats are multi-brooded, and females may initiate additional clutches while males take care of fledglings, so that male intruders may well gain reproductive benefits at this time. Moreover, in a study on closely related African stonechats

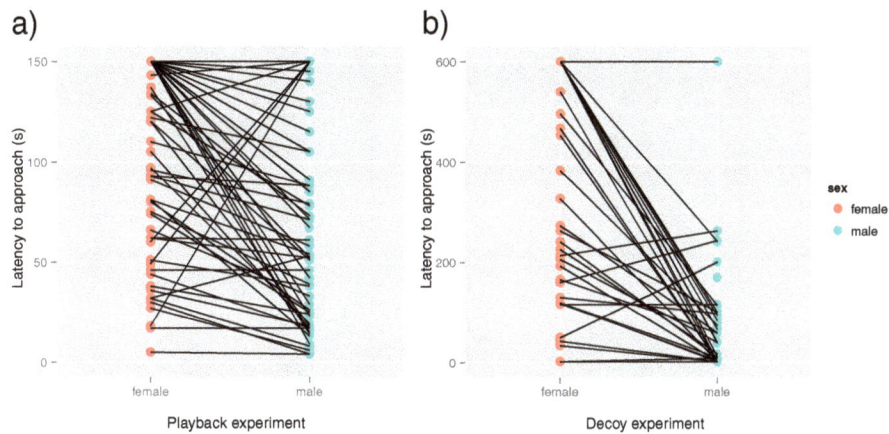

Figure 5 Response behaviour of female and male mates within pairs. Shown is the latency to approach within 5 m for females (red) and males (blue) within pairs (connected dots) for **(a)** the playback (n =15 pairs) and **(b)** decoy experiments (n =14 pairs) for all stimuli trials. Females and males differ significantly in their response behaviour, i.e. males approached the presented stimulus with lower latency.

with similar experimental designs, but conducted during simultaneous breeding stages for playback and decoy experiments, the birds' discrimination by song was more consistent than that by morphology (unpublished data by KGM). Overall, our data cautiously suggest that song may be indeed the stronger discriminatory signal for Stonechats.

In the chestnut-bellied flycatcher, *Monarcha castanei-ventris,* plumage colour played a greater role than song for the intensity of aggressive response by territory-owners, although both signals mattered [62]. Chestnut-bellied flycatchers display more variation in plumage colour than in song, which may indicate that plumage is more emphasised in sexual selection than song structure. The relative advantages of signalling with several modalities may be driven by the environment [63]. In general,

acoustic signals can be transmitted over long distances and are ideal for long-range communication, whereas visual signals can be more limited and therefore be more suitable for short-range communication [64]. For Stone-chats, which breed in open habitats, both signalling modes may be important.

Conclusions

Our study on Stonechats reveals geographic differenti-ation of sexually selected traits in a migratory songbird. Song traits differed significantly in populations of distinct geographic origin. Consistently, both sexes distinguished local morphological and especially acoustic phenotypes from those of foreign populations. These data demon-strate that variation in sexually selected traits of different modalities may contribute to geographic isolation over relatively short distances, and thereby aid local adaptation. The sexes had similar sensitivity to incipient behavioural barriers. Maintenance or further divergence in these phenotypic traits could lead to assortative mating, repro-ductive isolation, and potentially speciation, in migratory Stonechats.

Materials and methods
Subjects
Stonechats inhabit open habitats across a large extent of the Palearctic [65]. The study population of European stonechats, *Saxicola torquata rubicola,* is located in northwest Germany (51°N, 6°30'E) and overwinters in the Mediterranean region, predominantly in north Africa [40,66]. The study population has been observed, mea-sured and colour-banded for individual recognition since 1976. Stonechats arrive at the breeding grounds early in spring (late February/March), establish a territory, and form seasonal pair bonds with two to three broods per season

Table 3 Time spent within 5 m of the stimulus

	Fixed effects	Estimate	s.d.	CI 2.50%	CI 97.50%
(a) playback	local	0.84	0.24	0.37	1.29
	90 km	0.18	0.21	−0.22	0.62
	180 km	−0.28	0.21	−0.70	0.14
	African	−0.37	0.22	−0.80	0.05
	control	−1.21	0.31	−1.82	−0.63
(b) decoy	local	−1.93	0.45	−2.75	−0.97
	European	−2.33	0.38	−3.03	−1.55
	African	−1.97	0.36	−2.65	−1.24
	control	0.59	0.36	−0.14	1.29

Results of generalized linear mixed model with estimates, standard deviation, and credible intervals using WinBUGS for (a) playback and (b) decoy experiment. Stimulus is defined as random factor to compare paths of all stimuli and thus correct for multiple testing. A behavioural response differs significantly from the local population if its credible intervals do not include the mean of the local population. Significant results are shown in bold.

Table 4 Number of tail flips

	Fixed effects	Estimate	s.e.m	t	CI 2.50%	CI 97.50%
(a) playback	stimulus	−16.65	1.57	−10.58	−19.82	−13.45
	local	21.34	25.35	0.84	−30.85	73.32
	90 km	−37.68	6.18	−6.09	−49.77	−25.20
	180 km	−57.62	6.16	−9.35	−70.01	−45.05
	African	−54.67	6.17	−8.86	−67.06	−42.01
	control	−74.16	6.46	−11.48	−87.36	−60.74
	breeding stage	23.13	10.37	2.23	1.93	44.08
	trial order	−3.10	1.44	−2.16	−5.97	0.44
	date	2.75	1.45	1.89	−0.18	5.69
	time	0.49	0.50	0.97	−0.54	1.51
(b) decoy	stimulus	−18.43	14.13	−1.30	−48.35	11.16
	local	91.14	75.87	0.80	−102.65	224.57
	European	−10.80	43.36	−0.25	−106.25	83.18
	African	−52.38	43.82	−1.20	−146.92	40.28
	control	−48.50	44.02	−1.10	−144.07	46.80
	breeding stage	−2.78	20.12	−0.14	−46.04	39.67
	trial order	59.11	16.39	3.61	24.44	93.31
	date	−15.99	11.76	−1.36	−41.19	8.33
	time	−0.28	1.00	−0.28	−2.40	1.82

Results of general linear mixed model with estimates, standard error, t-value, and credible intervals fitted by maximum likelihood. Estimates for the stimulus locations refer to differences from the intercept estimates, which represent the number of tail flips of the local population. Subjects were included as random intercepts to control for repeated measures. A behavioural response differs significantly from the local population if its credible intervals do not include the mean of the local population. Significant results are shown in bold.

[40]. After the postnuptial moult they start migrating towards the wintering grounds in early autumn [41,67]. In the present study, all focal pairs were ringed. We conducted regular checks twice per week to monitor the breeding activity and to define the breeding stage of each pair.

To test the discrimination ability of the local population, we collected songs from the local population (Düffel) and two nearby European Stonechat populations at distances of 90 km (Heubach) and 180 km (Wahner Heide) from the study area. Furthermore, we used songs and decoys of African stonechats from Kenya [distance 4,000 km, [68]] and decoys from Stonechats from Austria [distance 1,000 km, 41]. Control species are explained below.

Recording method and song analysis
Stonechats, in common with most passerines, spend a higher proportion of their time singing just before dawn than at other times of day [69]. During the onset of the breeding season, we recorded the dawn song of a minimum of 28 individuals from each European stonechat population (n =3) for about ten minutes, using a Marantz PMD 661 solid state recorder (Osnabrück, Germany) and Sennheiser ME66/K6 directional microphones with windbreak (Georgsmarienhütte, Germany). To expand our set of stimuli, we also obtained 28 songs per species from African stonechats, *Saxicola torquata axillaris*, and

Winter wrens, *Troglodytes hiemalis*, from the Macaulay Library (www.macaulaylibrary.org). These song recordings were conducted in the Great Rift Valley (Kenya) for African stonechats and New York State (United States) for Winter wrens.

We analysed the songs of all five stimuli origins (sampling frequency: 44.1 kHz; resolution: 16 bit, Figure 6) with the software Avisoft Sound Analysis Pro, version 5.1.09 (Raimund Specht, Berlin, Germany). We examined the song duration, number of elements per song, element rate (number of elements per second), minimum and maximum frequency, peak frequency (frequency of the highest amplitude sound), and bandwidth for all populations (Additional file 1: Figure S3). With the automatic parameter measurements setup, we obtained the minimum and maximum frequency values measuring at a standard decibel threshold (here −20 dB, total option) below the peak in the power spectra [70].

To assess the song repertoire size, we analysed 100 consecutive songs of each male Stonechat (n =20) from the local population. In Stonechats, a song typically consists of a sequence of motifs, and these in turn each contain several consecutive elements (Additional file 1: Figure S3). They are stereotypically repeated at a constant rate, and thereby distinguishable from all other song types. In general, song motifs, rather than complete songs, are shared within a

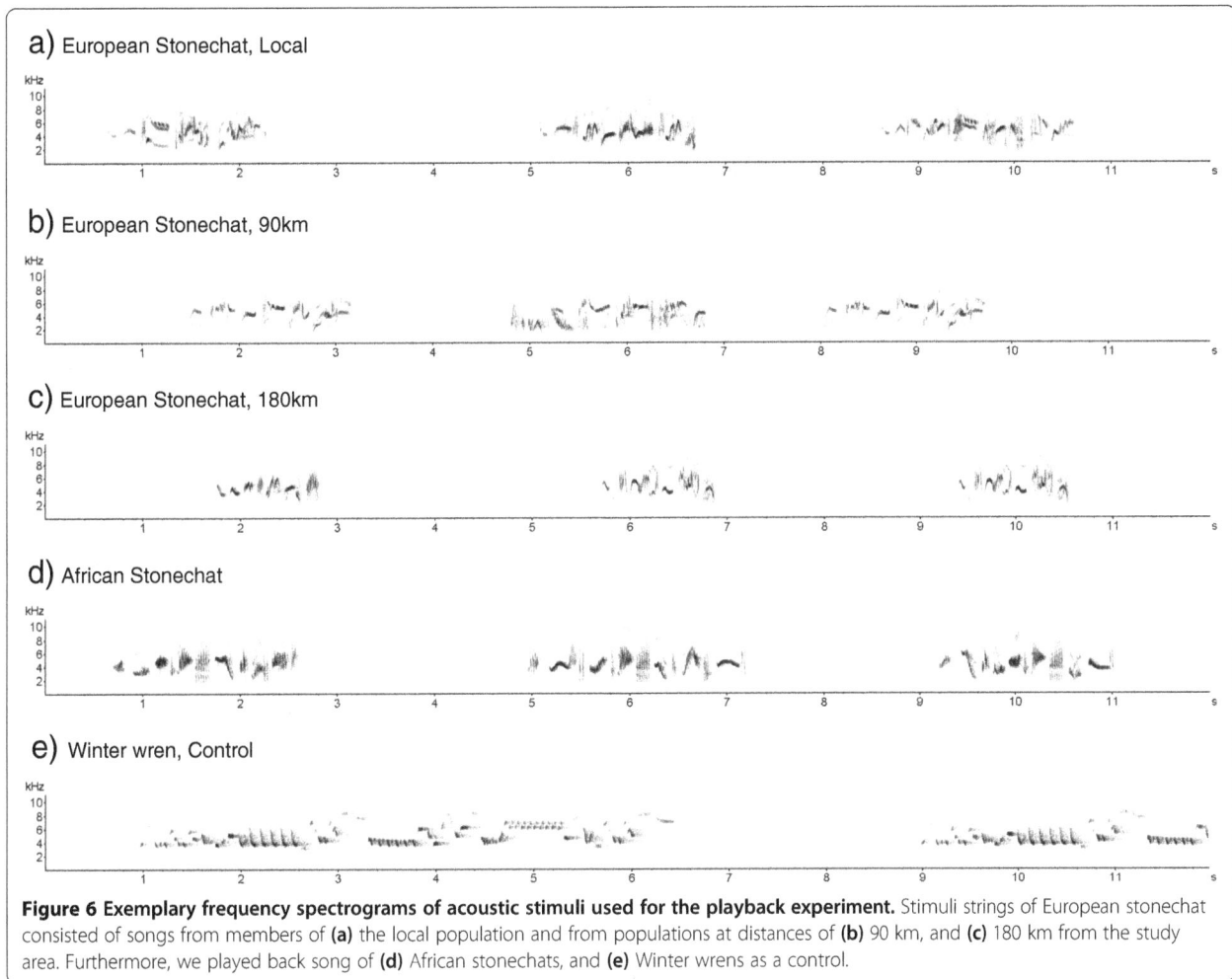

Figure 6 Exemplary frequency spectrograms of acoustic stimuli used for the playback experiment. Stimuli strings of European stonechat consisted of songs from members of **(a)** the local population and from populations at distances of **(b)** 90 km, and **(c)** 180 km from the study area. Furthermore, we played back song of **(d)** African stonechats, and **(e)** Winter wrens as a control.

population. The mean song repertoire consists of 16 ± 3.06 unique song types.

Playback experiment

To reveal behavioural responsiveness of male and female Stonechats to songs of different dialects, we performed a field-based playback experiment by simulating a territorial intrusion with songs of distinct dialects during the egg laying or incubation stage (13th – 26th April, 2011). Each subject received five playback trials with the sequence of exposure determined by a randomized block design created by Randlist 1.2 (DatInf GmbH, Tübingen, Germany) to account for bias by trial order effects. Stimuli strings consisted of songs from the three European and the single African Stonechat populations. As a control, we used song of heterospecific Winter wrens, *Troglodytes hiemalis*. This species was chosen following the rationale by Grant and Grant [6], using a species that is similar in note structure and frequency range, but has never been heard by the tested birds.

To avoid inclusion of rare motifs, we selected song types with defined common motifs, which are shared between members of a population, and thus are representative of

each population. However, this implies that the interpretation of song discrimination between populations should be taken with caution. The standardized stimuli selection of common and locally shared songs most likely excluded overlapping songs among populations. Their incorporation may have led to a slight decrease in the discrimination ability, and thus a potential overestimation of the responsiveness. To increase the number of independent samples, and thus improve the reliability and external validity, we generated a unique stimulus for each trial [71]. Each stimulus song was only used once for the entire study. We tested females and males simultaneously on their territories, and therefore each stimulus string was used once for both sexes (i.e. $28 \times 5 = 140$ unique playback stimuli). Each stimulus comprised song types from one individual's recording following the natural syntax of Stonechat song. We used 25 unique songs in total for each stimulus string, which were filtered (1 kHz high-pass filter) and normalised in peak amplitude (i.e. the amplitude of each song was adjusted to 75 % of the maximum amplitude). Songs were divided by pauses of 4.5 seconds. A trial comprised all five population stimuli played back consecutively in a random order, each with a duration of 150 seconds. Each stimulus string was followed by at least 150 seconds of silence. To ascertain a comparable behavioural response of the latency to approach for each stimulus, playback strings only started when the focal bird (males =28, females =15) was at a distance of at least 10 m from the caller (longest silence between consecutive strings 285 s). Hence, each trial was performed on an individual bird for a period of about 25 minutes in total depending on the start times of the consecutive playback stimuli. Stimuli were broadcasted with the caller Foxpro Scorpion X1B (digital game caller, FOXPRO Inc. Lewistown, USA), which could be operated with a remote control. It was mounted on top of a bush in the central area of a territory such that it was widely audible. Response songs were recorded during the entire trial. However, acoustic responses to the playback were rare, and thus were not included in further analysis.

Decoy experiment

We conducted a second experiment to test the responsiveness towards morphological traits by using a stuffed decoy simulating a territorial intrusion. During the nestling or fledgling stage we placed the decoy (male in full adult plumage protected by an inconspicuous cage) in the centre of respective territories for ten minutes in total for each trial. Decoy stimuli consisted of males from (a) local Stonechats, (b) European Stonechats from Austria [distance 1,000 km, 41], (c) African Stonechats from Kenya [distance 4,000 km, [68]], and as a control (d) European Robins, *Erithacus rubecula*. From extensive observations, we know that Stonechats aggressively chase off other small insectivorous passerines with similar feeding habits. European robins meet this criterion but their preference of deciduous wooded habitats limits their familiarity to Stonechats. To avoid pseudo-replication we randomly chose from five different decoys per stimulus for each trial. Each focal Stonechat (male =16, female =14) received all stimuli in a randomized and balanced order. We conducted each trial on a different date (5th-18th May, 2011) during morning hours with two days pause between trials.

Behavioural measurements

All behavioural responses were observed from a distance of about 30 m and were documented continuously by dictating to the Marantz PMD 661. To quantify behaviour, we used descriptors that are commonly used to measure responses to territorial intrusions and mate attraction [72], including studies in Stonechats [43]. Specifically, we measured the latency of a bird to approach the playback or decoy within 5 m; the time a bird spent within this 5 m zone; and the number of tail flips, which are defined as up- and downward movements of the entire tail and indicate agitation in Stonechats [65]. The descriptive statistics

of all behavioural responses can be found in the supplements (Additional file 1: Table S4).

Statistical analysis

All statistical analyses were performed with the software R v. 3.1.0 [73]. Tests were two-tailed and significance was accepted at $\alpha = 0.05$. We used principal component analyses (PCA, R package *FactoMineR* [74]) to compare song traits between groups (with and without both control groups) and then tested the first principal component in a general linear model (LM, R package *lme4* [75]) to identify the relationship between song traits and geographic distance of song origins from the local population. The latencies to approach within 5 m to the different stimuli were analysed using mixed-effects cox models (survival model) fitted by maximum likelihood accounting for breeding stage, randomized trial order, date and time (coxme, R package *survival* [76]). Subjects were included as random intercepts to control for repeated measures. A Spearman's correlation was run to determine the relationship of the behavioural response between paired females and males. For the time spent within 5 m we used a generalized linear mixed model with a beta distribution and stimulus as random factor using WinBUGS software 1.4 (GLMM, R package *R2WinBUGS* [77,78]). In WinBUGS we focussed exclusively on differences between stimuli. We defined stimulus as a random factor to compare paths of all stimuli, and thus correct for multiple testing. The response number of tail flips in males was analysed with a general linear mixed model fitted by maximum likelihood methods (LMMs, R package lme4 [75]) controlling for breeding stage, trial order, date and time. Subjects were included as random intercepts to control for repeated measures. Predictions from the general and generalized linear mixed models (Bayesian methods) were calculated as the median of their posterior distributions, and the 2.5 and 97.5% credible intervals (CI).

Additional file

Additional file 1: Figure S1. Geographic variation in the song of European stonechats as quantified by principal component analysis. **Table S1** Song traits of three European Stonechat populations. (a) Factor loadings of the principal component analysis for seven song traits. (b) Results of a general linear model testing whether the first principal component (PC1) differed between songs from different locations. **Figure S2** Female and male behavioural responses. The figures show the latency to approach within 5 m of a stimulus in females and males during (a) playback experiments and (b) decoy experiments. **Table S2** Behavioural response between pairs. **Table S3** Correlation of behavioural responses between females and males during playback and decoy experiment. **Figure S3** Spectrogram of an exemplary song in European stonechats. Indicated are typical measured song traits of Stonechats, i.e. song duration, total number of elements, minimum and maximum frequency, and bandwidth. **Table S4** Descriptive statistics of behavioural responses for a) playback and b) decoy experiment. **Figure S4** Trial order for the latency to approach during the playback experiment in (a) males and (b) females, and during the decoy experiment in (c) males and (d) females.

Competing interests

The authors declare that they have no competing interests.

Authors' contributions

KGM and BH conceived the study. KGM designed the experimental set-up. KGM and HF executed the experiments. KGM analysed the data and wrote the first draft of the manuscript. KGM and BH completed the manuscript. All authors read and approved the final manuscript.

Acknowledgements

All procedures follow NIH guidelines for the Care and Use of Experimental Animals and were conducted under the permission of the responsible authorities (West Münsterland, Kreis Borken, North-Rhine-Westphalia, Germany). The authors thank Martin Wikelski for providing equipment. We also thank Davide M. Dominoni, Fränzi Korner-Nievergelt, Beate Apfelbeck, Wolfgang Goymann, Manfred Gahr, Michaela Hau, Jim Caryl and three anonymous reviewers for providing valuable discussion and feedback on the manuscript. We are also very grateful for the splendid cover image of a male European stonechat by Beate Apfelbeck. KGM is a member of the International Max Planck Research School for Organismal Biology and is funded through the German Research Foundation (DFG grant HE3488/5-1).

Author details

[1]Department of Migration and Immuno-Ecology, Max Planck Institute for Ornithology, 78315 Radolfzell, Germany. [2]Department of Ornithology, University of Konstanz, 78457 Konstanz, Germany. [3]Am Kuhm 19, 46325 Borken, Germany. [4]Institute of Biodiversity, Animal Health and Comparative Medicine, University of Glasgow, G12 8QQ Glasgow, UK.

References

1. Coyne JA, Orr HA: *Speciation.* Sunderland MA: Sinauer Associates Inc; 2004:545.
2. Marler P: Specific distinctiveness in the communication signals of birds. *Behaviour* 1957, **11:**13–39.
3. Podos J: Acoustic discrimination of sympatric morphs in Darwin's finches: a behavioural mechanism for assortative mating? *Philos Trans R Soc B* 2010, **365:**1031–1039.
4. Edwards SV, Kingan SB, Calkins JD, Balakrishnan CN, Jennings WB, Swanson WJ, Sorenson MD: Speciation in birds: Genes, geography, and sexual selection. *Proc Natl Acad Sci U S A* 2005, **102:**6550–6557.
5. Price TD, Sol D: Introduction: Genetics of colonizing species. *Am Nat* 2008, **172:**S1–S3.
6. Grant BR, Grant PR: Simulating secondary contact in allopatric speciation: an empirical test of premating isolation. *Biol J Linn Soc* 2002, **76:**545–556.
7. Slabbekoorn H, Smith TB: Bird song, ecology and speciation. *Phil Trans R Soc Lond B* 2002, **357:**493–503.
8. Baker MC, Cunningham MA: The biology of bird-song dialects. *Behav Brain Sci* 1985, **8:**85.
9. Marler P, Tamura M: Song "dialects" in three populations of White-crowned Sparrows. *Condor* 1962, **64:**368–377.
10. Nottebohm F: The "critical period" for song learning. *Ibis (Lond 1859)* 1969, **111:**386–387.
11. Baker MC: Song dialects and genetic differences in white-crowned sparrows *(Zonotrichia leucophrys).* *Evolution (N Y)* 1975, **29:**226–241.
12. Searcy WA: Song repertoire and mate choice in birds. *Am Zool* 1992, **32:**71–80.
13. Baker MC: Vocal dialect recognition and population genetic consequences. *Am Zool* 1982, **22:**561–569.
14. Baker MC: The behavioral response of female nuttalis White-crowned sparrows to male song of natal and alien dialects. *Behav Ecol Sociobiol* 1983, **12:**309–315.
15. Searcy WA, Nowicki S, Hughes M, Peters S: Geographic song discrimination in relation to dispersal distances in song sparrows. *Am Nat* 2002, **159:**221–230.
16. Danner JE, Danner RM, Bonier F, Martin PR, Small TW, Moore IT: Female, but not male, tropical sparrows respond more strongly to the local song dialect: implications for population divergence. *Am Nat* 2011, **178:**53–63.

17. Uy JAC, Moyle RG, Filardi CE: Plumage and song differences mediate species recognition between incipient flycatcher species of the Solomon Islands. *Evolution (N Y)* 2009, **63:**153–164.
18. Seddon N, Botero CA, Tobias JA, Dunn PO, Macgregor HEA, Rubenstein DR, Uy JAC, Weir JT, Whittingham LA, Safran RJ: Sexual selection accelerates signal evolution during speciation in birds. *Proc R Soc B* 2013, **280:**20131065.
19. McGraw K, Stoehr A: Plumage redness predicts breeding onset and reproductive success in the house finch: a validation of Darwin's theory. *J Avian* 2001, **1:**90–94.
20. Stein A, Uy J: Unidirectional Introgression of a Sexually Selected Trait across an Avian Hybrid Zone: A Role for Female Choice? *Evolution (N Y)* 2006, **60:**1476–1485.
21. Price T: Sexual selection and natural selection in bird speciation. *Philos Trans R Soc B Biol Sci* 1998, **353:**251–260.
22. Kirschel ANG, Blumstein DT, Smith TB: Character displacement of song and morphology in African tinkerbirds. *Proc Natl Acad Sci U S A* 2009, **106:**8256–8261.
23. Baker MC, Mewaldt LR: Song dialects as barriers to dispersal in white-crowned sparrows, *Zonotrichia leucophrys nuttalli.* *Evolution (N Y)* 1978, **32:**712–722.
24. Patten MA, Rotenberry JT, Zuk M: Habitat selection, acoustic adaptation, and the evolution of reproductive isolation. *Evolution (N Y)* 2004, **58:**2144–2155.
25. Grant BR, Grant PR: Cultural inheritance of song and its role in the evolution of Darwins finches. *Evolution (N Y)* 1996, **50:**2471–2487.
26. Grant PR, Grant BR: The secondary contact phase of allopatric speciation in Darwin's finches. *Proc Natl Acad Sci U S A* 2009, **106:**20141–20148.
27. Kawecki TJ, Ebert D: Conceptual issues in local adaptation. *Ecol Lett* 2004, **7:**1225–1241.
28. Baldwin MW, Winklerà H, Organ CL, Helm B: Wing pointedness associated with migratory distance in common- garden and comparative studies of stonechats *(Saxicola torquata).* *J Evol Biol* 2010, **23:**1050–1063.
29. Helm B, Schwabl I, Gwinner E: Circannual basis of geographically distinct bird schedules. *J Exp Biol* 2009, **212:**1259–1269.
30. Benkman CW: Adaptation to single resources and the evolution of crossbill *(Loxia)* diversity. *Ecol Monogr* 1993, **63:**305–325.
31. Benkman CW: Divergent selection drives the adaptive radiation of crossbills. *Evolution (N Y)* 2003, **57:**1176–1181.
32. Smith JW, Benkman CW: A coevolutionary arms race causes ecological speciation in crossbills. *Am Nat* 2007, **169:**455–465.
33. Helbig AJ: *Evolution of Migration: A Phylogenetic and Biogeographic Perspective.* Heidelberg: Springer-Verlag Berlin Heidelberg; 2003:3–20.
34. Nelson DA: Ecological influences on vocal development in the white-crowned sparrow. *Anim Behav* 1999, **58:**21–36.
35. Paradis E, Baillie SR, Sutherland WJ, Gregory RD: Patterns of natal and breeding dispersal in birds. *J Anim Ecol* 1998, **67:**518–536.
36. Reed JM, Boulinier T, Danchin E, Oring LW: *Current Ornithology: Informal Dispersal: Prospecting by Birds for Breeding Sites.* New York: Kluwer Academic/Plenum Publishers; 1999:189–259.
37. Randler C: Avian hybridization, mixed pairing and female choice. *Anim Behav* 2002, **63:**103–119.
38. Nelson DA, Marler P, Morton ML: Overproduction in song development: an evolutionary correlate with migration. *Anim Behav* 1996, **51:**1127–1140.
39. Illera JC, Richardson DS, Helm B, Atienza JC, Emerson BC: Phylogenetic relationships, biogeography and speciation in the avian genus *Saxicola.* *Mol Phylogenet Evol* 2008, **48:**1145–1154.
40. Helm B, Fiedler W, Callion J: Movements of European Stonechats *(Saxicola torquata)* according to ringing recoveries. *Ardea* 2006, **94:**33–44.
41. Flinks H, Helm B, Rothery P: Plasticity of moult and breeding schedules in migratory European *Stonechats Saxicola rubicola.* *Ibis (Lond 1859)* 2008, **150:**687–697.
42. Canoine V, Gwinner E: The hormonal response of female European Stonechats to a territorial intrusion: the role of the male partner. *Horm Behav* 2005, **47:**563–568.
43. Canoine V, Gwinner E: Seasonal differences in the hormonal control of territorial aggression in free-living European stonechats. *Horm Behav* 2002, **41:**1–8.
44. Nelson DA, Soha JA: Male and female white-crowned sparrows respond differently to geographic variation in song. *Behaviour* 2004, **141:**53–69.
45. Sætre GP, Král M, Bureš S: Differential species recognition abilities of males and females in a flycatcher hybrid zone. *J Avian Biol* 1997, **28:**259–263.

46. Haavie J, Borge T, Bures S, Garamszegi LZ, Lampe HM, Moreno J, Qvarnström A, Török J, Saetre G-P: Flycatcher song in allopatry and sympatry - convergence, divergence and reinforcement. *J Evol Biol* 2004, **17**:227–237.

47. Balakrishnan CN, Sorenson MD: Song discrimination suggests premating isolation among sympatric indigobird species and host races. *Behav Ecol* 2006, **17**:473–478.

48. Nelson DA, Soha JA: Perception of geographical variation in song by male Puget Sound white-crowned sparrows, *Zonotrichia leucophrys pugetensis*. *Anim Behav* 2004, **68**:395–405.

49. MacDougall-Shackleton E, MacDougall-Shackleton S: Cultural and genetic evolution in mountain white-crowned sparrows: song dialects are associated with population structure. *Evolution (N Y)* 2001, **55**:2568–2575.

50. Soha JA, Nelson DA, Parker PG: Genetic analysis of song dialect populations in Puget Sound white-crowned sparrows. *Behav Ecol* 2004, **15**:636–646.

51. Kiefer S, Spiess A, Kipper S, Mundry R, Sommer C, Hultsch H, Todt D: First-Year Common Nightingales (*Luscinia megarhynchos*) Have Smaller Song-Type Repertoire Sizes Than Older Males. *Ethology* 2006, **112**:1217–1224.

52. Todt D, Geberzahn N: Age-dependent effects of song exposure: song crystallization sets a boundary between fast and delayed vocal imitation. *Anim Behav* 2003, **65**:971–979.

53. Kiefer S, Sommer C, Scharff C, Kipper S: Singing the popular songs? Nightingales share more song types with their breeding population in their second season than in their first. *Ethology* 2010, **116**:619–626.

54. Marler P, Peters S: Developmental overproduction and selective attrition: new processes in the epigenesis of birdsong. *Dev Psychobiol* 1982, **15**:369–378.

55. Ellers J, Slabbekoorn H: Song divergence and male dispersal among bird populations: a spatially explicit model testing the role of vocal learning. *Anim Behav* 2003, **65**:671–681.

56. Mortega KG, Horsburgh GJ, Illera JC, Dawson DA: Characterization of microsatellite markers for Saxicola species. *Conserv Genet Resour* 2014, 10.1007/s12686-014-0355-9.

57. Uy JAC, Borgia G: Sexual selection drives rapid divergence in bowerbird display traits. *Evolution (N Y)* 2000, **54**:273–278.

58. Irwin DE, Bensch S, Price TD: Speciation in a ring. *Nature* 2001, **409**:333–337.

59. Seddon N, Tobias JA: Song divergence at the edge of Amazonia: an empirical test of the peripatric speciation model. *Biol J Linn Soc* 2007, **90**:173–188.

60. Podos J: Correlated evolution of morphology and vocal signal structure in Darwin's finches. *Nature* 2001, **409**:185–188.

61. Baker MC, Baker AEM: Reproductive behavior of female buntings: Isolating mechanisms in a hybridizing pair of species. *Evolution (N Y)* 1990, **44**:332–338.

62. Uy JAC, Moyle RG, Filardi CE, Cheviron ZA: Difference in plumage color used in species recognition between incipient species is linked to a single amino acid substitution in the melanocortin-1 receptor. *Am Nat* 2009, **174**:244–254.

63. Rowe C, Guilford T: The evolution of multimodal warning displays. *Evol Ecol* 1999, **13**:655–672.

64. Bradbury JW, Vehrencamp SL: *Principles of Animal Communication.* Sinauer: Boston, MA; 1998.

65. Urquhart ED: *Stonechats. A Guide to the Genus Saxicola.* London: Christopher Helm; 2002.

66. Flinks H, Pfeifer F: Brutzeit, Gelegegröße und Bruterfolg beim Schwarzkehlchen (*Saxicola torquata*). *Charadrius* 1987, **23**:128–140.

67. Schwabl H, Flinks H, Gwinner E: Testosterone, reproductive stage, and territorial behavior of male and female European stonechats *Saxicola torquata*. *Horm Behav* 2005, **47**:503–512.

68. Dittami JP, Gwinner E: Annual cycles in the African stonechat *Saxicola torquata axillaris* and their relationship to environmental factors. *J Zool* 1985, **207**:357–370.

69. Zollinger SA, Podos J, Nemeth E, Goller F, Brumm H: On the relationship between, and measurement of, amplitude and frequency in birdsong. *Anim Behav* 2012, **84**:e1–e9.

70. Kroodsma DE: Suggested experimental designs for song playbacks. *Anim Behav* 1989, **37**:600–609.

71. McGregor PK: Quantifying responses to playback: one, many, or composite multivariate measures. In *Play Stud Anim Commun.* New York: Plenum; 1992:79–96.

72. R Core Team: *R: A language and environment for statistical computing.* Vienna: R Foundation for Statistical Computing; 2013:618–622.

73. Husson F, Josse J, Le S, Mazet J: *Multivariate Exploratory Data Analysis and Data Mining with R.* R Package Version 1; 2014:102–123.

74. Bates D, Mächler M, Bolker B, Walker S: Fitting linear mixed-effects models using lme4. R package version 1.0-6. *J Stat Softw* 2014, **55**:1–9.

75. Therneau T: coxme: mixed effects Cox models. R package version 2.2-3. 2011, See http://cran.r-project.org.

76. Sturtz S, Ligges U, Gelman A: R2WinBUGS: a package for running WinBUGS from R. *J Stat Softw* 2005, **12**:1–16.

77. Kacelnik A, Krebs JR: The dawn chorus in the great tit (*Parus major*): proximate and ultimate causes. *Behaviour* 1982, **83**:287–309.

78. Lunn DJ, Thomas A, Best N, Spiegelhalter D: WinBUGS - A Bayesian modelling framework: Concepts, structure, and extensibility. *Stat Comput* 2000, **10**:325–337.

10

FASconCAT-G: extensive functions for multiple sequence alignment preparations concerning phylogenetic studies

Patrick Kück[1]* and Gary C Longo[2]

Abstract

Background: Phylogenetic and population genetic studies often deal with multiple sequence alignments that require manipulation or processing steps such as sequence concatenation, sequence renaming, sequence translation or consensus sequence generation. In recent years phylogenetic data sets have expanded from single genes to genome wide markers comprising hundreds to thousands of loci. Processing of these large phylogenomic data sets is impracticable without using automated process pipelines. Currently no stand-alone or pipeline compatible program exists that offers a broad range of manipulation and processing steps for multiple sequence alignments in a single process run.

Results: Here we present FASconCAT-G, a system independent editor, which offers various processing options for multiple sequence alignments. The software provides a wide range of possibilities to edit and concatenate multiple nucleotide, amino acid, and structure sequence alignment files for phylogenetic and population genetic purposes. The main options include sequence renaming, file format conversion, sequence translation between nucleotide and amino acid states, consensus generation of specific sequence blocks, sequence concatenation, model selection of amino acid replacement with ProtTest, two types of RY coding as well as site exclusions and extraction of parsimony informative sites. Convieniently, most options can be invoked in combination and performed during a single process run. Additionally, FASconCAT-G prints useful information regarding alignment characteristics and editing processes such as base compositions of single in- and outfiles, sequence areas in a concatenated supermatrix, as well as paired stem and loop regions in secondary structure sequence strings.

Conclusions: FASconCAT-G is a command-line driven Perl program that delivers computationally fast and user-friendly processing of multiple sequence alignments for phylogenetic and population genetic applications and is well suited for incorporation into analysis pipelines.

Keywords: Multiple sequence alignment, Phylogenetic reconstruction, Sequence processing, Consensus sequence, Sequence translation, Sequence concatenation, File format conversion

Introduction

Phylogenetic and population genetic analyses commonly involve the manipulation and processing of multiple sequence alignments. For instance, concatenation of multiple gene alignments are common in rRNA analyses (e.g. [1-6]) and in 'mixed' nucleotide alignment analyses, combining rRNA genes like 18S and 28S as well as protein coding nucleotide sequences (e.g. [7-10]). Likewise, the ability to concatenate hundreds to thousands of nucleotide or amino acid single gene alignments has recently become an indispensable tool with the growth of phylogenomics (e.g. [11-24]). Sequence translation of nucleotide data (DNA/RNA) to protein coding sequences as well as RY coding [25] of nucleotide sequences are common practices to reduce the signal-to-noise ratio of underlying data in phylogenomic studies prior to tree reconstruction (e.g. [26-29]). In order to predict possible nucleotide sequences for a specified protein, researchers

*Correspondence: patrick_kueck@web.de
[1]Zoologisches Forschungsmuseum A. Koenig, Adenauerallee 160-163, 53113 Bonn, Germany
Full list of author information is available at the end of the article

often reverse translate amino acid sequences to nucleotide states (e.g. [30-33]). Another common analysis of multiple sequence alignments is consensus sequence generation, which is commonly used to identify and compare conserved and variable regions (e.g. [34-36]), design degenerated PCR primers for appropriate locations within the alignment, or to define operational taxonomic units using DNA barcode data for subsequent phylogenetic analysis (e.g. [37]). Consensus sequence generation has also become a valuable tool in large scale population genetic analyses that pool individuals as a cost effective method for determining population level data. Recent studies searching for genes potentially under selection among populations relied on identifying the most common allele at polymorphic sites as well as alleles fixed within populations [38], which can be accomplished through consensus generation.

Phylogenetic and population genetic analyses also commonly involve the tedious tasks of dealing with different sequence file formats and sequence renaming with the later becoming increasingly time-consuming when dealing with hundreds of sequences.

Although there are many scripts and online platforms that address these issues or manipulate sequence alignments with single processing steps, a software tool which enables combined processing steps in a single operation is lacking. Software like SequenceMatrix [39], TranslatorX [40], and CONCATENATOR [41] are pure concatenation tools which can be used only via graphical user interface or which are web server designed and therefore cannot be implemented in automatic process pipelines. 2matrix [42] is a pure concatenation tool as well but command line driven. SCaFoS [43] is a phylogenomic tool for selecting and concatenating sequences in large multigene and species datasets at either the amino acid or nucleotide level. Although SCaFoS is efficient at selecting orthologous sequences, creating chimerical sequences, and selecting genes according to their level of missing data, it lacks alignment processing options such as sequence translation, RY-coding, secondary structure handling, sequence renaming and consensus sequence generation.

With FASconCAT-G (FcC-G), we introduce a versatile software designed for processing and manipulating multiple sequence alignments. Conveniently, FcC-G allows for multiple processing steps in a single run and is easily implemented into pipeline analyses. FcC-G represents an advancement of FASconCAT [44], an already commonly used tool in phylogenetic studies (e.g. [45-53]).

Results and discussion

FASconCAT-G accepts multiple nucleotide, amino acid, and structure sequence alignment input files and can perform sequence renaming, file format conversion,

sequence translation of nucleotide and amino acid states, consensus generation of specific sequence blocks, sequence concatenation, RY coding, model selection of amino acid replacement using ProtTest [54], extraction of parsimony informative sites as well as generation of partitioned files for MrBayes [55] and RAxML analyses [56]. The process order of FcC-G allows for a wide range of optional process combinations (Figure 1), although, some process chains are not possible in a single process run. For instance, it is not possible to RY code nucleotide sequences before translating them to amino acid sequences or to build consensus sequences before the sequence translation process. For tasks of this nature, FcC-G has to be run twice. However, we hope the current process order of FcC-G is useful for most phylogenetic and population genetic applications. To avoid errors such as the exclusion of third nucleotide site positions before sequence translation to amino acid character states, FcC-G contains a hierarchical order of single file processing steps:

1. Sequence renaming
2. Sequence translation (nucleotide to amino acid sequences or vice versa)
3. Generation of consensus sequences of predefined sequence blocks
4. RY coding of nucleotide sequences
5. Exclusion of each third nucleotide site position
6. Sequence concatenation
7. Extraction of parsimony informative sites
8. Print out of edited sequences and additional sequence information

Sequence renaming

Sequence names are often coded during the sequencing process or, if downloaded from NCBI, extended with additional information and non-alphanumeric signs which are often not allowed in downstream analysis programs. Accordingly, FcC-G can rename defined sequence names prior to file processing by using a user supplied info file, which lists, in each row, the old name delimited from the new name by a tabstop. Sequences which are not listed in the user supplied info file remain unchanged. FcC-G will print additional information of the sequence renaming process to a new outfile.

Sequence translation

FcC-G can translate standard nucleotide sequence states to amino acid characters and vice versa. For sequence translation of nucleotide data FcC-G uses the standard IUPAC triplet codes for amino acid characters. When translating amino acid states to corresponding nucleotide characters, FcC-G uses compressed IUPAC codes. Conveniently, FcC-G can recognize and handle

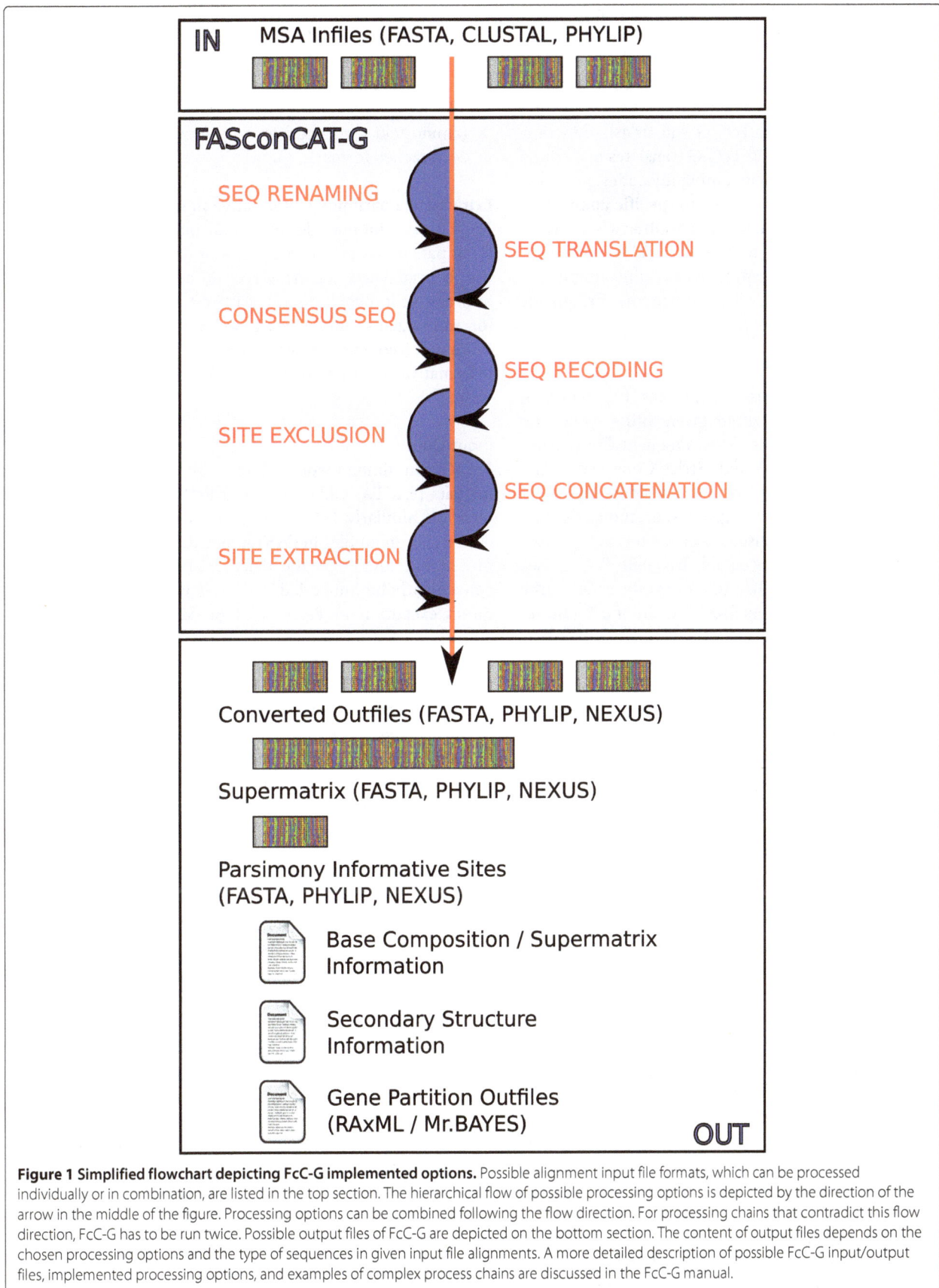

Figure 1 Simplified flowchart depicting FcC-G implemented options. Possible alignment input file formats, which can be processed individually or in combination, are listed in the top section. The hierarchical flow of possible processing options is depicted by the direction of the arrow in the middle of the figure. Processing options can be combined following the flow direction. For processing chains that contradict this flow direction, FcC-G has to be run twice. Possible output files of FcC-G are depicted on the bottom section. The content of output files depends on the chosen processing options and the type of sequences in given input file alignments. A more detailed description of possible FcC-G input/output files, implemented processing options, and examples of complex process chains are discussed in the FcC-G manual.

both amino acid and nucleotide data sets in a single processing run. Accordingly, FcC-G will only translate sequences of infiles which are suitable for a defined translation process. This makes it very easy to concatenate a mixture of different infile sequence types to one specific supermatrix sequence type FcC-G will translate incomplete nucleotide triplets to '?'. FcC-G translates nucleotide triplets even if triplets contain ambiguity codes, provided that the triplets are still assignable to specific amino acid characters (e.g. 'YTR' ↪ Leucine/L). Otherwise, unspecific triplets are translated to '?' (e.g. 'RCT' ↪ ?). FcC-G does not check for correctness of given reading frames but will print a warning in the terminal window if sequence lengths are not a multiple of three.

Consensus sequences

FcC-G can create consensus sequences for matching defined sequence blocks within given infiles using one of three consensus methods: 'Most Frequent Consensus', 'Majority Rule Consensus', and 'Strict Consensus'. The 'Most Frequent Consensus' option considers the most frequent character state at a given site among defined sequence blocks as the consensus character state. If two or more character states are equally frequent, FcC-G uses either the corresponding IUPAC ambiguity code as the consensus character state (nucleotide data) or a '?' (amino acid data and nucleotide data). The 'Majority Rule Consensus' option considers character states which occur at a given site position in more than 50% of sequences of a defined sequence block as consensus character state. Otherwise, FcC-G uses a '?' as the consensus character state (amino acid data and nucleotide data). The 'Strict Consensus' option considers all character states at a given site position to generate a strict consensus sequence for a defined sequence block using IUPAC ambiguity codes for nucleotide data and a 'X' for amino acid data. For nucleotide data, indel events (coded as '-') and missing data (coded as '?') are ignored using 'Strict Consensus' as long as a nucleotide character state exists for a specific site position. If a specific site on a defined sequence block consists of only indel events ('-') and missing data states ('?'), FcC-G will output a '?' as the consensus character state.

RY coding of nucleotide sequences

RY coding can be applied to each third nucleotide sequence position or to complete nucleotide sequences. The R code is used for purine states while the Y code is used for pyrimidines. Amino acid sequences are left unchanged unless the sequence translation option from amino acid to nucleotide states has been defined.

Sequence concatenation

FcC-G can concatenate sequence alignment infiles (nucleotide and amino acid as well as 'dot-bracket'

structure information) of identical taxa into a supermatrix file. It is also possible to concatenate amino acid and nucleotide alignments into one supermatrix. In the supermatrix file, taxon sequences which were missing from single files are encoded either by 'N' (nucleotide sequences), 'X' (amino acid sequences) or by ' . ' (dots structure strings in 'dot-bracket' format).

Extraction of parsimony informative sites

FcC-G can print out additional information file(s) identifying parsimony-informative sites of given infiles and/or the concatenated supermatrix. A site is parsimony-informative if it contains at least two types of nucleotides (or amino acids), and at least two of them occur with a minimum frequency of two. The file format of parsimony-informative alignment files depends on the chosen output format(s).

Input/Output

FcC-G can simultaneously handle three different infile formats (FASTA, CLUSTAL, and PHYLIP) in any combination. Similarly, FcC-G can print concatenated and/or edited alignment files in FASTA, NEXUS, and/or PHYLIP format but FASTA is the default. NEXUS outfiles can conveniently be imbedded with MrBayes commands for direct execution in PAUP [57] or MrBayes [58] (very convenient for partitioned or mixed DNA/RNA analyses) or output without any specific commands. Likewise, PHYLIP output files can be directly used for Maximum Likelihood tree reconstruction analyses with RAxML [56] or PhyML [59]. Additionally, our new software tool prints a file with useful information about alignment and sequence characteristics for the concatenated supermatrix as well as all single infiles. Information on this file includes single base compositions (including GC content), sequence types as well as sequence lengths and the number of taxa represented in each infile and the concatenated supermatrix. The file also contains information specific to the concatenation process, such as the position of each sequence fragment in the concatenated supermatrix as well as a list of all concatenated sequences and inserted replacement strings. However, the evaluation of this additional information often results in longer computation times depending on the size of data sets. Therefore, FcC-G offers an option to increase the overall computation speed by decreasing the information sampled and printed to the information file. If one or more infiles contain a secondary structure string, FcC-G will print another file with information about stem and loop character states and positions in both the concatenated supermatrix and infile(s). FcC-G can also print parsimony informative sites (sites which consist of at least two types of nucleotides, or amino acids, with a minimum frequency of two) from given infiles and/or the concatenated supermatrix to separate output files. Furthermore, FcC-G can optionally generate additional gene partition output files for the concatenated supermatrix which can be directly used for Maximum Likelihood analyses using RAxML [56] or for Bayesian analyses with MrBayes [58].

Model selection of amino acid replacement using ProtTest
FcC-G offers the option to generate the best-fit protein model for each amino acid gene partition in RAxML partition formatted supermatrices using the external software, ProtTest [54]. The ProtTest option can only be

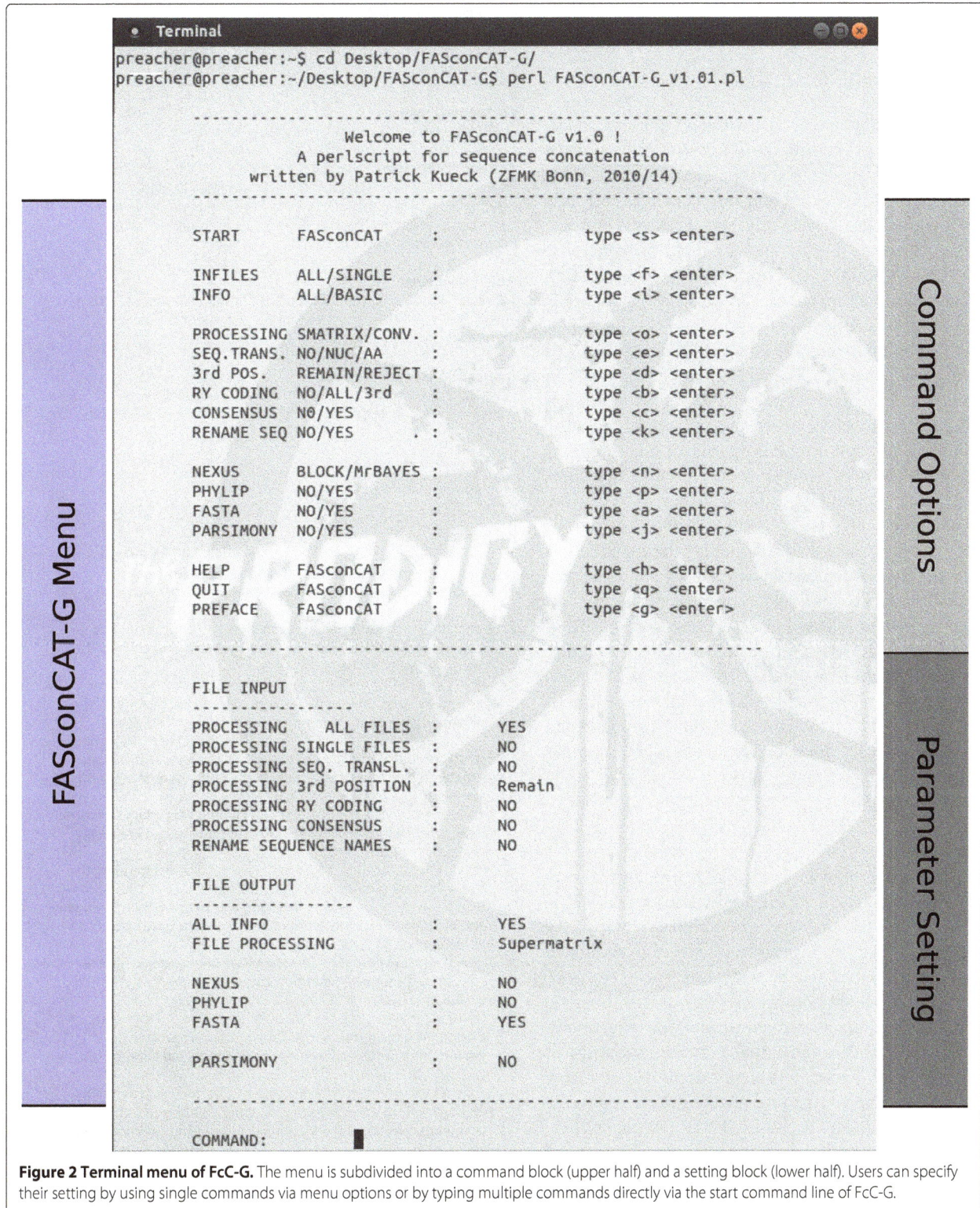

Figure 2 Terminal menu of FcC-G. The menu is subdivided into a command block (upper half) and a setting block (lower half). Users can specify their setting by using single commands via menu options or by typing multiple commands directly via the start command line of FcC-G.

executed with amino acid infiles or translated infiles and when sequence concatenation has been chosen together with the partition option ("-l"), but not for supermatrices in NEXUS format. FcC-G implements the default parameters for ProtTest version 3.3 and uses the ProtTest BIC criterium to select the best-fit model.

Conclusions

With FcC-G, we introduce an advanced editor to facilitate subsequent processing steps for multiple sequence alignments in phylogenetic and population genetic studies. Like its predecessor version, FASconCAT, FcC-G is easy to use, very fast (even with large data sets) and not limited in number of input files or input sequences. It facilitates data handling, it is time saving in generating and processing data matrices, and provides useful additional information about input sequences. FcC-G is implemented in Perl and runs on Windows PCs, Mac OS and Linux operating systems. FcC-G is command-line driven and well suited for incorporation into automatic process pipelines. Alternatively, the software tool can be operated and executed through interactive terminal menu options (Figure 2). Most processing options of FcC-G are combinable (Figure 1) and help is provided for every option. The executable source code (Additional file 1) as well as example test files and a detailed documentation of FcC-G are freely available at https://www.zfmk.de/en/research/research-centres-and-groups/fasconcat-g. The program is open-source and released under the terms of the GNU General Public License (GPL) 3.0. Detailed information and instructions are provided in the manual of FcC-G (Additional file 2). The manual also includes some practical examples, which demonstrate FcC-G is a suitable and user-friendly tool for complex phylogenetic and population genetic data processing.

Methods

FcC-G is implemented in Perl (Perl 5.0 or higher) and platform independent. Like the predecessor version FASconCAT, FcC-G can be used via command line or by terminal menu options. The terminal menu is subdivided into two parts, separated by a dashed line (Figure 2). The upper component constitutes of a list of all possible options and their associated commands for adjustment. The lower part shows the actual parameter settings of FcC-G. All default parameters can be optionally changed, and the new setting configuration will be displayed in the lower part of the menu. FcC-G is distributed under GNU GPL 3.0 and freely available from https://www.zfmk.de/en/research/research-centres-and-groups/fasconcat-g or upon request from the corresponding authors.

Additional files

Additional file 1: Executable Perl script of FASconCAT-G. FASconCAT-G is distributed under GNU GPL 3.0 and freely available. Windows users have to install a PERL interpreter on their operating system. Mac and Linux users can directly start FASconCAT-G via terminal options.

Additional file 2: FASconCAT-G manual. Detailed information and instructions and practical examples of FASconCAT-G. The pdf document can be opened with pdf readers like AdobeAcrobatReader, Xpdf, or DocumentViewer.

Competing interests
The authors declare that they have no competing interests.

Authors' contributions
PK and GCL designed FASconCAT-G. PK programmed FASconCAT-G. GCL performed the beta testing of FASconCAT-G. PK and GCL discussed and wrote the paper as well as the FASconCAT-G manual. Both authors read and approved the final manuscript.

Acknowledgements
We would like to thank all members of the Zoological Research Museum A. Koenig (Bonn, Germany) and the Center for Ocean Health (Santa Cruz, California) for inspiring discussions.

Author details
[1] Zoologisches Forschungsmuseum A. Koenig, Adenauerallee 160-163, 53113 Bonn, Germany. [2] Center for Ocean Health, 100 Shaffer Road, 95060 Santa Cruz, CA, USA.

References
1. Letsch HO, Greve C, Kück P, Fleck G, Stocsits RR, Misof B: **Simultaneous alignment and folding of 28S rRNA sequences uncovers phylogenetic signal in structure variation.** *Mol Phylogenet Evol* 2009, **53:**758–771.
2. Stocsits RR, Letsch HO, Hertel J, Misof B, Stadler PF: **Accurate and efficient reconstruction of deep phylogenies from structured RNAs.** *Nucelid Acids Res* 2009, **37:**6184–6193.
3. Keller A, Förster F, Müller T, Dandekar T, Schultz J, Wolf M: **Including RNA secondary structures improves accuracy and robustness in reconstruction of phylogenetic trees.** *Biol Direct* 2010, **5:**4.
4. Letsch HO, Kück P, Stocsits RR, Misof B: **The impact of rRNA secondary structure consideration in alignment and tree reconstruction: simulated data and a case study on the phylogeny of hexapods.** *Mol Biol Evol* 2010, **27**(11):2507–2521.
5. Murienne J, Edgecombe G, Giribet G: **Including secondary structure, fossils and molecular dating in the centipede tree of life.** *Mol Phylogenet Evol* 2010, **57:**301–313.
6. Wan Y, Kertesz M, Spitale RC, Segal E, Chang HY: **Understanding the transcriptome through RNA structure.** *Nat Rev Genet* 2011, **12:**641–655.
7. Dinapoli A, Klussmann-Kolb A: **The long way to diversity – Phylogeny and evolution of the Heterobranchia (Mollusca:Gastropoda).** *Mol Phyogenet Evol* 2010, **55:**60–76.
8. Hoppenrath M, Leander BS: **Dinoflagellate phylogeny as inferred from heat shock protein 90 and ribosomal gene sequences.** *PLoS ONE* 2010, **5**(10):e13220.
9. Goto R, Okamoto T, Ishikawa H, Hamamura Y, Kato M: **Molecular phylogeny of echiuran worms (phylum: annelida) reveals evolutionary pattern of feeding mode and sexual dimorphism.** *Plos ONE* 2013, **8**(2):e56809.
10. Lopez-Osorio F, Pickett KM, Carpenter JM, Ballif BA, Agnarsson I: **Phylogenetic relationships of yellowjackets inferred from nine loci (Hymenoptera: Vespidae, Vespinae, *Vespula and Dolichovespula*).** *Mol Phylogenet Evol* 2014, **73:**190–201.
11. Phillipe H, Lartillot N, Brinkmann H: **Multigene Analyses of Bilaterian Animals Corroborate the Monophyly of Ecdysozoa, Lophotrochozoa, and Protostomia.** *Mol Biol Evol* 2005, **22**(5):1246–1253.

12. Savard J, Tautz D, Richards S, Weinstock GM, Gibbs RA, Werren JH, Tettelin H, Lercher MJ: **Phylogenomic analysis reveals bees and wasps (Hymenoptera) at the base of the radiation of Holometabolous insects.** *Genome Res* 2006, **16:**1334–1338.

13. Dunn CW, Hejnol A, Matus DQ, Pang K, Browne WE, Smith SA, Seaver E, Rouse GW, Obst M, Edgecombe GD, Sorensen MV, Haddock SH, Schmidt-Rhaesa A, Okusu A, Kristensen RM, Wheeler W, Martindale MQ, Giribet G: **Broad phylogenomic sampling improves resolution of the animal tree of life.** *Nature* 2008, **452:**745–749.

14. Hejnol A, Obst M, Stamatakis A, Ott M, Rouse GW, Edgecombe GD, Martinez P, Baguna J, Bailly X, Jondellus U, Wiens M, Müller WEG, Seaver E, Wheeler WC, Martindale MQ, Giribet G, Dunn CW: **Assessing the root of bilaterian animals with scalable phylogenomic methods.** *Proc R Soc B* 2009, **276**(1677):4261–4270.

15. Simon S, Strauss S, von Haeseler A, Hadrys H: **A phylogenomic approach to resolve the basal pterygote divergence.** *Mol Biol Evol* 2009, **26**(12):2719–2730.

16. Meusemann K, von Reumont BM, Simon S, Roeding F, Kück P, Strauss S, Ebersberger I, Walzl M, Pass G, Breuers S, Achter V, von Haeseler A, Burmester T, Hadrys H, Wägele JW, Misof B: **A phylogenomic approach to resolve the arthropod tree of life.** *Mol Biol Evol* 2010, **27**(11):2451–2464.

17. Pick KS, Phillipe H, Schreiber F, Erpenbeck D, Jackson DJ, Wrede P, Wiens M, Alie A, Morgenstern B, Manuel M, Wörheide G: **Improved phylogenomic taxon sampling noticeably affects nonbilaterian relationships.** *Mol Biol Evol* 2010, **27**(9):1983–1987.

18. Rota-Stabelli O, Campbell L, Brinkmann H, Edgecombe GD, Longhorn SJ, Peterson KJ, Pisani D, Philippe H, Telford MJ: **A congruent solution to arthropod phylogeny: phylogenomics, microRNAs and morphology support monophyletic Mandibulata.** *Proc R Soc B* 2011, **278:**298–306.

19. Kocot KM, Cannon JT, Todt C, Citarella MR, Kohn AB, Meyer A, Santos SR, Schander C, Moroz LL, Lieb B, Halanych KM: **Phylogenomics reveals deep molluscan relationships.** *Nature* 2011, **477:**452–456.

20. Smith SA, Wilson NG, Goetz FE, Feehery C, Andrade SCS, Rouse GW, Giribet G, Dunn CW: **Resolving the evolutionary relationships of molluscs with phylogenomic tools.** *Nature* 2011, **480:**364–367.

21. Struck TH, Paul C, Hill N, Hartmann S, Hösel C, Kube M, Lieb B, Meyer A, Tiedemann R, Purschke G, Bleidorn C: **Platyzoan paraphyly based on phylogenomic data supports a noncoelomate ancestry of spiralia.** *Nature* 2011, **471:**95–98.

22. Rubin BER, Gee RH, Moreau CS: **Inferring phylogenies from RAD sequence data.** *PLoS ONE* 2012, **7**(4):e33394.

23. Wagner CE, Keller I, Wittwer S, Selz OM, Mwalko S, Greuter L, Sivasundar A, Seehausen O: **Genome-wide RAD sequence data provide unprecedented resolution of species boundaries and relationships in the Lake Victoria cichlid adaptive radiation.** *Mol Ecol* 2013, **22**(3):787–798.

24. Wheat CW, Wahlberg N: **Phylogenomic insights into the cambrian explosion, the colonization of land and the evolution of flight in arthropoda.** *Syst Biol* 2013, **62:**93–109.

25. Woese CR, Achenbach L, Rouvier P, Mandelco L: **Archaeal phylogeny: reexamination of the phylogenetic position of Archaeoglobus Fulgidus in light of certain composition-induced artifact.** *Syst Appl Microbiol* 1991, **14:**364–371.

26. Phillips MJ, Delsuc F, Penny D: **Genome-scale phylogeny and the detection of systematic biases.** *Mol Biol Evol* 2004, **21**(7):1455–1458.

27. Harshman J, Braun EL, Braun MJ, Huddleston CJ, Bowie RCK, Chojnowski JL, Hackett SL, Han KL, Kimball RT, Marks BD, Miglia KJ, Moore WS, Reddy S, Sheldon FH, Steadman DW, Steppan SJ, Witt CC, Yuri T: **Phylogenomic evidence for multiple losses of flight in ratite birds.** *Proc Natl Acad Sci U S A* 2008, **105**(36):13462–13467.

28. White NE, Phillips MJ, Gilbert TP, Alfaro-Nunez A, Willerslev E, Mawson PR, Spencer PBS, Bunce M: **The evolutionary history of cockatoos (Aves: Psittaciformes: Cacatuidae).** *Mol Phylogenet Evol* 2011, **59**(3):615–622.

29. Chen JN, Lopez A, Lavoue S, Miya M, Chen WJ: **Phylogeny of the Elopomorpha (Teleostei): Evidence from six nuclear and mitochondrial markers.** *Mol Phylogenet Evol* 2014, **70:**152–161.

30. Burger TD, Shao R, Beati L, Miller H, Barker SC: **Phylogenetic analysis of ticks (Acari: Ixodida) using mitochondrial genomes and nuclear rRNA genes indicates that the genus Amblyomma is polyphyletic.** *Mol Phylogenet Evol* 2012, **64:**45–55.

31. Liu GH, Wu CH, Song HQ, Wei SJ, Xu MJ, Lin RQ, Zhao GH, Huang SY, Zhu XQ: **Insect phylogenomics: results, problems and the impact of matrix composition.** *Mol Phylogenet Evol* 2012, **492:**110–116.

32. Lin RQ, Qiu LL, Liu GH, Wu XY, Weng YB, Xie WQ, Hou J, Pan H, Yuan ZG, Zou FC, Hu M, Zhu XQ: **Characterization of the complete mitochondrial genomes of five Eimeria species from domestic chickens.** *Mol Phylogenet Evol* 2012, **480:**28–33.

33. dos Reis M, Inoue J, Hasegawa M, Asher RJ, Donoghue PCJ, Yang Z: **Phylogenomic datasets provide both precision and accuracy in estimating the timescale of placental mammal phylogeny.** *Proc R Soc B* 2012, **279:**3491–3500.

34. Krüger M, Krüger C, Walker C, Stockinger H, Schüssler A: **Phylogenetic reference data for systematics and phylotaxonomy of arbuscular mycorrhizal fungi from phylum to species level.** *New Phytologist* 2012, **139:**970–984.

35. Waheed Y, Saeed U, Anjum S, Afzal MS, Ashraf M: **Development of global consensus sequence and analysis of highly conserved domains of the HCV NS5B protein.** *Hepat Mon* 2012, **12**(9):e6142.

36. Cotton M, Lam TT, Watson SJ, Palser AL, Petrova V, Grant P, Pybus OG, Rambaut A, Guan Y, Pillay D, Kellam P, Nastouli E: **Full-Genome Deep Sequencing and Phylogenetic Analysis of Novel Human Betacoronavirus.** *Emerg Infect Dis* 2013, **19**(5):736–742.

37. Blaxter M, Mann J, Chapman F, Thomas F, Whitton RF, Abebe E: **Defining operational taxonomic units using DNA barcode data.** *Phil Trans R Soc B* 2005, **360**(1462):1935–1943.

38. Rubin CJ, Zody MC, Eriksson J, Meadows RS, Sherwood E, Webster MT, Jiang L, Ingman M, Sharpe T, Ka S, Hallböök F, Besnier F, Carlborg O, Bed'hom B, Tixier-Boichard M, Jensen P, Siegel P, Lindblad-Toh K, Andersson L: **Whole-genome resequencing reveals loci under selection during chicken domestication.** *Nature* 2010, **464:**587–591.

39. Vaidya G, Meier R: **SequenceMatrix: concatenation software for the fast assembly of multi-gene datasets with character set and codon information.** *Cladistics* 2011, **27:**171–180.

40. Abascal F, Zardoya R, Telford MJ: **TranslatorX: multiple alignment of nucleotide sequences guided by amino acid translations.** *Nucl Acids Res* 2010, **38**(Web Server issue):W7—13. doi:10.1093/nar/gkq291.

41. Pina-Martins F, Paulo OS: **CONCATENATOR: sequence data matrices handling made easy.** *Mol Ecol Resour* 2008, **8:**1254–1255.

42. Salinas NR, Little DP: **2MATRIX: A utility for indel coding and phylogenetic matrix concatenation.** *Appl Plant Sci* 2014, **2:**1300083.

43. Roure B, Rodriguez-Ezpeleta N, Phillipe H: **SCaFoS: a tool for selection, concatenation and fusion of sequences for phylogenomics.** *BMC Evol Biol* 2007, **7:**S2.

44. Kück P, Meusemann K: **FASconCAT: Convenient handling of data matrices.** *Mol Phylogenet Evol* 2010, **56:**1115–1118.

45. Kück P, Hita-Garcia F, Misof B, Meusemann K: **Improved phylogenetic analyses corroborate a plausible position of Martialis Heureka in the ant tree of life.** *PLoS ONE* 2011, **6**(6):e21031.

46. Biswal KD, Debnath M, Kumar S, Tandon P: **Phylogenetic reconstruction in the Order Nymphaeales: ITS2 secondary structure analysis and in silico testing of maturase k *(matK)* as a potential marker for DNA bar coding.** *BMC Bioinformatics* 2012, **13:**16.

47. Boumans L, Baumann RW: **Amphinemura palmeni is a valid Holarctic stonefly species (Plecoptera: Nemouridae).** *Zootaxa* 2012, **3537:**59–75.

48. Kohn AB, Citarella MR, Kocot KM, Bobkova YV, Halanych KM, Moroz LL: **Rapid evolution of the compact and unusual mitochondrial genome in the ctenophore, Pleurobrachia bachei.** *Mol Phylogenet Evol* 2012, **63:**203–207.

49. McNulty SN, Mullin AS, Vaughan JA, Tkach VV, Weil GJ, Fischer PU: **Comparing the mitochondrial genomes of Wolbachia-dependent and independent filarial nematode species.** *BMC Genomics* 2012, **13:**145.

50. Young ND, Jex AR, Li B, Liu S, Yang L, Xiong Z, Li Y, Cantacessi C, Hall RS, Xu X, Chen F, Wu X, Zerlotini A, Oliveira G, Hofmann A, Zhang G, Fang X, Kang Y, Campbell BE, Loukas A, Ranganathan S, Rollinson D, Rinaldi G, Brindley PJ, Yang H, Wang J, Wang J, Gasser RB: **Whole-genome sequence of Schistosoma haematobium.** *Nat Genet* 2012, **44:**221–225.

51. Golombek A, Tobergte S, Nesnidal P, Purschke G, Struck T: **Mitochondrial genomes to the rescue – Diurodrilidae in the mystozomid trap.** *Mol Phylogenet Evol* 2013, **68**(2):312–326.

52. Larriba E, Jaime MDLA, Carbonell-Caballero J, Conesa A, Dopazo J, Nislow C, Martin-Nieto J, Lopez-Llorca LV: **Sequencing and functional analysis**

of the genome of a nematode egg-parasitic fungus, Pochonia
chlamydosporia. *Fungal Genet Biol* 2014, **65**:69–80.

53. Scheel BM, Hausdorf B: **Dynamic evolution of mitochondrial ribosomal proteins in Holozoa.** *Mol Phylogenet Evol* 2014, **76**:67–74.

54. Darriba D, Guillermo LT, Doallo R, Posada D: **ProtTest 3: fast selection of best-fit models of protein evolution.** *Bioinformatics* 2011, **27**:1164–1165.

55. Ronquist F, Huelsenbeck J: **MrBayes 3: Bayesian phylogenetic inference under mixed models.** *Bioinformatics* 2003, **19**(12):1572–1574.

56. Stamatakis A: **RAxML-VI-HPC: maximum likelihood-based phylogenetic analyses with thousands of taxa and mixed models.** *Bioinformatics* 2006, **22**(21):2688–2690.

57. Swofford D: *PAUP*: Phylogenetic Analysis Using Parsimony (*and Other Methods). Version 4.0.* Sunderland, MA: Sinauer Associates; 2003.

58. Huelsenbeck J, Ronquist F: **MrBayes: Bayesian inference of phylogenetic trees.** *Bioinformatics* 2001, **17**(8):754–755.

59. Guindon S, Dufayard JF, Lefort V, Anisimova M, Hordijk W, Gascuel O: **PhyML 3.0: New algorithms and methods to estimate maximum-likelihood phylogenies: assessing the performance of PhyML 3.0.** *Syst Biol* 2010, **59**(3):307–321.

11

Minding the gap: in-flight body awareness in birds

Ingo Schiffner[1*], Hong D Vo[1], Partha S Bhagavatula[1] and Mandyam V Srinivasan[1,2,3]

Abstract

Introduction: When birds fly in cluttered environments, they must tailor their flight to the gaps that they traverse. We trained budgerigars, *Melopsittacus undulatus,* to fly through a vertically oriented gap of variable width, to investigate their ability to perform evasive manoeuvres during passage.

Results: When the gap was wider than their wingspan, the birds passed through it without interrupting their flight. When traversing narrower gaps, however, the birds interrupted their normal flight by raising their wings or tucking them against the body, to prevent contact with the flanking panels. Our results suggest that the birds are capable of estimating the width of the gap in relation to their wingspan with high precision: a mere 6% reduction in gap width causes a complete transition from normal flight to interrupted flight. Furthermore, birds with shorter wingspans display this transition at narrower gap widths.

Conclusion: We conclude from our experiments that the birds are highly aware of their individual body size and use precise, anticipatory, visually based judgements to control their flight in complex environments.

Keywords: Birds, Flight, Vision, Body awareness, Obstacle avoidance

Introduction

When traversing cluttered environments at nearly cruising speeds, birds need to be constantly aware of the distances to oncoming obstacles and the spaces between them, in order to make split second decisions about whether a gap can be traversed, and to determine whether a change in the wing posture is necessary to facilitate an injury-free passage (e.g. [1]).

Do birds fly through passages that are narrower than their wingspan? If they indeed do so, what postural changes do they make to accommodate the passage? Are the wings held up, held down, held forward, or held behind, tucked close to the body? Furthermore, very little is known about a bird's ability to assess the width of a gap in relation to its own body size, and about how this assessment is made. In principle, there are a number of ways in which this could be accomplished. For example, the width of the gap could be judged from afar, by using vision to gauge the angular width of the gap and the viewing distance – from which the absolute width of the

gap can be estimated. Alternatively - as explained in the Discussion - this judgement could be made by measuring the rate at which the visual image of the gap expands when it subtends a prescribed visual angle. A third possibility is that the width of the gap is estimated just as it is being traversed, using vision to determine whether the wing tips are about to touch or clear the edges of the gap (For a general introduction to bird vision and its possible role in collision avoidance see [2,3]).

In the experiments presented here, seven budgerigars (*Melopsittacus undulatus*) were confronted with an aperture of variable width. We aimed to investigate their flight manoeuvres through the aperture and to enquire whether they display awareness of their body size while doing so. The aperture was a vertical slit, presented as a gap between two cloth panels (see Figure 1). A total of 560 flights were recorded, with 10 flights per bird in seven experimental and one control (unobstructed tunnel) condition.

Results

During the normal flapping flight mode – when the birds were not negotiating an aperture – the birds' wingbeat cycles proved to be very stable, with the duration of

---
* Correspondence: i.schiffner@uq.edu.au
[1]Queensland Brain Institute, University of Queensland, St Lucia QLD 4072, Australia
Full list of author information is available at the end of the article

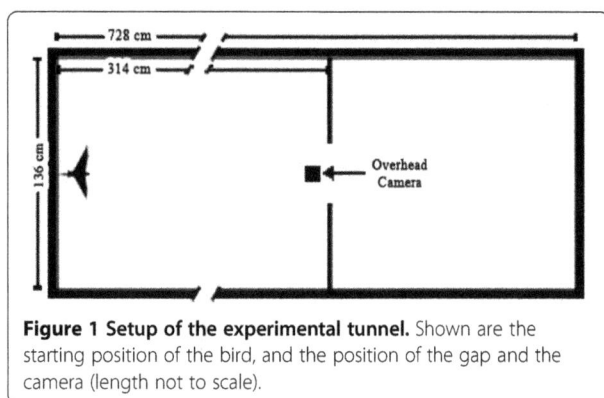

Figure 1 Setup of the experimental tunnel. Shown are the starting position of the bird, and the position of the gap and the camera (length not to scale).

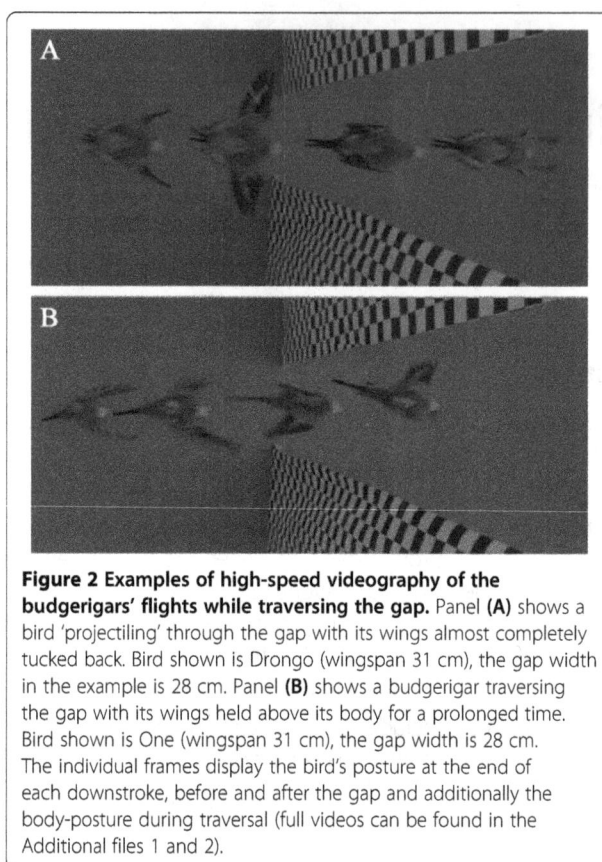

Figure 2 Examples of high-speed videography of the budgerigars' flights while traversing the gap. Panel **(A)** shows a bird 'projectiling' through the gap with its wings almost completely tucked back. Bird shown is Drongo (wingspan 31 cm), the gap width in the example is 28 cm. Panel **(B)** shows a budgerigar traversing the gap with its wings held above its body for a prolonged time. Bird shown is One (wingspan 31 cm), the gap width is 28 cm. The individual frames display the bird's posture at the end of each downstroke, before and after the gap and additionally the body-posture during traversal (full videos can be found in the Additional files 1 and 2).

the downstrokes being in the range of 38 ± 3 ms and the upstrokes in the range of 22 ± 2 ms, resulting in a mean wingbeat period of approximately 60 ms. Even though there was a slight variation in the duration of the downstroke and the upstroke, the ratio between the two remained almost constant at 0.59 ± 0.07. These figures were constant across individuals and across different experimental conditions.

As the width of the gap was reduced to approach the wingspan of each individual bird, the normal wingbeat cycle was interrupted during the actual passage through the gap. The duration of the upstroke was then longer than the duration of the downstroke, indicating that the birds actively held their wings in such position to avoid touching the panels during the passage. During traversal of the gap, the birds either held their wings in a position corresponding to the end of an upstroke, or tucked them in against the body (a behaviour very reminiscent of flap bounding, i.e. intermittent phases during which normal flapping flight is interrupted; see [4]). In either case, the birds closed their wings, 'projectiling' themselves through the gap, rather than actively flapping through it (see Figure 2 or the videos included as Additional files 1 and 2). The choice of the mode of traversal depended upon the duration of the passage: during longer traversals the wings were always tucked in completely. Only in one instance (out of a total of 490 narrow-gap traversals) did we observe a bird holding its wings pointing downwards, and once a bird holding one wing up and the other down. For simplicity, we shall refer to all of the 'projectiling' behaviours as 'wing closure'.

Mechanical contact with the cloth panel occurred in only 8% of the 490 gap traversals i.e. all trials excluding the flights in the unobstructed tunnel. No major collisions where observed in any of the flights. This result clearly indicated that budgerigars are capable of flying through passages that are narrower than their wingspan without injuring themselves.

We found that, as the gap was made narrower, the birds were more likely to interrupt their wingbeat cycle

and close their wings. The variation of the probability of wing closure [(No. of observed wing closures / 10) * 100] with gap width is shown in Figure 3A (note that the abscissa in this figure represents the *relative* gap width, i.e. the difference between the bird's wingspan and the width of the gap – a negative value means that the gap is narrower than the wingspan). Furthermore, the birds maintained wing closure for a longer duration as the gap was made narrower. This is illustrated in Figure 3B, which shows how the normalised duration of the upstroke varies with gap width at the moment of passage. More importantly, the birds are not simply adjusting the phase of their normal wingbeat cycle so as to ensure that wings are in the closed position when passing through the gap. They are definitely prolonging the period of wing closure during passage through the gap. This is evinced by the fact that the normalised duration of the upstroke is about 70% when passing through the narrowest gap, as compared to about 20% when passing through the widest gap (see Figure 3B).

Both the probability of wing closure (One Way ANOVA: $F_{(6,42)} = 8.539$; $p < 0.001$) and the normalised duration of the upstroke phase during passage vary strongly with gap width (One way ANOVA: $F_{(6,42)} = 5.002$ $p < 0.001$). Both measures increase significantly as the width of the gap is decreased from –1 to –3 cm (Wilcoxon signed rank test:

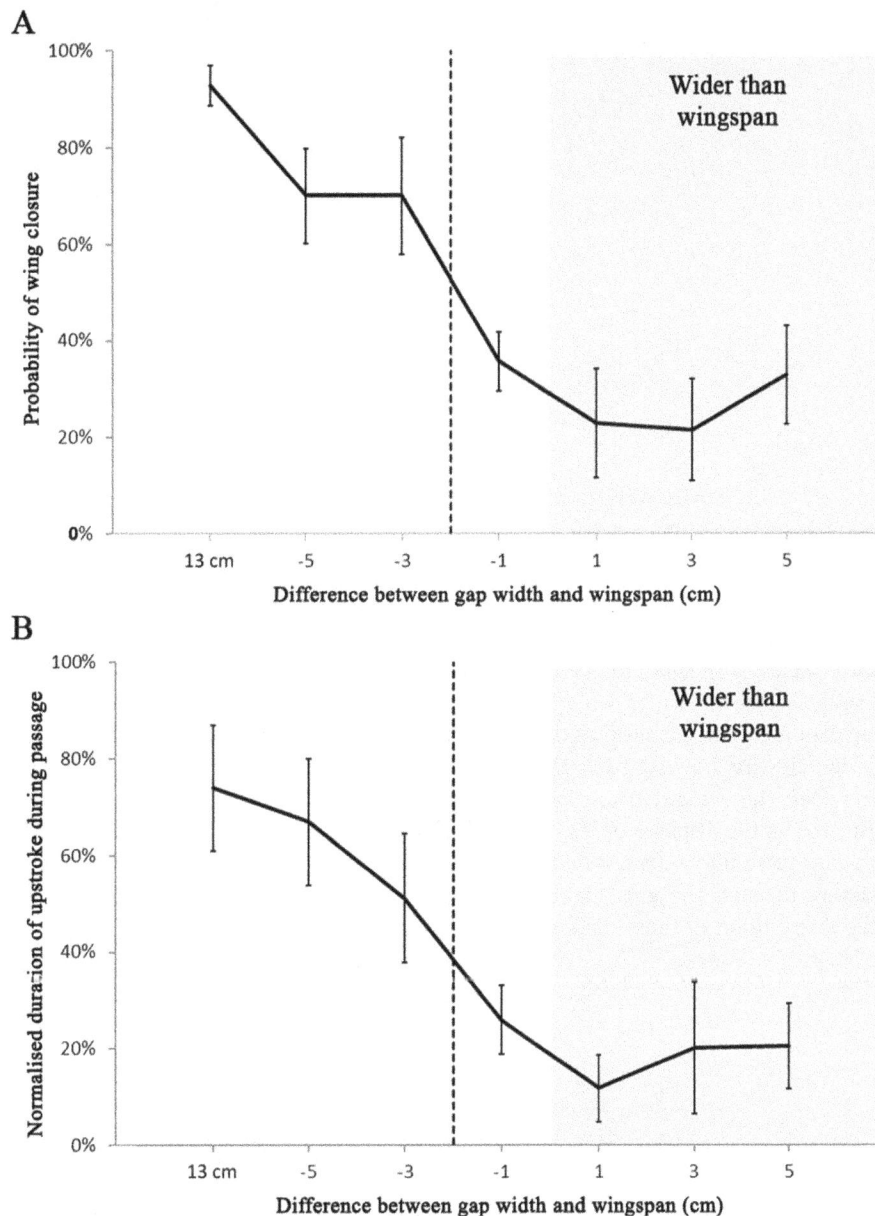

Figure 3 Mean probability of wing closure (A) and normalised mean duration of the up-stroke during passage (B) as a function of the difference between the gap width and the bird's wingspan, averaged over all 7 birds and including SEM. The percentage probability of wing closure for each bird was computed as 100* (No. of observed wing closures/10), from the 10 flights that were analysed for each bird, for each condition. The grey area represents the conditions in which the gap was wider than the bird's wingspan.

$p < 0.05$), indicating a transition from uninterrupted to interrupted flight when the width of the gap is reduced to a value that is 2 cm narrower than the bird's wingspan, as shown by the dashed lines in Figure 3A and B. Thus, the birds are aware of their own wingspan with a precision of +/− 2 cm. Inspection of the data with the birds divided into 3 groups according to their wingspans (29, 31 and 33 cm) as shown in Figure 4, suggests that the larger birds (33 cm wingspan) show a propensity to close their wings even when the gap is larger than their wingspan.

Furthermore, all three groups of birds exhibit a significant increase in the probability of wing closure when the gap becomes narrower than their respective wingspans (Figure 4).

In addition, we examined the probabilities of wing closure for the three groups of birds at each of the three critical gap widths (28, 30 and 32 cm), as shown in Figure 5. For any given gap width, larger birds are more likely to close their wings than smaller birds (Two way ANOVA: $F_{(2,12)} = 4.742$; $p = 0.0304$), and close their wings for longer times (Two way ANOVA: $F_{(2,12)} = $

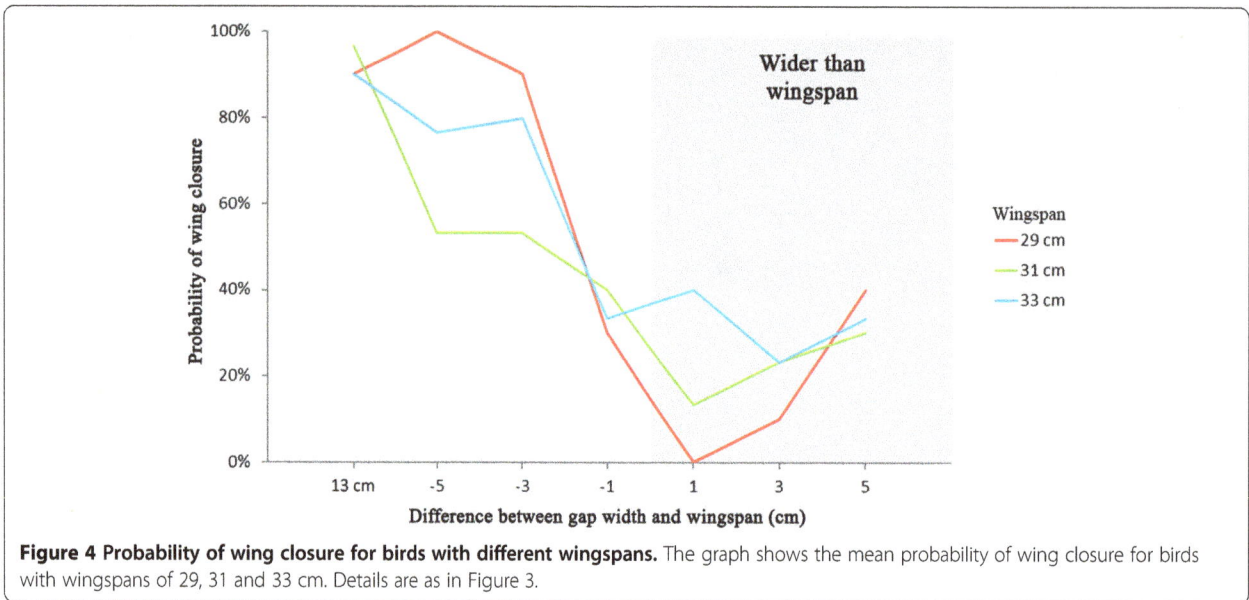

Figure 4 Probability of wing closure for birds with different wingspans. The graph shows the mean probability of wing closure for birds with wingspans of 29, 31 and 33 cm. Details are as in Figure 3.

8.722; p = 0.00458). Thus, it would appear that each bird possesses a body template that is specific to its own wingspan: it does not simply close its wings when the width of the gap is lower than a certain absolute value.

When a gap induces wing closure, how far back along the approach trajectory does the wing closure commence? This can be estimated as the distance of the bird from the gap when the wings were fully outstretched for the last time before passing through the gap. Figure 6 shows a histogram of the distribution of these distances,

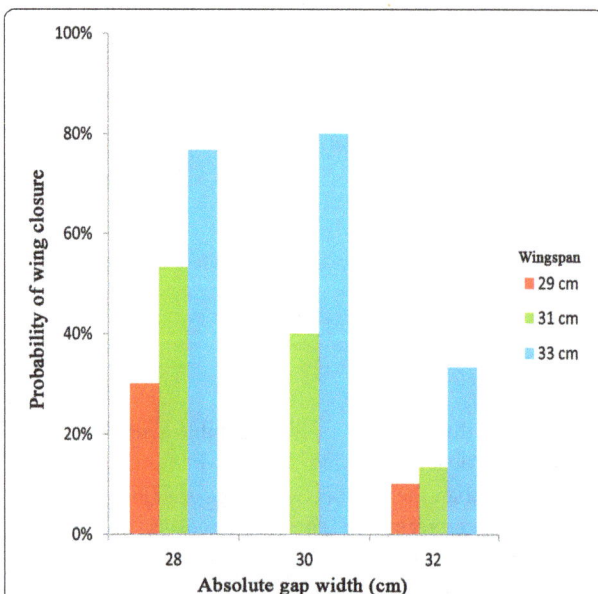

Figure 5 Probability of wing closure at the critical gap widths. The graph shows the mean probability of wing closure for birds with wingspans of 29, 31 and 33 cm at each of the three critical gap widths: 28, 30 and 32 cm.

obtained by pooling data from all flights in which the birds closed their wings – as indicated by a longer up-stroke duration—assuming a constant flight speed of 4 m/s. This histogram shows that, for these flights, the final wing extension occurred most frequently at a distance of about 220 mm from the gap. Furthermore, the cumulative histogram in Figure 6 shows that only 20% of the final wing extensions occur at distances that are closer than 120 mm from the gap; 80% of these extensions occur at greater distances.

For comparison, Figure 7 shows the histogram of the distances of final wing extensions for gaps that did not induce wing closure. Here the histogram of final wing-extension distances is approximately uniformly distributed over the entire period of an average, uninterrupted wing beat cycle. This wing-stroke period is approximately 60 ms, as mentioned above, and is indicated by the vertical arrow on the time scale. The fact that the timing of the final wing extension is distributed more or less uniformly over the duration of an entire wing beat cycle indicates that this extension occurs independently of the distance of the bird from the aperture, and that it is a part of the normal wing beat cycle of the bird during uninterrupted flight. This is in contrast with the data in Figure 6, which show a non-uniform distribution of wing extensions over the period of a wing beat cycle, indicating that the narrower gaps are clearly affecting the wing beat cycle during the approach to the gap.

Discussion

The observation that birds close their wings when negotiating narrow gaps is not, in itself, surprising. This phenomenon would be apparent to anyone who watches a bird weave its way through dense foliage, and it is

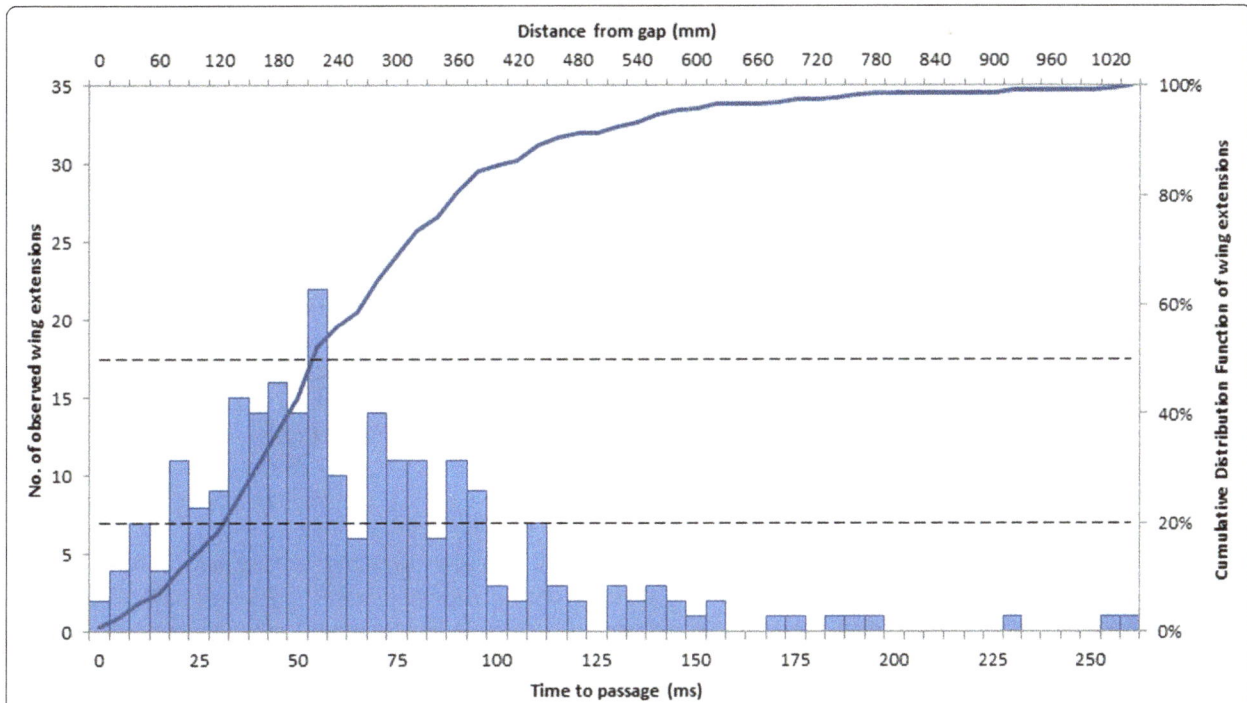

Figure 6 Histogram and cumulative distribution function of all flights where birds closed their wings. The blue bars depict a histogram of the frequency of occurrence of the last wing extension prior to passage through the gap, along a time scale (lower abscissa) and a distance scale (upper abscissa), assuming an average flight speed of 4 m/s, as explained in the text. The blue curve shows the cumulative distribution function. The dashed horizontal lines show cumulative probability levels of 20% and 50%.

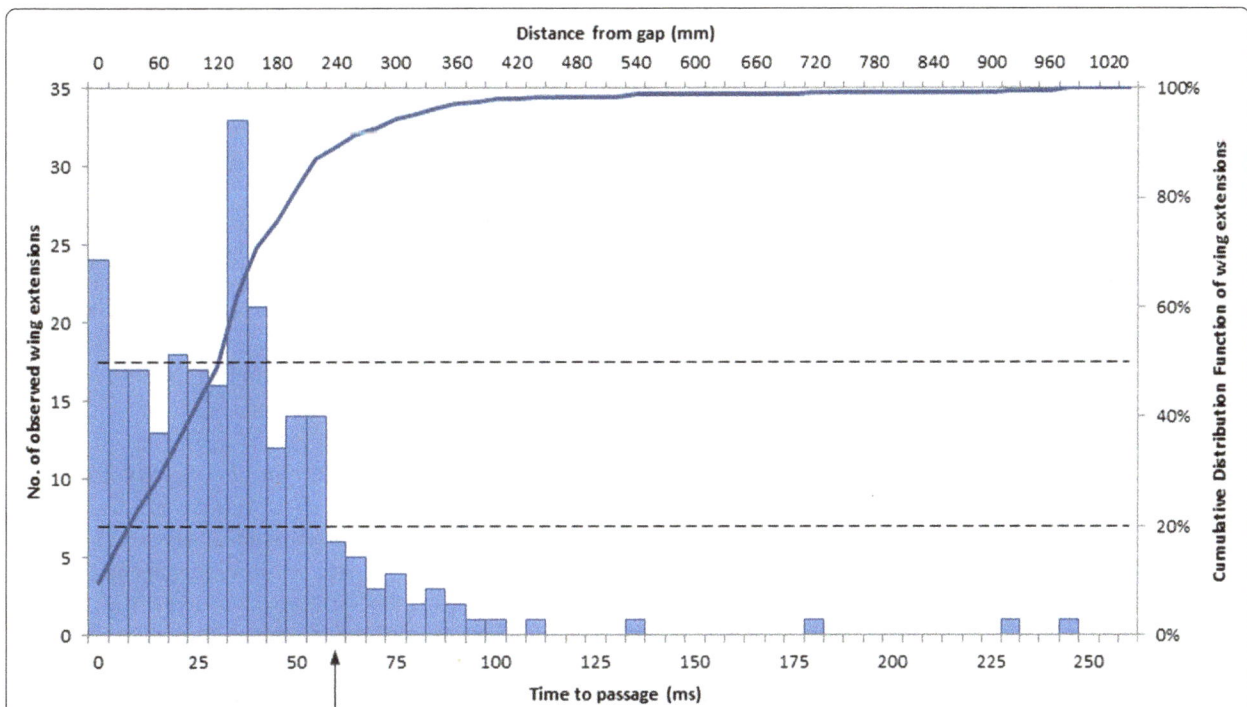

Figure 7 Histogram and cumulative distribution function of all flights where birds did not close their wings. The blue bars depict a histogram of the frequency of occurrence of the last wing extension prior to passage through the gap, along a time scale (lower abscissa) and a distance scale (upper abscissa), assuming an average flight speed of 4 m/s, as explained in the text. The blue curve shows the cumulative distribution function. The dashed horizontal lines show cumulative probability levels of 20% and 50%. The vertical arrow on the time scale represents the average duration of one uninterrupted wing beat cycle (60 ms).

obviously a sensible thing to do to avoid injury. What is surprising, however, is the precise tuning of this behaviour to the width of the gap. Our results reveal that wing closure occurs only when the width of the gap approaches a value that is within 6% of the bird's wingspan. This means that the wings are never folded back unless it is absolutely necessary. Why don't birds adopt a safer, more conservative approach by closing their wings at larger gap widths? The answer may be that wing closure causes the bird to lose (a) altitude and (b) manoeuvrability. We have observed that the birds in fact lose altitude when they are projectiling [5]. Furthermore, we have also observed that when the birds approach a narrow gap they often increase their altitude prior to entering the gap, presumably compensating for the loss of altitude that is incurred during the passage in advance [5]. Thus, it may be important to ensure that the wings are closed only when absolutely necessary; and that the closure period is as brief as possible. In order to achieve this, birds need to be accurately aware of the relationship between their wingspan and the width of the gap.

Our study cannot answer the question as to whether birds are 'consciously' aware of their wingspans. Nevertheless, we have demonstrated that birds are able to tailor their behaviour in an accurate way to facilitate safe passage through narrow gaps. This must mean that precise information about body size is incorporated into their behaviour.

Awareness of body size has been studied mainly in humans (e.g. when walking through a narrow or short doorway, humans rotate their shoulders sideways if the doorway is too narrow, or duck if it is not high enough) and they appear to be able to predict in advance whether such accommodative movements are necessary [6-9]. Humans are also able to estimate, with an accuracy of about a centimetre, whether a hand will pass through an aperture [10]. Toads decide to either pass through a gap or to detour around it depending upon whether the gap is wider or narrower than 3 cm, which is approximately the width of the head [11]. To the best of our knowledge, body size awareness has not been investigated in any other animal. Most studies in this general area have focused on the sensory aspects of obstacle detection, but not on how this information is linked to body size to facilitate collision-free navigation.

Our experiments reveal that birds are capable of estimating their own body size (wingspan) with a precision of +/– 2 cm. Furthermore, birds are individually aware of their body size, with larger birds commencing wing closure at larger gap widths. It has been observed that humans perform accommodative shoulder rotations when preparing to walk through doorways that are narrower than 130% of their shoulder width, which allows for a considerable margin of error [6]. Our birds, on the

other hand, begin to close their wings only when the gap is 2 cm (or ca. 6%) narrower than the wingspan. The reasons for this less cautious behaviour in birds may be (i) a lower chance of serious injury or blockage of passage, as the wings are compliant structures; and (ii) with a free-running wingbeat cycle, the chances that the wings are fully extended exactly at the time of passage through the gap are rather low – the time-averaged distance between the wingtips would be approximately half the maximum wingspan, although the exact figure would depend upon the details of the wing kinematics.

In our experiments, the only information available to the birds to estimate the width of the gap are the visual cues carried by the checkerboard patterns on the panels that flank the gap. Boundary airflow effects near the edges of the gap are unlikely to be the triggers of wing closure, as the closure usually occurs well before the gap is traversed. Furthermore, collisions with the flanking panels occur substantially more frequently when they are devoid of visual texture (white). For flights through a 17 cm gap, the collision rate was 0% when both panels carried the checkerboard, and 23% when both panels were white (data from 4 birds and a total of 70 flights; results taken from [5]).

What could be the nature of these visual cues? For humans walking through doorways, it has been suggested that intrinsic knowledge of the height of the eye above the ground, coupled with measurement of the visual angle subtended by the gap at any particular distance from the doorway, and the assumption of walking on a horizontal plane, should provide enough trigonometrical information to derive the distance to the doorway, and hence the width of the gap – see for example, [8]. However, it is unlikely that birds use the same strategy because their flights are of variable height and the ground cannot always be approximated by a horizontal plane. Alternatively, birds may use stereo information to determine the distance to the gap, and to combine this with information on the angular subtense of the gap to estimate the physical width of the gap. While this is theoretically possible, good stereo vision is unlikely to be present in budgerigars, given the low acuity and small region of binocular overlap in the frontal visual fields of most non-predatory birds [2].

Another possibility to be considered is that the birds are directly comparing the size of their wingspans with the width of the gap, just prior to traversing the gap. However, this is very unlikely because, during the passage through the gap, each wingtip and the corresponding gap edge would be in the lateral field of view, and would therefore be viewed by one eye only. This would preclude the acquisition of stereo cues. With monocular vision there would be no distance cues, consequently it would be impossible to gauge whether a wingtip extends beyond the edge of the gap, even when these two features are closely juxtaposed in the retinal image. Furthermore, as we have mentioned above, under the conditions in which the birds close their wings whilst flying through the gap, only 20% of the final wing extensions occur at distances that are closer than 120 mm from the gap; and 80% of these wing extensions occur at distances greater than 120 mm from the gap. These figures make it very improbable that the relationship between wingspan and gap width is gauged by a direct visual comparison, because in most of the instances, the bird is too far away from the gap to be able to make such a comparison.

It is possible that the birds are able to extract information about the width of the gap from the rate at which the image of the gap widens as the gap is approached. It can be shown that, if the approach speed (V) of the bird is known or estimable, the width w of the gap can be inferred from the relationship

$$\frac{d\theta}{dt} = \left(\frac{4V}{w}\right)\sin^2\left(\frac{\theta}{2}\right) \tag{1}$$

where θ is the angular width of the gap on the retinal image, and $\frac{d\theta}{dt}$ is the observed rate of change of this angular width.

Figure 8, based on Equation (1) shows the way in which $\frac{d\theta}{dt}$ is expected to vary with θ, for a number of

different gap widths (w). In addition, the figure shows that if the speed of approach is known (it is assumed to be 4 m/s in this example), the width of the gap can be estimated by monitoring the profile of the rate of expansion of the image of the gap as it is approached, or, alternatively, by sensing the rate of expansion of the image when it has a particular angular width. For example, when the angular width of the gap is 90 degrees, as shown by the dashed line, an expansion rate of 1240 degrees/second would indicate that the gap has a physical width of 36 cm; an expansion rate of 1500 degrees/second would indicate a gap width of 30 cm; and so on. If a bird with a wingspan of 34 cm experiences an image expansion of 1600 degrees/second, it would have to close its wings because the indicated gap width is 28 cm.

In principle, the relationships shown in Figure 8 can be used to determine if closure is necessary and when to close the wings. Let us assume that each bird carries within its nervous system a representation of the characteristic function that corresponds to a gap of width equal to its wingspan. If the bird approaches a gap that is narrower than its wingspan, the rate of change of gap width will be consistently higher than that carried in the internal representation, at each visual angle. The "confidence" of the hypothesis (gap < wingspan) can therefore be built up by integrating the differences that are sensed

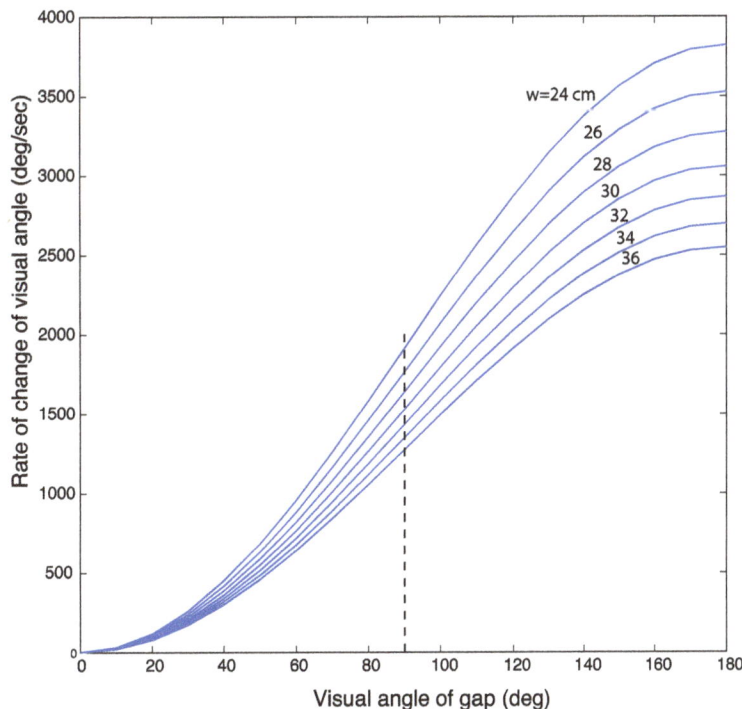

Figure 8 Determining whether, and, if so, when, to close the wings. The plot shows the expected rate of expansion of the image of a gap as it is approached, as a function of the instantaneous visual angle subtended by the gap, with the physical width of the gap as a parameter. The vertical dashed line represents an angular width of 90 deg. These curves are plotted assuming an approach speed of 4 m/s, based on analysis of video sequences (Vo, Schiffner and Srinivasan, unpublished data).

at successive visual angles as the gap is approached. A decision to close the wings can be made when this integrated difference exceeds an internally set threshold. With such a scheme, wing closure will occur at a relatively small visual angle (i.e. at a relatively large distance from the gap) when the gap is much narrower than the bird's wingspan, at a larger visual angle (i.e. at a relatively small distance from the gap) when the gap is only marginally narrower than the wingspan, and never when the gap is wider than the wingspan.

This strategy for estimating gap width and triggering wing closure would require information about the speed of approach. It is not yet known whether birds are able to estimate the speed of their flight. In principle, flight speed can be estimated by sensing the air speed, or calibrating it in terms of the parameters of the wing stroke (such as the amplitude, frequency, plane and wing articulation). One way to test the visual expansion strategy would be to examine whether wing closure occurs at larger distances for narrower gaps, as predicted above. Another way would be to film the birds' behaviour when the width of the gap is varied dynamically during the approach.

Conclusion

In summary, our findings indicate that budgerigars are very precise at estimating the widths of passages and assessing their navigability. Birds are individually aware of their own wingspan, and are capable of combining this "body image" with visual information from the environment to orchestrate their flight through narrow passages.

Materials and methods
Ethics statement

All experiments were carried out in accordance with the Australian Law on the protection and welfare of laboratory animals and the approval of the Animal Experimentation Ethics Committees of the University of Queensland, Brisbane, Australia (QBI/303/12/ARC).

Subjects

The subjects were seven adult male wild type budgerigars (*Melopsittacus undulatus*) between three and five years old. They were purchased from a pet shop at an age of approximately one month and were housed in an outdoor aviary. The aviary was 4.0 m in length, 2.0 m in width and 2.2 m in height, and had large mesh-screened windows, which provided natural light and a diurnal light cycle. Perches were provided in areas that presented sunshine, as well as in areas that offered protection from inclement weather. The experiments were conducted at The University of Queensland's Pinjarra Hills field station. On the days in which experiments were conducted (usually two to three days per week),

the birds were kept in groups of up to four in smaller cages of 47 cm length, 34.5 cm width and 82 cm height for a period not exceeding eight hours, to permit easy access to individual birds for experimentation. At the end of a day's experimentation the birds were moved back to the aviary.

Training

The birds were trained to take off at one end of a tunnel from a perch, which was either held by an experimenter or fixed on the wall, and to fly to the other end of the tunnel. To provide proper motivation for the birds, a cage with other familiar birds was placed at the other end of the tunnel. Further into the experiment this motivational stimulus proved to be unnecessary, but the procedure was continued in order to preserve uniformity across all of the experiments.

Apparatus and implementation

The experiments took place indoors in a tunnel that was 7.28 m long, 1.36 m wide and 2.44 m high, with white walls and ceiling, and a grey floor (see Figure 1). The end walls were covered with black cloth, to standardise visual cues. In the middle of the tunnel the birds were presented with an aperture of varying width, formed by two sheets of cloth suspended from the ceiling. The sheets carried a checkerboard pattern (check size 4 cm) making the aperture and the flanking panels clearly visible to the birds.

The width of the gap was varied in a pseudo random fashion, in separate flight trials. This width ranged from 24 to 38 cm, and was chosen to bracket the birds' wingspans, which were 29(1 bird), 31(3 birds) and 33(3 birds) cm (measurements were rounded to the nearest cm). Consequently, for all birds, the absolute differences between gap width and wingspan varied at least over the range −5 to +5 cm. In one additional trial, the gap had a width of 13 cm, which was narrow enough that all of the birds would be forced to move both wings out of the way while traversing the gap.

Recording

The birds' flights were filmed using a high-speed camera (DRS data & imaging systems, Inc., Oakland, NJ). Camera operation and video acquisition were controlled by special-purpose software (MiDAS 2.0 (Xcitex, Inc., Cambridge, MA)). The camera was mounted on the ceiling, 50 cm in front of the gap, with its optical axis pointing vertically downwards (Figure 1). The camera had a visual field of 110 degrees along the tunnel axis and 93 degrees in the transverse direction, which enabled flight trajectories to be filmed from about 2.0 m ahead of the gap to about 1.0 m beyond it, the exact figures depending up the height of the bird's flight. All flights were recorded at

200 frames per second, which provided sufficient temporal resolution to investigate the wingbeat cycle before, during and after passage through the gap. From these recordings we determined the times when each upstroke and each downstroke had been completed.

Statistics

In order to check for significant differences in the birds' behaviour across the different gap widths, we used both one and two way ANOVA with repeated measurements with a factorial design to test for differences in probability of wing-closure and normalized duration of the upstroke. The independent factors were the size of the gap (one way ANOVA) and the wingspan of the bird (additionally for the two way ANOVA). Each bird was treated as a repeated measure within each wingspan-category. When the ANOVA indicated significance, we used the Tukey HSD test –correcting for multiple comparisons - to verify the findings of the ANOVA and later the Wilcoxon signed Rank test for paired data to determine the critical gap width at which the birds started to fold their wings. For each ANOVA we ensured beforehand that neither normality, homogeneity of variances, nor sphericity had been violated using, Shapiro-Wilks test for normality, Levene's Test for homogeneity of variance and Mauchly's test for sphericity respectively.

Additional files

Additional file 1: Examples of high-speed videography of the budgerigars' flights while traversing the gap. Video shows a bird 'projectiling' through the gap with its wings almost completely tucked back. Bird shown is Drongo (wingspan 31 cm), the gap size in the example is 28 cm.

Additional file 2: Examples of high-speed videography of the budgerigars' flights while traversing the gap. Video shows a budgerigar traversing the gap with its wings held above its body for a prolonged time. Bird shown is One (wingspan 31 cm), the gap size is 28 cm.

Competing interests
The authors declare that they have no competing interests.

Authors' contributions
IS participated in the design of the study, carried out the experiments, performed the statistical analysis, and drafted the manuscript. HDV helped to carry out the experiments and analysed the raw video footage. PSB participated in the design of the study, helped to carry out the experiments, and analysed the raw video footage. MVS participated in the design of the study, and helped to draft the manuscript. All authors read and approved the final manuscript.

Acknowledgements
This research was supported by ARC Discovery Grant DP0559306, ARC Centre of Excellence in Vision Science Grant CE0561903, HFSP Research Grant RGP0003/2013, and by a Queensland Smart State Premier's Fellowship. We thank Lena Heeman and Brianna Hey for their help in conducting the experiments.

Author details
[1]Queensland Brain Institute, University of Queensland, St Lucia QLD 4072, Australia. [2]School of Information Technology and Electrical Engineering, University of Queensland, St Lucia QLD 4072, Australia. [3]ARC Centre of Excellence in Vision Science, University of Queensland, St Lucia QLD 4072, Australia.

References
1. Peregrine Falcon and Gos Hawk in flight. In [http://www.youtube.com/watch?v=p-_RHRAzUHM]
2. Martin GR: Visual fields and their functions in birds. *J Ornithol* 2007, **148**:547–562.
3. Martin GR: Understanding bird collisions with man-made objects: a sensory ecology approach. *Ibis* 2011, **153**:239–254.
4. Tobalske BW: Morphology, velocity, and intermittent flight in birds. *Am Zool* 2001, **41**:177–187.
5. Bhagavatula P: *Visually guided flight in birds using the Budgerigar (Melopsittacus undulatus) as a model system. Ph. D. Thesis.* Australian National University; 2011.
6. Warren WH Jr, Whang S: Visual guidance of walking through apertures: Body-scaled information for affordances. *J Exp Psychol Human* 1987, **13**:371.
7. Wilmut K, Barnett AL: Locomotor behaviour of children while navigating through apertures. *Exp Brain Res* 2011, **210**:185–194.
8. Fath AJ, Fajen BR: Static and dynamic visual information about the size and passability of an aperture. *Perception* 2011, **40**:887.
9. Franchak JM, Celano EC, Adolph KE: Perception of passage through openings depends on the size of the body in motion. *Exp Brain Res* 2012, **223**:1–10.
10. Ishak S, Adolph KE, Lin GC: Perceiving affordances for fitting through apertures. *J Exp Psychol Human* 2008, **34**:1501.
11. Lock A, Collett T: The three-dimensional world of a toad. *Proc R Soc B* 1980, **206**:481–487.

12

Temperature tolerance of different larval stages of the spider crab *Hyas araneus* exposed to elevated seawater PCO_2

Melanie Schiffer[1]*, Lars Harms[2], Magnus Lucassen[1], Felix Christopher Mark[1], Hans-Otto Pörtner[1] and Daniela Storch[1]

Abstract

Introduction: Exposure to elevated seawater PCO_2 limits the thermal tolerance of crustaceans but the underlying mechanisms have not been comprehensively explored. Larval stages of crustaceans are even more sensitive to environmental hypercapnia and possess narrower thermal windows than adults.

Results: In a mechanistic approach, we analysed the impact of high seawater CO_2 on parameters at different levels of biological organization, from the molecular to the whole animal level. At the whole animal level we measured oxygen consumption, heart rate and activity during acute warming in zoea and megalopa larvae of the spider crab *Hyas araneus* exposed to different levels of seawater PCO_2. Furthermore, the expression of genes responsible for acid–base regulation and mitochondrial energy metabolism, and cellular responses to thermal stress (e.g. the heat shock response) was analysed before and after larvae were heat shocked by rapidly raising the seawater temperature from 10°C rearing temperature to 20°C. Zoea larvae showed a high heat tolerance, which decreased at elevated seawater PCO_2, while the already low heat tolerance of megalopa larvae was not limited further by hypercapnic exposure. There was a combined effect of elevated seawater CO_2 and heat shock in zoea larvae causing elevated transcript levels of heat shock proteins. In all three larval stages, hypercapnic exposure elicited an up-regulation of genes involved in oxidative phosphorylation, which was, however, not accompanied by increased energetic demands.

Conclusion: The combined effect of seawater CO_2 and heat shock on the gene expression of heat shock proteins reflects the downward shift in thermal limits seen on the whole animal level and indicates an associated capacity to elicit passive thermal tolerance. The up-regulation of genes involved in oxidative phosphorylation might compensate for enzyme activities being lowered through bicarbonate inhibition and maintain larval standard metabolic rates at high seawater CO_2 levels. The present study underlines the necessity to align transcriptomic data with physiological responses when addressing mechanisms affected by an interaction of elevated seawater PCO_2 and temperature extremes.

Keywords: *Hyas araneus*, Larvae, Ocean acidification, Climate change, Thermal tolerance, Gene expression

* Correspondence: melanie.schiffer@awi.de
[1]Integrative Ecophysiology, Alfred-Wegener-Institute for Polar and Marine Research, Am Handelshafen 12, 27570 Bremerhaven, Germany
Full list of author information is available at the end of the article

Introduction

The surface waters of the worlds' ocean are affected by anthropogenic warming and accumulating atmospheric CO_2. Sea surface temperatures are predicted to reach 1.5 to 8°C above preindustrial values by the year 2300 [1], and the concentration of atmospheric CO_2 may reach levels of 2000 ppm by 2300, leading to a drop in surface water pH by up to 0.8 pH units [2]. Marine organisms will thus have to cope with concomitant changes in seawater temperature and pH. Combined or interactive effects of these environmental factors on the physiology of marine organisms can result from the same physiological mechanisms being affected by both factors [3].

To address the question of how organisms deal with thermal challenges, the concept of oxygen and capacity limited thermal tolerance (OCLTT) has been developed [4]. The observations supporting the concept include those in temperate zone crustaceans, among others, and led to the hypothesis that a mismatch between oxygen demand and oxygen supply results from limited capacity of ventilatory and circulatory systems at temperature extremes. The resulting limits in aerobic performance are the first lines of limitation in thermal tolerance [5]. These earliest, ecologically relevant, thermal tolerance limits are called pejus temperatures (T_p). Beyond the pejus range critical temperatures (T_c) indicate the transition to anaerobic metabolism. Within the pejus temperature range, heartbeat and ventilation increase with temperature supporting the rising oxygen demand in the warmth [5] as well as a scope for aerobic performance such as growth. Beyond the T_p, haemolymph oxygen partial pressure decreases as a result of limited capacities of ventilation and circulation indicating a progressive mismatch between oxygen demand for maintenance and oxygen supply. In warm temperate species, hypoxia occurs on both flanks of the thermal performance curve and, finally, anaerobic metabolism sets in at the critical temperature. Survival beyond the T_c is time-limited [5]. At the upper end of the thermal tolerance window, denaturation temperature might elicit a loss of protein function, the heat shock response and oxidative stress [4].

The interactions of elevated seawater PCO_2 and temperature extremes have been proposed to cause a narrowing of the thermal tolerance window of an organism exposed to high CO_2 levels [6]. With rising seawater CO_2 concentration, upper thermal tolerance limits have been observed to be lowered by several °C in adult crustaceans and coral reef fishes [7-9]. Zittier et al. [10] found elevated seawater PCO_2 and heat stress to act synergistically reducing the righting response in the spider crab *Hyas araneus*.

To understand the synergistic effects of increasing seawater PCO_2 and temperature at population level, it is important to include the most vulnerable life cycle stages.

Early developmental stages are suggested to be most sensitive to environmental hypercapnia [11] and to possess narrow thermal windows [12,13]. They might, thus, be a bottleneck for successful survival and viability of a species in a warm and high CO_2 ocean. Embryos of the Sydney rock oyster, *Saccostrea glomerata* yielded in a reduced number of D-veligers with a greater percentage of abnormalities as well as reduced size when exposed to high CO_2 and high temperature during both fertilization and embryonic development compared to embryos that were exposed to the treatments for embryonic development only [14]. In temperate sea urchin larvae concomitant exposure to high temperature and high PCO_2 reduced larval metabolism and led to a down-regulation of histone encoding genes [15]. However, in tropical sea urchin larvae during concomitant exposure to elevated temperature and PCO_2 effects of acidification on larval size were dominant [16]. Additive effects of increased temperature and CO_2 were recorded for survival, development, growth, and lipid synthesis of larvae and juveniles of Northwest Atlantic bivalves [17]. At ambient temperature, elevated CO_2 (3100 ppm) resulted in increased mortality and prolonged developmental time accompanied with a decrease in oxygen consumption rates of developing zoea I of *Hyas araneus*, when they were exposed to CO_2 during their embryonic development [18,19]. So far, there is limited data available on the thermal tolerance of larval stages exposed to elevated seawater PCO_2.

The aim of the paper is to investigate the effect of elevated seawater PCO_2 on the heat tolerance of the three larval stages of the spider crab *Hyas araneus*. *Hyas araneus* is a benthic shelf species and has a wide distribution range from temperate to Arctic waters [20]. Larvae go through two zoea stages and one megalopa stage before settling into the adult habitat. In a mechanistic approach, we analysed parameters on different levels of functional hierarchy, from the whole animal to the molecular level. As temperature tolerance of adult *Hyas araneus* has been shown to be reduced by high CO_2 [8] and larvae are supposed to be more sensitive to synergistic effects of CO_2 and temperature [12], larvae were exposed to high seawater CO_2 of 3300 µatm and temperature extremes (10°C above rearing temperature) to study mechanisms affected by both factors and the interaction between these factors. For the identification of affected mechanisms it is necessary to use high levels of CO_2 and high temperatures followed by subsequent studies of these mechanisms at intermediate levels of physico- chemical parameters [6]. At the whole organism level, we measured active metabolic rate, heart rate and larval activity during continuous warming in the three larval stages reared at different seawater PCO_2 to identify differences in heat tolerance between CO_2 treatments and stages.

CO_2 and temperature induced shifts in gene expression were studied in batches of larvae of each stage by sampling directly from the different CO_2 treatments and after exposure to short term heat shock. Expression levels of genes responsible for cellular stress phenomena including the heat shock response as a protection process, for acid–base regulation as an important energy consuming process [21] and for mitochondrial energy metabolism as an energy supplying process, were analysed. These processes are hypothesized to be of central importance for a limitation in thermal tolerance during hypercapnic exposure. Previous studies reported differential responses of heat shock protein expression in larval and adult marine ectotherms. Responses ranged from a reduced expression [22,23], to an up-regulation of heat shock protein expression at low pH [24].

The capacities to regulate hypercapnia-induced blood acid–base disturbances by means of ion transporters might prevent strong acid–base disturbances that could lead to reduced protein function and lower temperature tolerance. Systemic hypercapnia also causes metabolic depression by lowering pH [25] accompanied by increasing gas partial pressure gradients [26] and will reduce the organisms' capacity to increase its rate of aerobic energy turnover [3]. Metabolic depression may also be reflected at gene expression level. Hypoxia caused the repression of genes of the mitochondrial citric acid cycle and the electron transport system in gills of adult zebrafish [27].

With our data, we have been able to align whole organism performance to molecular responses and to reveal mechanisms affected by the combined action of elevated CO_2 and temperature levels.

Results
Larval mortality
There was no significant difference in larval mortality between the treatments for both zoea I and zoea II larvae. Mortality of zoea I larvae was $15.5 \pm 5\%$ in larvae exposed

to 420 µatm and $21.6 \pm 6\%$ in larvae exposed to 3300 µatm (t-test, $p = 0.413$), while zoea II larvae showed $14.7 \pm 11\%$ mortality in the control treatment and $32.3 \pm 13\%$ in the high CO_2 treatment (t-test, $p = 0.320$).

Determination of the larval thermal tolerance window
Oxygen consumption
Oxygen consumption of zoea I larvae increased significantly with temperature, while no effect of seawater CO_2 concentration on metabolic rate was detected (2 way-ANOVA, Table 1, Figure 1A). At high temperature extremes a posteriori tests identified peaks in oxygen consumption at 25°C in control larvae (2.3 ± 0.3 µO$_2$ mg DW^{-1*} h^{-1}) and at 22°C in high CO_2 larvae (2.2 ± 0.4 µO$_2$ mg DW^{-1*} h^{-1} (Figure 1A). At 28°C larval oxygen consumption showed a significant decrease even below values observed at 10°C for control and CO_2 treatments (Figure 1A). Oxygen consumption was significantly lower under high seawater PCO_2 at 25°C in comparison to oxygen consumption of control larvae.

Oxygen consumption patterns of zoea II revealed a significant interaction between temperature and CO_2 levels (Table 1, Figure 1B). Oxygen consumption of control larvae increased between 10°C and 19°C, remained constant between 19°C and 25°C followed by a significant decrease at 28°C (Figure 1B). In contrast, oxygen consumption of larvae reared at elevated CO_2 increased between 10°C and 22°C and showed a sharp decrease already at 25°C.

There was also a significant interaction between temperature and CO_2 in the oxygen consumption rates of megalopa larvae (Table 1). A posteriori tests found an increase in respiration rates between 10°C and 22°C and a significant decrease between 22°C and 28°C for megalopa kept under control conditions (Figure 1C). Under high PCO_2 oxygen consumption increased only between 10°C and 13°C and was significantly lower at 22°C than in control larvae. The highest oxygen consumption was found at 22°C in control larvae (1.9 ± 0.5

Table 1 Results of two-way repeated measures ANOVAs

Stage	Response variable	CO$_2$ effect			Temperature effect			Interaction		
		F	df	p	F	df	p	F	df	p
Zoea I	Oxygen consumption	0.723	1	0.411	32.838	6	< 0.001	1.709	6	0.133
Zoea II	Oxygen consumption	5.229	1	**0.041**	51.374	6	< 0.001	11.025	6	**0.001**
Megalopa	Oxygen consumption	0.965	1	0.345	7.595	6	< 0.001	2.769	6	**0.022**
Zoea I	Heart rate	0.000	1	0.978	32.255	6	< 0.001	0.712	6	0.642
Zoea II	Heart rate	0.001	1	0.974	41.980	6	< 0.001	18.755	6	**< 0.001**
Megalopa	Heart rate	0.136	1	0.723	28.959	6	< 0.001	0.419	6	0.862
Megalopa	Maxilliped beat rate	1.277	1	0.295	4.789	6	< 0.001	0.433	6	0.852
Zoea II	Maxilliped beat rate	1.109	1	0.333	18.092	6	< 0.001	1.289	6	0.288

ANOVAs were conducted to investigate effects of CO_2 and temperature on oxygen consumption (Figure 1A-C), heart rate (Figure 2A-C) and maxilliped beat rate (Figure 3A and B) of *Hyas araneus* zoea and megalopa larvae. Bold values indicate statistical significance.

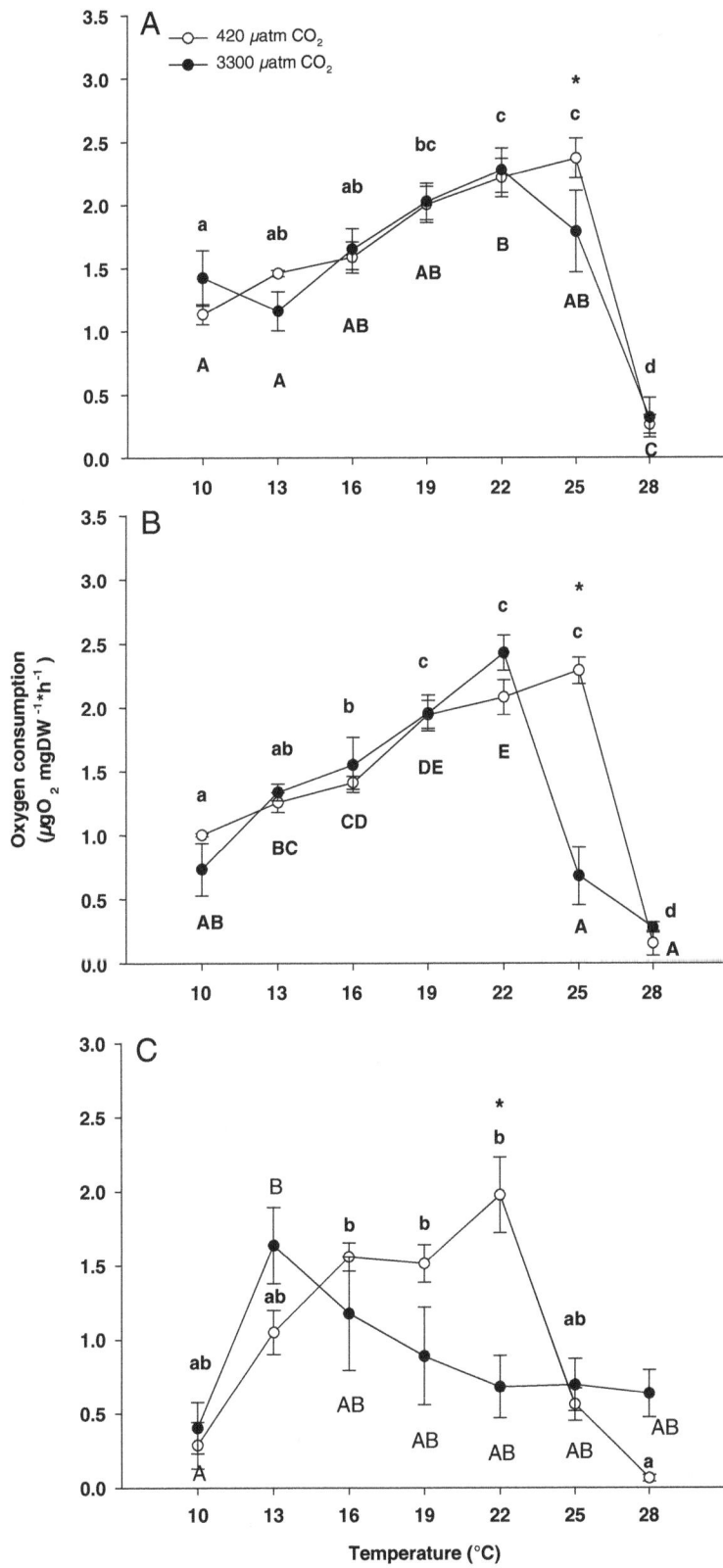

Figure 1 (See legend on next page.)

(See figure on previous page.)
Figure 1 Temperature dependent oxygen consumption of zoea I (A), zoea II (B) and megalopa larvae (C) of *Hyas araneus*. Larvae were reared at two different seawater PCO_2 (open circle: controls, 420 µatm CO_2; closed circle: 3300 µatm CO_2; Mean ± SE, N = 5-8). Asterisks indicate significant differences between treatments at the same experimental temperature. Different letters indicate significant differences between temperatures within one treatment (lowercase letters: 420 µatm CO_2; uppercase letters: 3300 µatm CO_2).

μO_2 mg $DW^{-1}*h^{-1}$) and at 13°C under elevated PCO_2 (1.6 ± 0.7 μO_2 mg $DW^{-1}*h^{-1}$).

Heart rate

The heart rate of zoea I larvae was significantly affected by temperature, but not by CO_2 (two-way-ANOVA, Table 1). Heart rate of zoea I reared under control conditions increased between 10°C and 25°C with highest heart rates at 25°C (418 ± 14 beats min^{-1}, Figure 2A). A similar increase between 10°C and 25°C could be seen under high PCO_2 with highest rates of 353 ± 54 beats min^{-1} at 25°C. Upon further warming to 28°C there was a significant decrease of heart rate in both treatments to 133 ± 70 beats min^{-1} under control and 153 ± 46 beats min^{-1} under high PCO_2 conditions.

There was a significant interaction between temperature and CO_2 in zoea II larvae (Table 1). A posteriori tests identified a significant increase of zoea II heart rates between 10°C and 25°C in control larvae and between 10°C and 22°C in high CO_2 larvae, respectively. Subsequently, heart rates decreased at 28°C in zoea II kept at control seawater PCO_2, whereas a significant decrease of heart rates already occurred at 25°C in larvae reared at elevated PCO_2, followed by a further decrease at 28°C (Figure 2B). Larvae showed higher heart rates at 16°C and 19°C and lower heart rates at 25°C and 28°C when kept at high CO_2 (Figure 2B).

The heart rate of megalopa larvae was significantly affected by temperature, but not by CO_2 (two-way-ANOVA, Table 1). Heart rates remained constant between 10°C and 22°C followed by a significant decrease between 22°C and 28°C in control larvae and larvae from the high CO_2 treatment (Figure 2C). At 28°C no heart beat could be detected at either treatment.

Maxilliped beat rate

The maxilliped beat rate of zoea I larvae was significantly affected by temperature, but not by CO_2 (two-way-ANOVA, Table 1). A posteriori Tukey tests revealed constant maxilliped beat rates between 10°C and 25°C and a decrease upon further warming to 28°C, which was significant between 16°C and 28°C in zoea I larvae reared under control conditions and between 19°C and 28°C at high CO_2. There was no significant difference between maxilliped beat rates of zoea I larvae reared at control or high CO_2 level (Figure 3A).

A two-way ANOVA revealed a significant effect of temperature but not of CO_2 on maxilliped beat rates of zoea II larvae (Table 1). Rates decreased upon warming in both, control and CO_2 treatments (a posteriori Tukey tests) (Figure 3B). There was no significant difference between rates at 22°C and 25°C in control larvae, whereas a significant drop occurred in larvae reared at elevated seawater PCO_2. All zoea II stopped maxilliped beating at 28°C under control conditions while beating ceased already at 25°C at high seawater PCO_2.

Gene expression patterns

For the purpose of clarity only significant changes in gene expression of proteins involved in the cellular stress/heat shock response, acid–base regulation and mitochondrial energy metabolism are reported and discussed (Tables 2, 3 and 4). Significant differences of heat shock refer to the gene expression in larvae kept at 10°C compared to those exposed to a heat shock at 20°C of control and CO_2 treatments. Significant effects of CO_2 refer to the gene expression of larvae reared at control and high CO_2 levels within each temperature treatment (10°C (control larvae) and 20°C (heat shocked larvae)). We presumed a combined effect when both factors, heat shock and seawater CO_2, significantly affected larval gene expression (up- or down-regulation) (Tables 2, 3 and 4).

Cellular stress/heat shock response

Seven different sequences, identified as heat shock proteins (HSP) by Blastx (E-Value cut-off of $1E^{-3}$), were selected to investigate effects of elevated seawater CO_2 and heat shock on HSP gene expression in the different larval stages, among those were 4 representatives of the HSP 70 family (HSP70_1-4), 1 HSP 90, 1 HSP 26 and 1 HSP 60.

Exposure to heat shock (20°C) for 5 h affected the gene expression of HSP70_1-4, HSP90 and HSP26 in larvae of *Hyas araneus* (Table 2). Significant interactions were detected between heat shock and seawater CO_2 concentration for the expression of HSP70_1 in zoea II larvae on day 3 and day 15, HSP70_4 in zoea II on day 3 and in megalopa larvae on day 3 as well as for HSP90 in zoea II larvae on day 15 and in megalopa larvae on day 3 (Table 5).

On day 0 expression of HSP70_1 (unpaired t-test: $p = 0.020$), HSP_2 ($p = 0.016$) and HSP_3 ($p = 0.010$) in zoea I larvae was twice or even for times higher after a heat shock of 20°C. A strong increase in the gene expression of HSP70_1-4 and HSP90 (Table 2) could also be observed on day 15 after heat shock in control and

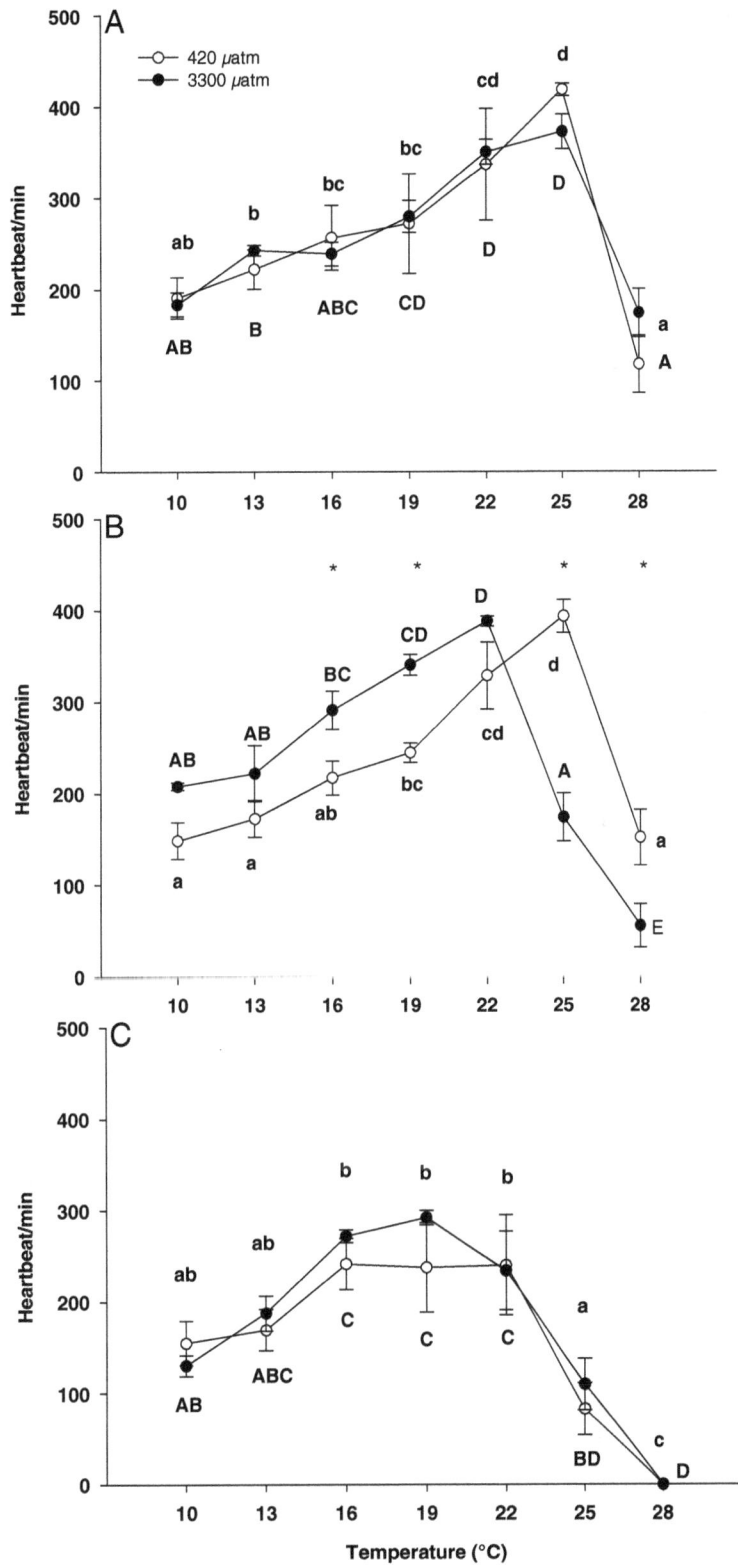

Figure 2 Temperature dependent heart rate of zoea I (A), zoea II (B) and megalopa larvae (C) of *Hyas araneus*. Larvae were reared at two different seawater PCO_2 (open circle: controls, 420 µatm CO_2; closed circle: 3300 µatm CO_2; Mean ± SE, N = 5-6). Different letters indicate significant differences between temperatures within one treatment (lowercase letters: 420 µatm CO_2; uppercase letters: 3300 µatm CO_2).

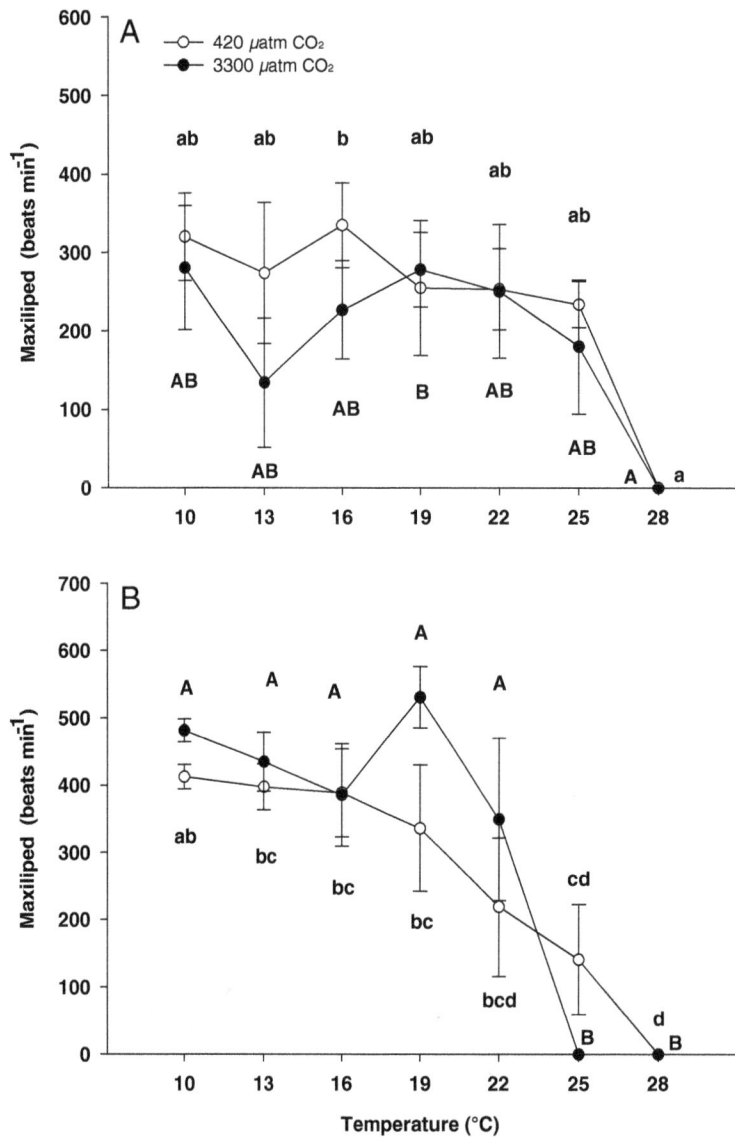

Figure 3 Temperature dependent maxilliped beat rate of zoea I (A) and zoea II (B) of *Hyas araneus*. Larvae were reared at two different seawater PCO_2 (open circle: controls, 420 µatm CO_2; closed circle: 3300 µatm CO_2. Mean ± SE, N = 4-6). Asterisks indicate significant differences between treatments at the same experimental temperature. Different letters indicate significant differences between temperatures within one treatment (lowercase letters: 420 µatm CO_2; uppercase letters: 3300 µatm CO_2).

CO_2 treatments (p <0.05, a posteriori analysis). The strongest increase in HSP expression of all stages was observed in zoea I on day 15. HSP 70_1 expression increased from 0.4 to 2.0 within the CO_2 group.

In zoea II larvae HSP70_1 and HSP70_4 expression doubled after heat shock in control and CO_2 treatments on day 3 (Table 2). A similar pattern could be observed for the gene expression of heat shock protein 70_3 and 90 (Table 5) with higher expression at 20°C in control larvae and high CO_2 larvae and in those reared at elevated PCO_2, respectively (p <0.05, a posteriori analysis). In zoea II on day 15 and in the megalopa stage on day 3 expression of HSP70_1-4 and HSP90 was strongly up-

regulated after 5 h heat shock of 20°C independent of PCO_2 (p <0.05, a posteriori analysis) (Table 2).

HSP26 was the only heat shock protein, which was significantly down-regulated after heat shock. A Tukey test revealed a significantly lower gene expression in zoea II on day 15 independent of PCO_2.

Considered as isolated factor, hypercapnia affected HSP gene expression only at 10°C in zoea I on day 15. Hypercapnic exposure doubled the gene expression of HSP70_4 and HSP26 (Table 5, Table 2).

In several larval stages of *Hyas araneus*, a combined effect of high CO_2 and heat shock on the gene expression of heat shock proteins could be observed (Tables 2,

Table 2 Gene expression analysis: gene expression (quantities) in zoea and megalopa larvae of *Hyas araneus* at different time points in development classified according to their function in cellular stress/heat shock response

Cellular stress/heat shock response

Gene	Zoea I day 0		Zoea I day 15				Zoea II day 3				Zoea II day 15				Megalopa day 3			
	C 10°C	C 20°C	C 10°C	CO₂ 10°C	C 20°C	CO₂ 20°C	C 10°C	CO₂ 10°C	C 20°C	CO₂ 20°C	C 10°C	CO₂ 10°C	C 20°C	CO₂ 20°C	C 10°C	CO₂ 10°C	C 20°C	CO₂ 20°C
HSP 70_1	0.39 ±0.02	1.85 ±0.29	0.36 ±0.03	0.42 ±0.06	1.72 ±0.10	2.09 ±0.20	0.77 ±0.07	0.66 ±0.10	1.31 ±0.08	1.76 ±0.07	0.27 ±0.03	0.25 ±0.01	1.00 ±0.16	1.51 ±0.11	0.19 ±0.01	0.24 ±0.03	1.04 ±0.10	0.94 ±0.21
HSP 70_2	0.92 ±0.14	2.14 ±0.38	0.84 ±0.07	1.01 ±0.11	1.97 ±0.15	2.19 ±0.08	0.47 ±0.01	0.40 ±0.02	1.11 ±0.13	0.88 ±0.06	0.69 ±0.05	0.83 ±0.04	1.46 ±0.14	1.91 ±0.11	0.71 ±0.05	0.73 ±0.04	1.59 ±0.08	1.40 ±0.09
HSP 70_3	0.39 ±0.10	1.19 ±0.19	0.56 ±0.07	1.03 ±0.25	1.55 ±0.25	1.87 ±0.29	0.85 ±0.03	0.77 ±0.10	1.18 ±0.13	1.08 ±0.07	0.64 ±0.12	0.71 ±0.04	1.15 ±0.04	1.56 ±0.12	0.44 ±0.05	0.64 ±0.03	1.05 ±0.17	1.17 ±0.16
HSP 70_4	0.94 ±0.27	2.17 ±0.41	0.84 ±0.05	1.56 ±0.24	1.87 ±0.17	2.43 ±0.19	0.75 ±0.04	0.81 ±0.10	1.23 ±0.15	1.81 ±0.11	0.69 ±0.04	0.88 ±0.07	1.31 ±0.03	1.58 ±0.13	0.51 ±0.03	0.89 ±0.10	1.71 ±0.23	1.41 ±0.18
HSP 90	1.01 ±0.32	1.82 ±0.29	0.94 ±0.09	1.10 ±0.16	1.90 ±0.22	2.13 ±0.19	1.26 ±0.05	1.11 ±0.22	1.41 ±0.16	1.84 ±0.04	0.81 ±0.04	0.69 ±0.03	1.19 ±0.04	1.63 ±0.07	0.55 ±0.02	0.66 ±0.07	1.89 ±0.19	1.40 ±0.11
HSP 26	1.18 ±0.24	2.16 ±0.37	1.08 ±0.18	2.00 ±0.29	0.97 ±0.22	1.48 ±0.36	1.47 ±0.09	1.68 ±0.37	0.80 ±0.07	1.16 ±0.09	2.23 ±0.10	2.03 ±0.13	1.69 ±0.08	1.61 ±0.12	1.28 ±0.08	1.57 ±0.05	1.22 ±0.13	1.28 ±0.09
HSP 60	1.40 ±0.10	1.98 ±0.43	1.56 ±0.12	2.08 ±0.43	1.94 ±0.05	2.10 ±0.18	1.32 ±0.15	1.25 ±0.24	0.90 ±0.08	0.99 ±0.18	1.11 ±0.13	1.53 ±0.12	1.24 ±0.02	1.61 ±0.18	1.08 ±0.05	0.91 ±0.08	1.40 ±0.31	1.01 ±0.12

Larvae were reared at control PCO₂ (C) and high PCO₂ (CO₂) at control temperature (10°C) or exposed to a heat shock for 5 h at 20°C. Arrow direction indicates significantly higher (upwards) or lower (downwards) gene expression between CO₂ treatments at the same temperature or between temperatures within the same CO₂ treatment. Black arrows: CO₂ effect at the same temperature (10°C or 20°C). White arrows: heat shock effect. White/Black arrows in one direction indicate a combined effect of CO₂ and heat shock.

Table 3 Gene expression analysis: gene expression (quantities) in zoea and megalopa larvae of *Hyas araneus* at different time points in development classified according to their function in acid-base regulation

	Acid-base regulation																		
	Zoea I day 0		Zoea I day 15				Zoea II day 3				Zoea II day 15				Megalopa day 3			
	C	C	C	CO₂	C	CO₂	C	CO₂	C	CO₂	C	CO₂	C	CO₂	C	CO₂	C	CO₂
Gene	10°C	20°C	10°C	10°C	20°C	20°C	10°C	10°C	20°C	20°C	10°C	10°C	20°C	20°C	10°C	10°C	20°C	20°C
CA	1.59 ±0.26	2.06 ±0.39	1.50 ±0.14	1.78 ±0.24	1.52 ±0.18	1.65 ±0.11	1.31 ±0.10	1.58 ±0.07 ◀	1.10 ±0.08	1.23 ±0.06 ⇨	2.20 ±0.09	1.55 ±0.08 ▶	1.74 ±0.08 ⇨	1.62 ±0.06	1.35 ±0.06	0.99 ±0.10 ▶	1.26 ±0.07	1.24 ±0.05 ⇦
NaK	1.30 ±0.31	1.74 ±0.34	1.64 ±0.09	1.51 ±0.25	1.55 ±0.15	1.26 ±0.10	1.35 ±0.07	1.46 ±0.09	0.96 ±0.05	0.95 ±0.03 ⇨	1.79 ±0.10	1.34 ±0.14 ▶	1.59 ±0.11	1.43 ±0.12	1.27 ±0.09	1.32 ±0.08	1.16 ±0.10	1.17 ±0.06
NBC	1.24 ±0.30	2.04 ±0.37	1.82 ±0.08	1.44 ±0.20	1.87 ±0.15	1.24 ±0.10 ▶	1.25 ±0.09	1.65 ±0.04 ◀	1.15 ±0.05	1.51 ±0.04 ◀	1.90 ±0.34	0.89 ±0.05 ▶	1.02 ±0.18 ⇨	0.88 ±0.04	1.29 ±0.05	1.49 ±0.06 ◀	1.33 ±0.03	1.37 ±0.04
NKCC	0.92 ±0.28	1.77 ±0.44	1.27 ±0.07	1.97 ±0.43	1.56 ±0.12	2.05 ±0.21	1.19 ±0.15	1.08±0.33	0.93 ±0.16	1.02 ±0.20	1.31 ±0.15	1.50 ±0.12	1.28 ±0.02	1.51 ±0.11	1.18 ±0.08	1.46 ±0.05 ◀	1.10 ±0.08	0.96 ±0.00 ⇦

Larvae were reared at control $P\text{CO}_2$ (C) and high $P\text{CO}_2$ (CO_2) at control temperature (10°C) or exposed to a heat shock for 5 h at 20°C. Arrow direction indicates significantly higher (upwards) or lower (downwards) gene expression between CO_2 treatments at the same temperature or between temperatures within the same CO_2 treatment. Black arrows: CO_2 effect at the same temperature (10°C or 20°C). White arrows: heat shock effect. White/Black arrows in one direction indicate a combined effect of CO_2 and heat shock.

Table 4 Gene expression analysis: gene expression (quantities) in zoea and megalopa larvae of *Hyas araneus* at different time points in development classified according to their function in mitochondrial energy metabolism

Mitochondrial energy metabolism

Gene	Zoea I day 0 C 10°C	C 20°C	Zoea I day 15 C 10°C	CO2 10°C	C 20°C	CO2 20°C	Zoea II day 3 C 10°C	CO2 10°C	C 20°C	CO2 20°C	Zoea II day 15 C 10°C	CO2 10°C	C 20°C	CO2 20°C	Megalopa day 3 C 10°C	CO2 10°C	C 20°C	CO2 20°C
PDH	1.29 ±0.29	1.33 ±0.29	1.76 ±0.09	1.76 ±0.35	1.20 ±0.23	1.47 ±0.13	1.51 ±0.04	1.49 ±0.12	0.98 ±0.14	1.35 ±0.09	1.74 ±0.18	1.76 ±0.08	1.78 ±0.18	1.65 ±0.10	1.28 ±0.08	1.05 ±0.01	1.26 ±0.11	0.91 ±0.04
IDH	1.52 ±0.20	2.13 ±0.37	1.70 ±0.04	2.75 ±0.37	1.79 ±0.09	2.25 ±0.07	2.02 ±0.12	1.89 ±0.07	1.83 ±0.15	1.66 ±0.06	1.45 ±0.10	1.53 ±0.05	1.17 ±0.06	1.64 ±0.05	1.45 ±0.08	1.39 ±0.04	1.54 ±0.06	1.29 ±0.07
NAD	1.19 ±0.15	1.34 ±0.16	0.98 ±0.14	1.34 ±0.26	1.05 ±0.05	1.83 ±0.35	0.83 ±0.02	0.48 ±0.11	0.73 ±0.18	0.37 ±0.09	0.20 ±0.04	1.80 ±0.15	1.17 ±0.08	1.69 ±0.21	0.82 ±0.27	1.46 ±0.08	0.98 ±0.11	1.26 ±0.19
SDH	1.55 ±0.15	2.07 ±0.41	1.69 ±0.10	2.31 ±0.29	1.84 ±0.12	2.14 ±0.14	1.37 ±0.08	1.58 ±0.06	1.11 ±0.09	1.34 ±0.03	1.77 ±0.15	1.54 ±0.05	1.60 ±0.06	1.60 ±0.06	1.29 ±0.02	1.35 ±0.09	1.38 ±0.04	1.31 ±0.07
CCR	1.27 ±0.28	2.04 ±0.37	1.49 ±0.15	2.12 ±0.32	1.68 ±0.13	1.72 ±0.11	1.43 ±0.12	1.36 ±0.12	0.99 ±0.07	1.56 ±0.05	1.28 ±0.12	1.46 ±0.05	1.65 ±0.14	1.64 ±0.08	1.43 ±0.11	1.37 ±0.08	1.57 ±0.12	1.16 ±0.05
COX	1.38 ±0.27	2.07 ±0.38	1.73 ±0.11	2.47 ±0.45	0.85 ±0.13	1.33 ±0.10	1.38 ±0.12	1.18 ±0.08	0.53 ±0.04	1.00 ±0.10	1.37 ±0.21	2.36 ±0.09	1.60 ±0.19	1.56 ±0.07	1.16 ±0.15	1.53 ±0.14	1.00 ±0.09	0.79 ±0.04
atpA	1.44 ±0.27	2.10 ±0.28	1.81 ±0.03	2.68 ±0.38	2.72 ±0.04	2.09 ±0.09	1.49 ±0.06	1.49 ±0.04	1.14 ±0.05	1.56 ±0.08	1.93 ±0.13	2.12 ±0.09	1.89 ±0.14	1.79 ±0.05	1.62 ±0.06	1.61 ±0.08	1.65 ±0.06	1.52 ±0.11

Larvae were reared at control PCO_2 (C) and high PCO_2 (CO_2) at control temperature (10°C) or exposed to a heat shock for 5 h at 20°C. Arrow direction indicates significantly higher (upwards) or lower (downwards) gene expression between CO_2 treatments at the same temperature or between temperatures within the same CO_2 treatment. Black arrows: CO_2 effect at the same temperature (10°C or 20°C). White arrows: heat shock effect. White/Black arrows in one direction indicate a combined effect of CO_2 and heat shock.

Table 5 Results of two-way ANOVAs

Gene	Stage	Day	Heat shock effect			CO$_2$ effect			Interaction		
			F	df	p	F	df	p	F	df	p
Cellular stress/heat shock response											
HSP70_1	Zoea I	15	165.2	1	**<0.001**	3.3	1	0.088	1.8	1	0.198
	Zoea II	3	104.8	1	**<0.001**	4.2	1	0.056	12.1	1	**0.003**
	Zoea II	15	127.2	1	**<0.001**	7.4	1	**0.017**	8.8	1	**0.011**
	Megalopa	3	42.8	1	**<0.001**	0.05	1	0.823	0.4	1	0.535
HSP70_2	Zoea I	15	114.3	1	**<0.001**	3.2	1	0.091	0.02	1	0.870
	Zoea II	15	106.5	1	**<0.001**	10.6	1	**0.006**	2.9	1	0.108
	Megalopa	3	136.8	1	**<0.001**	1.5	1	0.232	2.5	1	0.129
HSP70_3	Zoea I	15	15.6	1	**0.001**	2.9	1	0.104	0.1	1	0.740
	Zoea II	3	11.4	1	**0.004**	0.8	1	0.369	0.009	1	0.922
	Zoea II	15	43.8	1	**<0.001**	5.4	1	**0.038**	2.67	1	0.128
	Megalopa	3	23.2	1	**<0.001**	1.8	1	0.192	0.09	1	0.758
HSP70_4	Zoea I	15	29.5	1	**<0.001**	13.4	1	**0.002**	0.1	1	0.668
	Zoea II	3	45.0	1	**<0.001**	8.7	1	**0.009**	5.4	1	0.333
	Zoea II	15	51.5	1	**<0.001**	6.5	1	**0.024**	0.1	1	0.674
	Megalopa	3	30.0	1	**<0.001**	0.05	1	0.813	4.6	1	**0.046**
HSP90	Zoea I	15	33.2	1	**<0.001**	1.2	1	0.282	0.03	1	0.846
	Zoea II	3	7.7	1	**0.015**	0.7	1	0.388	3.4	1	0.085
	Zoea II	15	171.4	1	**<0.001**	10.1	1	**0.007**	30.4	1	**<0.001**
	Megalopa	3	79.8	1	**<0.001**	2.6	1	0.123	6.4	1	**0.022**
HSP26	Zoea I	15	1.2	1	0.283	6.4	1	**0.022**	0.5	1	0.477
	Zoea II	15	15.2	1	**0.002**	1.3	1	0.267	0.2	1	0.660
	Megalopa	3	3.5	1	0.080	3.6	1	0.074	1.5	1	0.229
HSP60	Zoea I	15	0.6	1	0.430	1.9	1	0.182	0.5	1	0.473
	Zoea II	3	3.8	1	0.068	0.003	1	0. .955	0.1	1	0.662
	Zoea II	15	0.5	1	0.491	7.2	1	**0.019**	0.03	1	0.857
	Megalopa	3	1.4	1	0.246	2.6	1	0.126	0.3	1	0.537
Acid–base regulation											
CA	Zoea I	15	0.1	1	0.746	1.3	1	0.257	0.1	1	0.670
	Zoea II	3	12.9	1	**0.002**	6.5	1	**0.02**	0.8	1	0.383
	Zoea II	15	6.1	1	**0.028**	22.8	1	**<0.001**	11.2	1	**0.005**
	Megalopa	3	1.1	1	0.302	6.9	1	**0.018**	5.0	1.	**0.039**
NaK	Zoea I	15	1.1	1	0.304	1.7	1	0.209	0.2	1	0.628
	Zoea II	3	49.2	1	**<0.001**	0.6	1	0.437	0.8	1	0.370
	Zoea II	15	0.2	1	0.662	5.9	1	**0.031**	1.3	1	0.272
	Megalopa	3	2.5	1	0.127	0.1	1	0.737	0.04	1	0.829
NBC	Zoea I	15	0.3	1	0.579	12.8	1	**0.003**	0.7	1	0.393
	Zoea II	3	3.7	1	0.074	35.8	1	**<0.001**	0.06	1	0.800
	Zoea II	15	6.2	1	**0.027**	10.3	1	**0.007**	5.8	1	**0.031**
	Megalopa	3	0.7	1	0.403	6.5	1	**0.022**	3.1	1	0.095
NKCC	Zoea I	15	0.5	1	0.471	5.6	1	**0.030**	0.1	1	0.670
	Zoea II	3	0.4	1	0.490	0.003	1	0.955	0.1	1	0.666
	Zoea II	15	0.002	1	0.963	2.9	1	0.109	0.03	1	0.862

Table 5 Results of two-way ANOVAs *(Continued)*

	Megalopa	3	18.3	1	**<0.001**	1.0	1	0.333	9.7	1	**0.007**
Mitochondrial energy metabolism											
PDH	Zoea I	15	3.2	1	0.093	0.3	1	0.578	0.3	1	0.582
	Zoea II	3	9.0	1	**0.009**	2.5	1	0.134	3.0	1	0.101
	Zoea II	15	0.07	1	0.794	0.1	1	0.69	0.3	1	0.577
	Megalopa	3	1.1	1	0.297	14.3	1	**0.002**	0.6	1	0.438
IDH	Zoea II	3	4.9	1	**0.045**	2.4	1	0.144	0.03	1	0.862
	Zoea II	15	1.2	1	0.285	13.8	1	**0.002**	6.9	1	**0.019**
	Megalopa	3	0.01	1	0.915	5.7	1	**0.029**	2.0	1	0.170
NAD	Zoea I	15	1.4	1	0.247	6.0	1	**0.025**	0.8	1	0.379
	Zoea II	3	0.7	1	0.388	9.1	1	**0.008**	0.003	1	0.953
	Zoea II	15	6.8	1	**0.021**	42.9	1	**<0.001**	11.2	1	**0.005**
	Megalopa	3	0.01	1	0.919	6.3	1	**0.023**	0.9	1	0.348
SDH	Zoea I	15	0.005	1	0.943	6.6	1	**0.020**	0.7	1	0.388
	Zoea II	3	14.5	1	**0.002**	10.7	1	**0.005**	0.03	1	0.848
	Zoea II	15	0.3	1	0.565	1.6	1	0.221	1.6	1	0.221
	Megalopa	3	0.1	1	0.704	0.02	1	0.881	1.2	1	0.276
CCR	Zoea I	15	0.2	1	0.604	2.9	1	0.105	2.2	1	0.155
	Zoea II	3	1.7	1	0.210	7.2	1	**0.016**	11.2	1	**0.004**
	Zoea II	15	8.6	1	**0.011**	0.7	1	0.405	1.0	1	0.314
	Megalopa	3	0.1	1	0.721	6.1	1	**0.025**	3.4	1	0.082
COX	Zoea I	15	16.7	1	**<0.001**	6.1	1	**0.025**	0.3	1	0.589
	Zoea II	3	32.2	1	**<0.001**	2.0	1	0.168	13.7	1	**0.002**
	Zoea II	15	4.2	1	0.060	11.6	1	**0.005**	13.6	1	**0.003**
	Megalopa	3	15.9	1	**0.001**	0.4	1	0.491	6.4	1	**0.022**
atpA	Zoea II	3	5.0	1	**0.041**	11.9	1	**0.003**	12.5	1	**0.003**
	Zoea II	15	3.4	1	0.088	0.2	1	0.661	2.1	1	0.169
	Megalopa	3	0.1	1	0.710	0.7	1	0.387	0.5	1	0.454

ANOVAs were conducted to investigate the effects of heat shock and seawater CO_2 on the gene expression of *Hyas araneus* zoea and megalopa larvae (significant differences are indicated by arrows in Tables 2, 3 and 4). Data for the expression of HSP70_2 and HSP26 in the zoea II larvae (day 3) and IDH and atpA in the zoea I larvae (day 15) were excluded as they did not meet the assumptions for a two-way ANOVA. Bold values indicate statistical significance.

3 and 4 see arrows of column CO_2/20°C). Expression of HSP70_1 and HSP70_4 were significantly higher in zoea I on day 15 and zoea II on day 3 after heat shock in larvae reared at elevated PCO_2. On day 15 of high CO_2 exposure in zoea II larvae expression of HSP70_1-3 and HSP90 was also higher than in controls.

Acid–base regulation

Four different sequences, among them carbonic anhydrase (CA), sodium potassium ATPase (NaK), sodium bicarbonate cotransporter (NBC) and sodium potassium chloride cotransporter (NKCC) were down-regulated under heat shock (Table 3). In zoea II larvae on day 3 CA expression was lower in high CO_2 treatment and NaK was down-regulated in both heat-shocked control and high CO_2 zoea II larvae (Table 3). On day 15 a

significantly lower CA expression was observed after the heat shock in control zoea II larvae. In megalopa larvae expression of NKCC decreased from 1.1 to 0.9, while CA was up-regulated in heat-shocked larvae at elevated seawater PCO_2.

A stronger response in gene expression of transporters relevant for acid–base regulation was found at high CO_2 in comparison to the levels found after heat shock (Table 5, Table 3). Expression of NBC was reduced in zoea I larvae on day 15 (Table 3) after heat shock at elevated PCO_2. On day 3, CO_2 caused significantly increased NBC expression (Table 3) in high CO_2 larvae at 10°C and 20°C and higher CA expression in high CO_2 zoea II at 10°C (Table 3). On day 15, lower CA and NaK expression in the high CO_2 treatment was found at 10°C in zoea II larvae (Table 3). In megalopa larvae CA expression was down-regulated, while NKCC and NBC expression was up-regulated at 10°C in larvae exposed to elevated PCO_2 (p <0.05, a posteriori analysis).

In all larval stages, fold-changes of acid–base relevant genes were smaller than that of cellular response and no combined effect of high CO_2 and heat shock became obvious.

Mitochondrial energy metabolism

Seven different sequences were identified as enzymes of the mitochondrial energy metabolism by Blastx (E-Value cut-off of $1E^{-3}$), among them pyruvate dehydrogenase (PDH), isocitrate dehydrogenase (IDH), NADH dehydrogenase (NAD), succinate dehydrogenase (SDH), cytochrome c reductase (CCR), cytochrome c oxidase (COX) and ATP synthase (atpA).

Larvae of *Hyas araneus* responded to the heat shock mainly with a down-regulation of genes relevant for mitochondrial energy metabolism (Table 4). After heat shock the expression of COX in zoea I larvae on day 15 (Table 5) was reduced regardless of CO_2 concentration (p <0.5, a posteriori analysis). The strongest response was observed in zoea II larvae on day 3. Five of seven investigated genes were down-regulated in larvae reared at control PCO_2. However, only SDH was significantly down-regulated by heat shock at high seawater PCO_2. (Table 4). A contrary pattern was recorded in zoea II on day 15. In control larvae a higher CCR, NAD and IDH expression was found in heat shocked larvae at   20°C compared to expression at 10°C (Table 4). COX expression was down-regulated in high CO_2 larvae exposed to a heat shock of 20°C as it could be observed in megalopa larvae (Table 4).

In contrast to heat shock, exposure to elevated seawater PCO_2 led mainly to an up-regulation of genes relevant for mitochondrial energy metabolism (Table 4). NAD, SDH and COX genes were up-regulated in zoea I larvae on day 15 (Table 5). NAD expression was higher at 20°C, while SDH and COX were up-regulated in high CO_2 zoea I larvae at 10°C (Table 4).

Again, the strongest response could be observed in zoea II on day 3 with six out of seven genes responding to a treatment with high seawater PCO_2. However, changes in gene expression were only recorded after heat shock. While PDH, SDH, CCR, COX and atpA were slightly up-regulated at 20°C, NAD expression decreased in larvae exposed to high PCO_2. On day 15, NAD expression was elevated 9-fold from control to hypercapnic conditions. Higher NAD and COX expression due to elevated seawater CO_2 at 10°C and an up-regulation of IDH and NAD in heat shocked high CO_2 zoea II larvae was recorded (Table 4).

In the megalopa stage seawater CO_2 concentration influenced PDH, IDH, NAD and CCR expression significantly, but differentially (Table 5). PDH, IDH and CCR expression levels were lowered by elevated PCO_2 in heat shocked larvae, while NAD expression was higher in high CO_2 megalopa at the control temperature of 10°C (p <0.5, a posteriori analysis) (Table 4).

Discussion
Determination of the larval thermal tolerance window

In the present study, the concept of oxygen and capacity limited thermal tolerance [4] was applied to determine the thermal tolerance and putatively synergistic effects of elevated seawater PCO_2 in different larval stages of the crustacean *Hyas araneus*. We could show that the three different larval stages of *Hyas araneus* display different upper critical thermal tolerance limits, 25°C in zoea I and zoea II and 22°C in megalopa larvae. According to Frederich and Pörtner [5], limited capacities of ventilation and circulation lead to a progressive mismatch between oxygen supply and oxygen demand for maintenance and finally lead to hypoxemia and anaerobic metabolism beyond the upper critical limit. Upon further warming, standard metabolic rate and heart beat rate decreased. A corresponding decrease in heart rate and oxygen consumption could also be observed in *Hyas araneus* larvae with maximal values for both parameters detected at 25°C in zoea I and zoea II larvae and at 22°C in megalopa larvae and a decrease at 28°C in zoea and at 25°C, in megalopa larvae, respectively. The sharp drop in oxygen consumption of the zoea larvae was correlated with ceased maxilliped beating rates. The concomitant decrease in heart rates of zoea larvae strongly suggests synchronous limitation or onset of failure of both ventilatory and circulatory systems. Different optimum temperature ranges in different larval stages have also been reported for the kelp crab *Taliepus dentatus* with the narrowest window found in the megalopa [13]. The high sensitivity of megalopae to environmental stressors suggests that this larval stage is a physiologically sensitive bottleneck within the life cycle of decapod crustaceans [13,28].

Exposure to elevated seawater PCO_2 constrained the thermal tolerance of zoea stages of *Hyas araneus* and resulted in a downward shift of upper thermal limits that was less pronounced in zoea I larvae than in zoea II larvae. In zoea I larvae, a decrease in thermal tolerance involves a higher oxygen consumption rate reached in control larvae at 25°C than in those under elevated CO_2 indicating an earlier metabolic depression under elevated CO_2. Oxygen supply (O_2 concentration in the hemolymph) was not measured, however, the collapse in respiration in high CO_2 zoea I larvae was not accompanied by significantly lower heart rates and maxilliped beating rates. Increasing heart rates at concomitantly decreasing oxygen consumption rates, could also be seen in warming larval stages of two populations of the kelp crab *Taliepus dentatus* and were attributed to a progressive mismatch between oxygen demand and oxygen supply [13]. Such pattern of limitation was more pronounced in the second zoea stage. The two-way ANOVA detected a significant interaction of CO_2 concentration and experimental temperature for the second zoea stage. At both CO_2 concentrations oxygen consumption increased with increasing temperature. This pattern changed at 25°C with a strong drop of oxygen consumption rates of larvae reared at 3300 µatm CO_2 leading to a significant interaction. When larvae were reared at control CO_2 concentration, oxygen consumption increased until 25°C suggesting a reduced thermal tolerance with increase of CO_2 concentration. The drop in oxygen consumption between 22°C and 25°C was accompanied by an earlier decrease in heart rate and ceased maxilliped beating rate at 25°C. In zoea I larvae maxilliped beating rates did not stop until 28°C. Albeit not statistically significant, a higher resilience of zoea I than in zoea II also becomes visible under hypercapnia where mortality of zoea II larvae doubled compared to controls while differences in mortality were less pronounced in zoea I larvae (Additional file 1: Figure S1). In the study by Walther et al. [8] on thermal tolerance of adult *Hyas araneus* under elevated seawater PCO_2, a CO_2 induced rise in Q_{10} values of heart rate has been proposed to cause the narrowing of thermal window under CO_2. Our data are in line with those findings, showing a steep rise in the Q_{10} values of respiration between rearing and critical temperatures in zoea II larvae exposed to high CO_2. Higher tissue oxygen demands with increasing temperature might be compensated for to some extent by higher heart rates (albeit not statistically significant), observed in zoea II larvae under elevated CO_2.

The two-way ANOVA also detected a significant interaction of CO_2 concentration and experimental temperature for the megalopa stage. Patterns of oxygen consumption with increasing temperature were dependent on seawater CO_2 concentration. Oxygen consumption of megalopa reared at 3300 µatm started to decrease at 16°C while oxygen consumption of control megalopa continued to increase leading to the significant interaction. In megalopa larvae significant higher oxygen consumption rates in larvae under control compared to high CO_2 conditions were seen at 22°C. These patterns indicate a downward shift of the upper thermal limit at high seawater CO_2 at even lower temperatures than found for the zoea stages. This is emphasized by the finding that maximum oxygen consumption rates were reached at 22°C in untreated megalopa but already at 13°C under CO_2. However, no significant difference between respiration rates of control and high CO_2 megalopa was seen across temperatures below the critical temperature. Here elevated seawater PCO_2 affected oxygen consumption but not heart rate, reflecting the CO_2 induced mismatch between the two processes.

Gene expression patterns

The three physiological parameters (oxygen consumption, heart rate and maxilliped beat rate) were measured in 9-13-day old zoea I and zoea II larvae and can, thus, be tentatively aligned with the gene expression data for physiological processes like cellular stress/heat shock response, acid–base regulation and mitochondrial energy metabolism studied in zoea I and zoea II at day 15. The gene expression data measured on other developmental days support a comparison of CO_2 responses during the time course of development in the different larval stages.

Cellular stress/heat shock response

A 5 h heat shock of 20°C caused a strong upregulation of heat shock proteins HSP 70 (1–4) and heat shock protein 90 in all three larval stages at any developmental time point. However, there was a stronger response to thermal stress in 15 day old than in 3 day old zoea I and zoea II larvae. Around 70% of all investigated HSPs were up-regulated after the heat shock in zoea I and zoea II on day 15, while around 50% were up-regulated on day 3 in zoea II. These findings indicate that larvae in the early stage phase might be less responsive to the stress than in the late stage phase (Table 2), possibly reflecting a narrowing of thermal tolerance or improved resilience with progressive development (see below). Heat shock proteins help to prevent denaturation of proteins and to refold denatured proteins. The high degree of up-regulation of HSP 70 and HSP 90 in *Hyas araneus* larvae suggests that 20°C is close to the upper thermal limits seen in the physiological data of both zoea stages.

There was a combined effect of temperature and PCO_2 on HSP70 and HSP90 expression in both zoea stages of *Hyas araneus*, resulting in higher HSP expression at 20°C in larvae reared at high CO_2 (Table 2). This finding was more pronounced in the second zoea stage on day 15 in comparison to the first zoea stage on day 15. In the first zoea stage, high CO_2 in synergy with a heat shock of 20°C resulted in an up-regulation of around 30% of all investigated HSPs, while around 60% of all HSPs were up-regulated in the second zoea stage. This reflects the CO_2 induced downward shift in the upper thermal limit seen in oxygen consumption and heart rate data (see above).

Although there might be a difference between HSP transcription and translation, we assume that the strong increase in HSP expression in *Hyas araneus* should result in increased protein levels. HSP70 is an ATP-dependent chaperone and the prevention of heat-induced protein denaturation is a highly ATP-demanding process. Increased expression of HSPs starts at some temperature (T_{on} or threshold temperature) above the acclimation temperature and increases until a maximum is reached (T_{peak}) and expression starts to drop (T_{off}) [29]. In marine organisms,

T_{on} was found to be close to the upper pejus temperature at which mortality starts to rise [30], while T_{off} was close to the upper critical temperature at which survival was strongly compromised [29]. There was a correlated decrease of heart rate and HSP expression in three decapod crustaceans [31]. The heat shock response and threshold temperature for HSP induction is highly plastic responding to acclimation and habitat [29,32]. Higher threshold temperatures for heat shock protein production as found in warm-acclimatized or summer animals reflect the shifted limits of thermal tolerance and also a trade-off between costs for passive thermal tolerance and costs of thermal denaturation of the protein pool at low HSP levels [32,33]. Again, the synergistic effects of elevated PCO_2 and heat shock leading to higher HSP expression in high CO_2 zoea larvae could indicate a left shift of the three key characteristics of the heat-shock response, T_{on}, T_{peak} and T_{off}, equivalent to the left-shift of the OCLTT thresholds.

An up-regulation of HSP70 in response to more alkaline or acidic seawater conditions than experienced in their natural environment could also be seen in the Antarctic bivalve *Laternula elliptica* [34], indicating a central role of HSPs in stabilizing enzymes outside their pH optimum. This function might become especially evident when pH changes are extreme or occur together with other stressors. CO_2 sensitivities of different marine taxa seem to be highly dependent on their capacities to regulate blood acid–base disturbances at elevated seawater PCO_2 [35]. The capacity to regulate acid–base disturbances might become limited when organisms are exposed to temperature extremes. As elevated seawater CO_2 and temperature concomitantly affect the acid–base status, strong acid–base disturbances leading to reduced protein function may be responsible for an up-regulation of HSP at high CO_2 and elevated temperature.

There was no combined effect of elevated seawater PCO_2 and heat shock on the gene expression of heat shock proteins in megalopa larvae. Previous studies already suggested a stronger response of the megalopa stage of Arctic *Hyas araneus* to thermal stress than to enhanced CO_2 levels [28]. CO_2 effects also tend to vanish in *Hyas araneus* megalopa larvae from a temperate population around Helgoland (North sea) [28]. The narrow thermal window of the megalopa indicates distinct stenothermy of this larval stage, which might prevent further narrowing under hypercapnia-exposure or reduce the possibility to detect any small differences in its thermal tolerance. High thermal sensitivity of the megalopa under control conditions is then paralleled by the limited capacity of stress response mechanisms to shift thermal limits or enhance the capacity for passive thermal tolerance, emphasizing the inflexibility or bottleneck characteristics of this larval stage (Figure 4).

Neither exposure to 20°C nor high CO_2 concentration induced an elevated expression of HSP60 in all *Hyas araneus* larval stages. HSP60 is a mitochondrial matrix protein and is involved in the folding of polypeptides into complex mitochondrial enzymes [36]. In crustaceans, HSP60 was found to respond to bacterial infections and contaminant exposures [37,38] and might play a more important role in the immune response than during heat stress. Heat shock protein 26 was the only heat shock protein down regulated at increased temperature. These findings are in line with those of Al-Fageeh et al. [39] and Colinet et al. [40] who found that HSP26 was induced by cold in *Drosophila melanogaster* and mammalian cells. This indicates a greater significance of HSP 26 during cold exposure.

Acid–base regulation and mitochondrial energy metabolism

Hyas araneus larvae displaying limited thermal tolerance at elevated seawater PCO_2 mirrors findings in adult specimens of edible and spider crabs [7,8]. This limitation might be attributed to the elevation in CO_2 levels or an incomplete compensation of extracellular acidosis. It is known that elevated seawater PCO_2 leads to decreasing

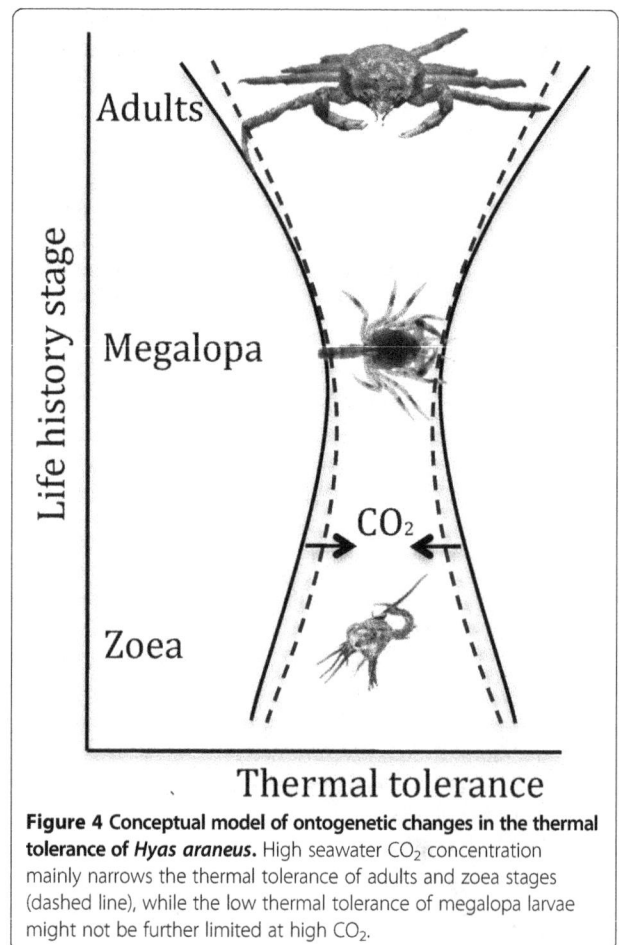

Figure 4 Conceptual model of ontogenetic changes in the thermal tolerance of *Hyas araneus*. High seawater CO_2 concentration mainly narrows the thermal tolerance of adults and zoea stages (dashed line), while the low thermal tolerance of megalopa larvae might not be further limited at high CO_2.

extracellular pH in *Hyas araneus* adults [10] which might cause metabolic depression in tissues and cells as found in invertebrates and fish [25,41]. Metabolic depression might concomitantly decrease the capacity to increase aerobic energy turnover at increasing temperatures. In our study, gene expression of acid–base related transporters and enzymes as well as enzymes from mitochondrial pathways were examined to determine whether or not acid–base regulation and/or metabolism respond to elevated seawater PCO_2 at a transcriptomic level.

Acid–base regulation under elevated CO_2 mainly involves active ion transporters like H^+-ATPase or transporters (sodium potassium chloride cotransporter NKCC; sodium bicarbonate co-transporter NBC), depending on the ion gradient maintained by sodium potassium ATPase (NaK). Carbonic anhydrase (CA) facilitates the formation of bicarbonate [35,42]. Transcript sequences related to ion and acid–base regulation and responding to thermal stress (CA, NaK, NKCC), were down regulated in *Hyas araneus* larvae from both CO_2 treatments (Table 3), reflecting thermal compensation as higher specific enzyme activities in the warmth might allow for reduced gene expression. This is in line with findings by Edge et al. [43] on coral gene expression associated with stressful temperature conditions. Coral carbonic anhydrase also showed a decrease in expression at elevated temperatures. Conversely, fish responded to cold acclimation by enhancing Na^+K^+-ATPase gene expression [44]. In *Hyas araneus* thermal compensation takes priority over CO_2 acclimation as all larval stages showed no strong response in the expression of transporters and enzymes to high seawater CO_2 levels.

In gills of adult zebrafish metabolic depression observed during hypoxia was indicated by the repression of genes in the citric acid cycle and in the electron transport system [27]. In *Hyas araneus* larvae gene expression of various genes from the citric acid cycle and the electron transport system gave no indication of CO_2 induced metabolic depression. Again the down-regulation of genes was mainly associated with an increase in temperature and in line with effects typically seen in warm acclimated eurytherms [45].

The majority of genes from mitochondrial metabolic pathways responding to CO_2 stress were up-regulated (Table 4). Up-regulation of enzymes of the electron transport system and the citric acid cycle in larvae reared at elevated seawater PCO_2 could indicate compensation for elevated demand on mitochondrial energy or compensation for reduced mitochondrial capacities under elevated CO_2 levels. The latter seems to be the case. An increased energy demand in high CO_2 larvae should be reflected in higher metabolic rates, which was not observed in *Hyas araneus*. It seems that a larger number of enzymes was necessary for the maintenance of standard metabolism in the high CO_2 treatment,

possibly caused by lowered enzyme activities at elevated seawater PCO_2. Strobel et al. [46] reported lower cytochrome c oxidase activity in the Antarctic fish *Notothenia rossii* exposed to seawater CO_2 levels of 2000 µatm. Furthermore, bicarbonate inhibits citrate synthase in mouse kidney mitochondria [47] and activates adenylyl cyclase, which produces the second messenger cAMP involved in enzyme regulation by phosphorylation and also transcription factor regulation [48,49]. As bicarbonate levels rise in parallel to rising CO_2 levels in intracellular as well as extracellular compartments in marine organisms [46,50], it might inhibit enzymes of the mitochondrial metabolic pathways. An up-regulation of these enzymes, as we found in *Hyas araneus* larvae, could be a compensatory measure to maintain standard metabolic rates and aerobic scope at high seawater CO_2 levels.

Interestingly, these regulatory shifts in ion transport and metabolism were mainly seen in 3 day old zoea II, paralleled by a lower heat shock response than in 15 day old zoea II. 63% and 27% of the corresponding genes were up-regulated in zoea II reared at high CO_2 on day 3 and day 15, respectively. This may again indicate a lower resilience in the earlier developmental stages and a lower capacity to maintain cellular homeostasis. A decreased heat shock response results in a lower protection of proteins, thus the identified decrease in gene expression of the analysed corresponding proteins might also indicate a destruction of the proteins, which cannot be seen in the 15 day old zoea II. Alternatively and more likely, 3 day old zoea II may be more thermally tolerant and still within their thermal window such that they are able to compensate for high seawater CO_2 levels by the up and down regulation of enzymes supporting cellular homeostasis and/or metabolic pathway fluxes. In contrast, less thermally tolerant 15 day old zoea II would already be forced to protect their proteins by an increased HSP response for passive survival indicating that they are beyond the temperature where regulatory mechanisms can maintain cellular functioning. Further research needs to test these alternative hypotheses.

Conclusion

Our findings reveal differences in thermal tolerance between the three larval stages of the spider crab *Hyas araneus* with the narrowest window found in the megalopa. Exposure to elevated seawater PCO_2 narrowed the thermal tolerance window of zoea larvae causing a breakdown in respiration and heart rate at a lower temperature than under control conditions. The distinct stenothermy of the megalopa stage might prevent further limitation of thermal tolerance during hypercapnic exposure.

In previous studies, effects of elevated seawater PCO_2 on thermal tolerance of marine organisms focused on whole animal performance, showing synergistic effects of high CO_2 and high temperature [8-10]. However, our knowledge of mechanisms affected by both factors and shaping sensitivities of an organism to ocean acidification and warming is far from complete and further studies are necessary. In the present study, we were able to unravel mechanisms at the molecular level that are affected by high temperature, high CO_2 and the combined action of both factors. In different larval stages of the spider crab *Hyas araneus*, we found a strong CO_2 effect with an up-regulation of genes involved in oxidative phosphorylation indicating potential compensation for enzyme activities being limited by bicarbonate inhibition. A strong increase in HSP expression in zoea stages of *Hyas araneus* under heat stress and CO_2 reflects an exacerbation of thermal stress and the capacity to adjust tolerance at the edges of the thermal window. Our study underlines the importance of integrative approaches to link molecular and cellular to whole organism responses to understand the biological consequences of ocean warming and acidification.

Methods
Larval collection and maintenance
Ovigerous females of *Hyas araneus* were collected by local fishermen in Gullmarsfjorden (west coast of Sweden, at 32 PSU and 15°C) in September 2010 and transferred to the Alfred Wegener Institute in Bremerhaven. They were maintained in flow-through aquaria at 10°C, 32 PSU and a constant dark: light cycle (12 h: 12 h). During larval hatching, which started in June 2011, twelve females were placed individually in 2 l aquaria to collect larvae of each female separately. Equal numbers of newly hatched larvae of the twelve females were pooled and subsequently transferred into 0,5 l enclosed culture vessels at a density of 30 individuals per vessel for the zoea larvae. The density was reduced to 15 larvae for the bigger megalopa stage. All experiments were conducted with larvae that had hatched within 24 h. They were reared in enclosed culture vessels filled with seawater of different CO_2 concentrations at a constant temperature of 10.0 ± 0.5°C and a salinity of 31.8 PSU (450 µatm: control treatment; 3300 µatm: high CO_2 treatment). Zoea I that moulted into the zoea II stage or zoea II that moulted into the megalopa stage at the same day were pooled together into another culture vessel at a maximum of 30 zoea larvae or 15 megalopa larvae, respectively. Seawater was provided from reservoir tanks (60 l) at 10.0 ± 0.5°C and a salinity of 31.8 PSU, continuously bubbled with an air/CO_2 mixture using a mass flow controller (HTK

Hamburg GmbH, Germany). Seawater in culture vessels and food (freshly hatched *Artemia* sp. nauplii, Sanders Brine Shrimp Company, Ogden, Utah, USA) were changed daily and dead larvae and moults were removed. Water physicochemistry was monitored by measuring temperature, salinity and pH (NBS scale, pH_{NBS}, corrected by Dixon buffered seawater) and the collections of water samples for the determination of dissolved inorganic carbon (DIC). Water PCO_2 was calculated from DIC, pH_{NBS}, temperature and salinity using the program CO_2SYS [51] (Table 6).

Larval mortality
About 200 zoea I and 120 zoea II larvae per treatment were used for investigating the effect of elevated CO_2 on the larval mortality. Mortality (number of dead zoea) were recorded on a daily basis until all larvae were either dead or moulted into the zoea II. Dead larvae and zoea II were removed. Larval total mortality were calculated and expressed as percentage.

Determination of the larval thermal tolerance window
All experiments were conducted during the middle of larval development with 9-13-day old zoea I and zoea II larvae and 14-18-day old megalopa larvae as thermal tolerance might change with development time. Measurements started at the rearing temperature of 10°C. After each measurement temperature was increased to the next experimental temperature by 3°C in 30 min. Experimental temperatures were 10°C, 13°C, 16°C, 19°C, 22°C, 25°C and 28°C. At each temperature, oxygen consumption, heart rate and maxilliped beat rate were measured in the various larval stages.

Oxygen consumption
Oxygen consumption rates of individual larvae were measured in closed, double-walled respiration chambers (OXY041 A, Collotec Meßtechnik GmbH, Niddatal, Germany). Chambers were connected via tubing to a thermostatted water bath to control temperature. Oxygen saturation was recorded by oxygen micro-optodes (NTH-PSt1-L5-TF-NS*46/0,80-YOP, PreSens GmbH, Regensburg, Germany), connected to a Microx TX3 oxygen meter (PreSens GmbH, Regensburg, Germany).

For measurements, the larvae were transferred into the respiration chamber. After each measurement, the next experimental temperature was established within half an hour. Between each measurement during adjustment of the new experimental temperature, larvae were maintained

Table 6 Seawater parameters measured during incubation

Incubation	Temperature (C°)	pH_T	DIC (µmol/kg)	PCO_2 (µatm)	Salinity (PSU)
Control	10.0 ± 0.5	8.04 ± 0.03	2277 ± 25	428 ± 35	31.8 ± 0.3
High CO_2	10.0 ± 0.5	7.18 ± 0.03	2473 ± 67	3390 ± 169	31.8 ± 0.3

Values are given in mean ± SD. N = 5 pH_T: pH total scale; DIC: dissolved inorganic carbon; PCO_2: partial pressure of CO_2.

in culture vessels containing seawater of the corresponding CO_2 concentration, which were placed in the thermostatted water bath to increase the temperature according to the experimental protocol. Afterwards larvae were allowed to acclimate for half an hour before being transferred to the respiration chamber. The plunger of the chamber lid was inserted and the volume of the chamber was reduced to 150 µl. The needle of the micro-sensor was inserted into the chamber through a hole in the lid and the sensitive tip of the optode was placed in the middle of the chamber. Respiration measurements were carried out for thirty minutes. Before each measurement, blanks were run to consider bacterial oxygen consumption. Larval oxygen consumption was expressed as μgO_2 * mg DW^{-1} *h^{-1} to allow for treatment-specific differences in larval dry weight. For all larval stages, at least six larvae from each CO_2 treatment were used to measure oxygen consumption. Individual larvae were measured at each experimental temperature.

After respiration measurements at the highest experimental temperature of 28°C, larvae were removed from the chamber and briefly rinsed with deionized water and blotted dry. For dry weight determination, larvae were stored at –20°C in pre-weighed tin cartridges, freeze-dried over night and subsequently weighed on a high precision balance (Mettler Toledo AG, Greifensee, CH-8606, CH).

Heart rate and maxilliped beat rate

Heart rates of individual larvae were measured according to Storch et al. [13]. Heart rate was recorded using a digital camera (AxioCam MRm, Carl Zeiss, Mikroimaging GmbH, Göttingen, Germany) mounted onto a microscope (Axio Observer A1, Carl Zeiss). Larvae were placed under the scope in a temperature-controlled flow-through microchamber (built at Alfred Wegener Institute, Bremerhaven, Germany) filled with seawater of the corresponding CO_2 concentration, which allowed changing the temperature according to the experimental protocol without disturbing the larvae. Temperature controlled seawater (10°C, 32PSU) was provided from a reservoir vessel placed in the thermostatted water bath and was pumped through the chamber with a flow rate of 5 ml/min to avoid a decrease in oxygen concentration due to larval respiration. Before closing the chamber, larvae were positioned in the centre of the micro-chamber by gluing the carapace to a thin glass spine, which itself was attached to a glass table. Larvae were left for 1 h to recover from handling stress and were videotaped for 1 min. Afterwards temperature was changed according to the protocol described above and at each experimental temperature the larvae were videotaped for 1 min. The video sequence was analysed for heart and maxilliped beat rates, respectively, by counting the beats min^{-1}. The beating heart can easily be seen through the transparent carapace. Heart rate and maxilliped beat rate was calculated for each larva as the mean number of beats min^{-1} ± SE from three 10s intervals. For all larval stages, five larvae from each CO_2 treatment were used to measure heart rates. The same five individual larvae were used to calculate maxilliped beat rates. Individual larvae were measured at each experimental temperature. Unfortunately, no data on pleopod beat rate of the megalopa stage could be obtained, as pleopod beating was too inconsistent for calculations.

Gene expression patterns
Sampling
Samples were taken on day 0 and day 15 post hatching in zoea I larvae, on day 3 and day 15 post moulting in zoea II and on day 3 in megalopa larvae. These time points were chosen to analyse hypercapnia-induced changes in gene expression at different time point within the larval development as CO_2 sensitivities might change with development time. On day 15, gene expression can be aligned to whole organism performance. Unfortunately, no data on gene expression could be obtained for the megalopa stage on day 15 due to loss of samples during RNA isolation. At each time point batches of 15 to 20 larvae (depending on larval stage) were transferred into 1.5 ml Eppendorf tubes containing RNAlater (Ambion, Austin, TX) and stored at –80°C. One batch of larvae from each CO_2 treatment was sampled directly from the culture vessel, while a second batch from each CO_2 treatment was heat shocked by transferring the larvae from the rearing temperature of 10°C into a 2 l glass jar containing seawater of 20°C and the corresponding CO_2 concentration. The glass jar was placed in a thermostatted water bath to keep the temperature constant. After 5 hours at 20°C larvae were sampled and frozen as described above. Each treatment (control/high CO_2 concentration at 10°C and 20°C, respectively) was replicated five times and for each replicate isolation of RNA and real-time PCR was conducted.

Isolation of RNA
Frozen samples were thawed and larvae were transferred from RNAlater into homogenisation buffer (Qiagen, Hilden, Germany). Larvae were homogenized in a Precellys homogenizer (Bertin Technologies, France) using 2 ml homogenisation tubes. Afterwards total RNA was extracted using the RNeasy kit (Qiagen, Hilden, Germany) following the manual. Extracted RNA was solubilized in 0.1 mM EDTA and 10 mM Tris and RNA purity and concentration were determined using a Thermal Scientific Nanodrop 2000 spectrometer.

Quantitative real-time PCR
10 µg of total RNA was treated with DNAse (Turbo DNA-free, Ambion) in order to digest genomic DNA remnants

Table 7 Primers of 21 genes used for RT-qPCR

Primer ID	Protein description	Primer sequence	Primer efficiency
Cellular stress/heat shock response			
HSP 26	Heat shock protein 26	F_AGGCAAGAGGCCGACAGA	1.98
		B_AAGCGGCGGTTGAAACG	
HSP 60	Heat shock protein 60	F_GCCAACGGCACCTTCGT	1.99
		B_CCTTGGTGGGATCGATGATT	
HSP 70_1	Heat shock protein 70	F_AGCACTTCGTCGGCTAAGGA	2.01
		B_CCTGGGCAGATGATGAAAGAG	
HSP 70_2	Heat shock protein 70	F_GTGTGGGCGTGTTCAAGAATG	2.03
		B_CGGTTGCCCTGGTCGTT	
HSP 70_3	Heat shock protein 70	F_CAACGTGCTCATCTTCGATCTG	1.98
		B_CTCGATGGTCAGGATGGATACA	
HSP 70_4	Heat shock protein 70	F_CCAACATGTCGGGAGAGATGA	2.00
		B_CATGAGCGTTCCCCTAGGAA	
HSP 90	Heat shock protein 90	F_GACACATCCACCATGGGATACA	2.03
		B_TGCTGTGGTCTGGGTTGATC	
Acid–base regulation			
CA	Carbonic anhydrase	F_TACGTGTCGGCCGATAGCA	1.92
		B_AAAGTCCGACCCGCTTCAC	
NBC	Sodium bicarbonate cotransporter	F_CCGCCGTCATTGTCAACAG	2.01
		B_TGGTATCCGCCACCCTTCT	
NKA	Sodium potassium ATPase	F_CCCCGAGAGGATCCTTGAAC	2.06
		B_AGGCTTCTCCTCGCCATTC	
NKCC	Sodium potassium chloride cotransporter	F_GGGCAAGGACATCAGAAAGG	1.96
		B_TCTACTTCACGGCGGAGCTT	
NHE	Sodium hydrogen exchanger	F_GCGGAGACCTGCTGGCTAT	1.99
		B_CGACTTGGCTAACACGTATTGG	
VA	V1-ATPase	F_CACCCCATCCCCGATCTC	1.96
		B_CTGCCGCTCCACGTAGATTT	
Mitochondrial energy metabolism			
atpA	ATP synthase	F_GGTGAATACTTCCGCGACAAC	1.96
		B_TGCTTGGACAGATCGTCGTAGA	
CCR	Cytochrome C reductase	F_GATCAGACCCAGACCAGTCCTT	1.99
		B_CATAGCGCCGGAGGTGTT	
COX	Cytochrome C oxidase	F_CGCTGCAGATGTTATTCACTCAT	1.99
		B_TCCAGGGATAGCATCAGCTTTT	
IDH	Isocitrate dehydrogenase	F_TGGCTCAAAAGAGGACCTATGCA	1.95
		B_CCACCACCGGGTTCTTCAC	
NAD	NADH dehydrogenase	F_CCCATAATTAACATCTCGGCAA	1.98
		B_CTGCCCACATTGATTTAGCTTTT	
SDH	Succinate dehydrogenase	F_CTCCGAGGAGAGGCTCAAGA	2.02
		B_GGTGTGGCAGCGGTATACG	

Table 7 Primers of 21 genes used for RT-qPCR *(Continued)*

PDH	Pyruvate dehydrogenase	F_CTGGACGAGGAGACCATCGT	2.02
		B_TCCACCGTCACCAGGTTGT	
Potential housekeeping			
Tub	Tubulin	F_GAGGACGCGGCCAACA	2.03
		B_GACAATTTCCTTGCCGATGGT	

Genes were classified according to their function into cellular stress/heat shock response, acid–base regulation and mitochondrial energy metabolism.

and RNA concentration was measured again (NanoDrop). Subsequently, 0.4 μg total RNA was subject to cDNA synthesis using the High capacity cDNA Reverse Transcriptase kit (Applied Biosystems, Darmstadt, Germany). Expression of 20 important genes involved in mitochondrial energy metabolism, acid–base regulation and stress/heat shock response, were analysed because they were assumed to be affected by synergistic effects of temperature and CO_2. Furthermore, tubulin was chosen as potential housekeeping candidate. Primers (Table 7) for 21 genes were designed using the Primer Express software for real-time PCR (version 3.0, Applied Biosystems, Darmstadt, Germany). Sequences were obtained from the recently in our lab established transcriptome of Arctic *Hyas araneus* [52]. The PCR was performed using a 7300 Real Time PCR System (Applied Biosystems) and the Sybr green qPCR master mix (Fermentas). For all primers, PCR efficiency was examined by using six different dilutions of template cDNA (1:20; 1:40; 1: 100; 1:200; 1:1000; 1:2000). Efficiency was estimated by $10^{(-1/slope)}$ and was for all primers >1.9 (Table 7). Primer concentrations were 300 nM and all genes were amplified with 1 ng of template cDNA. After amplification, a melting curve was acquired to verify the specific amplification of fragments.

C_t values of all genes were transformed into quantities (Q) using the formula $Q = E^{(-Ct)}$ with E being the reaction specific efficiency and C_t being the C_t value determined by the 7300 Real Time PCR System. The analysis by geNorm plus suggested VATPase (VA) and sodium hydrogen exchanger (NHE) as the most stable expressed genes in all treatments (VA M = 0.310; NHE M = 0.325). A normalization factor was determined by calculating the geometric mean of these two reference genes. Quantities of each sample for each gene were divided by the appropriate normalisation factor. The initially intended house keeping gene tubulin was one of the genes responding most to PCO_2 and temperature changes and could, thus, not be used as house keeping gene.

Statistical analysis

Results were analysed using SigmaPlot (Version 12, Systat Software, Inc., San Jose, California). All data were checked for outliers by Nalimov's test [53]. A two-way repeated measures ANOVA was used to investigate effects of CO_2 concentration and temperature on larval oxygen consumption, heart rate and maxilliped rate. Tukey's multiple comparison tests were used for a posteriori analysis.

An unpaired t-test was conducted to analyse the effect of temperature on gene expression in zoea I on day 0. When data did not meet assumptions for an unpaired t-test, a Mann–Whitney Rank Sum test was run. Two-way ANOVAs were applied to analyse the effect of CO_2 concentration and temperature on gene expression for each time point within the different larval stages. Tukey's multiple comparison tests were used for a posteriori analysis.

Ethics statement

This study was carried out according to the ethics and guidelines of German law and did not involve endangered or protected species. According to §8 Tierschutzgesetz (18.05.2006; 8081. I p. 1207), the experiments on crustaceans in this study do not require formal approval, but have been indicated nonetheless to the ethics committee of the Senatorin für Arbeit, Frauen, Gesundheit, Jugend und Soziales, Abt. Veterinärwesen, Lebensmittelsicherheit und Pflanzenschutz, Bahnhofsplatz 29, 28195 Bremen, Germany.

Additional file

Additional file 1: Figure S1. Mortality (%) of zoea I (black) and zoea II larvae (grey) of *Hyas araneus* reared under 420 μatm and 3300 μatm CO2.

Competing interests
The authors declare that they have no competing interests.

Authors' contributions
MS developed the concept of this study, carried out the experiments and the molecular genetic studies and drafted the manuscript. LH carried out the experiments and participated in the laboratory work. ML participated in the gene expression analysis. FCM participated in the study design, coordination of the study and substantially contributed to writing the manuscript. HOP participated in the coordination of the study and contributed to revising the manuscript. DS participated in the study design, coordination of the study and substantially contributed to writing the manuscript. All authors read and approved the final version of the manuscript.

Acknowledgements
Financial support was provided by Federal Ministry of Education and Research (BMBF), within phase I of the BIOACID research programme to MS, LH, ML, FCM, HOP and DS (FKZ 03F0608B WP 2.2.1 to PI D. Storch). We thank T. Sandersfeld and R.E. Dinner for technical assistance. We wish to thank J.H. Stillman for his technical and intellectual support.

Author details
[1]Integrative Ecophysiology, Alfred-Wegener-Institute for Polar and Marine Research, Am Handelshafen 12, 27570 Bremerhaven, Germany. [2]Scientific Computing, Alfred-Wegener-Institute for Polar and Marine Research, Am Handelshafen 12, 27570 Bremerhaven, Germany.

References
1. Meinshausen M, Smith SJ, Calvin K, Daniel JS, Kainuma MLT, Lamarque JF, Matsumoto K, Montzka SA, Raper SCB, Riahi K, Thomson A, Velders GJM, van Vuuren DPP: **The RCP greenhouse gas concentrations and their extensions from 1765 to 2300.** *Climate Change* 2011, 109(Suppl1-2):213–241.
2. Caldeira K, Wickett ME: **Ocean model predictions of chemistry changes from carbon dioxide emissions to the atmosphere and ocean.** *J Geophys Res Oceans* 2005, 110:C09S04.
3. Pörtner HO, Langenbuch M, Michaelidis B: **Synergistic effects of temperature extremes, hypoxia, and increases in CO_2 on marine animals: from earth history to global change.** *J Geophys Res* 2005, 110:C09S10.
4. Pörtner HO: **Climate change and temperature-dependent biogeography: oxygen limitation of thermal tolerance in animals.** *Naturwissenschaften* 2001, 88:137–146.
5. Frederich M, Pörtner HO: **Oxygen limitation of thermal tolerance defined by cardiac and ventilatory performance in spider crab, *Maja squinado*.** *Am J Physiol Regul Integr Comp Physiol* 2000, 279:R1531–R1538.
6. Pörtner HO: **Ecosystem effects of ocean acidification in times of ocean warming: a physiologist's view.** *Mar Ecol Prog Ser* 2008, 373:203–217.
7. Metzger R, Sartoris FJ, Langenbuch M, Pörtner HO: **Influence of elevated CO_2 concentrations on thermal tolerance of the edible crab *Cancer pagurus*.** *J Therm Biol* 2007, 32:144–151.
8. Walther K, Sartoris FJ, Bock C, Pörtner HO: **Impact of anthropogenic ocean acidification on thermal tolerance of the spider crab *Hyas araneus*.** *Biogeosciences* 2009, 6:2207–2215.
9. Munday PL, Crawley NE, Nilsson GE: **Interacting effects of elevated temperature and ocean acidification on the aerobic performance of coral reef fishes.** *Mar Ecol Prog Ser* 2009, 388:235–242.
10. Zittier ZMC, Hirse T, Pörtner HO: **The synergistic effects of increasing temperature and CO_2 levels on activity capacity and acid–base balance in the spider crab, *Hyas araneus*.** *Mar Biol* 2013, 160:2049–2062.
11. Kurihara H: **Effects of CO_2-driven ocean acidification on the early developmental stages of invertebrates.** *Mar Ecol Prog Ser* 2008, 373:275–284.
12. Pörtner HO, Farrell AP: **Physiology and climate change.** *Science* 2008, 322:690–692.
13. Storch D, Fernández M, Navarrete SA, Pörtner HO: **Thermal tolerance of larval stages of the Chilean kelp crab *Taliepus dentatus*.** *Mar Ecol Prog Ser* 2001, 429:157–167.
14. Parker LM, Ross PM, O'Connor WA: **The effect of ocean acidification and temperature on the fertilization and embryonic development of the Sydney rock oyster *Saccostrea glomerata* (Gould 1850).** *Glob Change Biol* 2009, 15(Suppl9):2123–2136.
15. Padilla-Gamiño JL, Kelly MW, Evans TG, Hofmann GE: **Temperature and CO_2 additively regulate physiology, morphology and genomic responses of larval sea urchins, *Strongylocentrotus purpuratus*.** *Proc R Soc B* 2013, 280(Suppl1759). doi:10.1098/rspb.2013.0155.
16. Brennand HS, Soars N, Dworjanyn SA, Davis AR, Byrne M: **Impact of ocean warming and ocean acidification on larval development and calcification in the sea urchin *Tripneustes gratilla*.** *Plos One* 2010, 5(Suppl6):1–7.
17. Talmage SC, Gobler CJ: **Effects of elevated temperature and carbon dioxide on the growth and survival of larvae and juveniles of three species of Northwest Atlantic bivalves.** *Plos One* 2011, 6(Suppl10):1–12.
18. Schiffer M, Harms L, Pörtner HO, Lucassen M, Mark FC, Storch D: **Tolerance of *Hyas araneus* zoea I larvae to elevated seawater PCO_2 despite elevated metabolic costs.** *Mar Biol* 2013, 160(Suppl8):1943–1953.
19. Schiffer M, Harms L, Pörtner HO, Mark FC, Storch D: **Pre-hatching seawater PCO_2 affects development and survival of the zoea stages of the Arctic spider crab *Hyas araneus*.** *Mar Ecol Pro Ser* 2014, doi:10.3354/meps10687.
20. Christiansen ME: **Crustacea Decapoda Brachyura.** In *Marine invertebrates of Scandinavia*, Volume 2. Oslo: Universitetsforlaget; 1969.
21. Pörtner HO, Bock C, Reipschläger A: **Modulation of the cost of pHi regulation during metabolic depression: a 31P-NMR study in invertebrate (*Sipunculus nudus*) isolated muscle.** *J Exp Biol* 2000, 203:2417–2428.
22. Todgham AE, Hofmann GE: **Transcriptomic response of sea urchin larvae *Strongylocentrotus purpuratus* to CO_2-driven seawater acidification.** *J Exp Biol* 2009, 212:2579–2594.
23. O'Donnell MJ, Hammond LM, Hofmann GE: **Predicted impact of ocean acidification on a marine invertebrate: elevated CO_2 alters response to thermal stress in sea urchin larvae.** *Mar Biol* 2009, 156:439–446.
24. Chapman R, Mancia A, Beal M, Veloso A, Rathburn C, Blair A, Holland AF, Warr GW, Didinato G, Sokolova IM, Wirth EF, Duffy E, Sanger D: **The transcriptomic responses of the eastern oyster, *Crassostrea virginica*, to environmental conditions.** *Mol Ecol* 2011, 20(Suppl 7):1431–1449.
25. Reipschläger A, Pörtner HO: **Metabolic depression during environmental stress: the role of extracellular versus intracellular pH in *Sipunculus nudus*.** *J Exp Biol* 1996, 199:1801–1807.
26. Pörtner HO, Reipschlager A, Heisler N: **Acid–base regulation, metabolism and energetics in *Sipunculus nudus* as a function of ambient carbon dioxide level.** *J Exp Biol* 1998, 201:43–55.
27. van der Meer DLM, van den Thillart GEEJM, Witte F, de Bakker MAG, Besser J, Richardson MK, Spaink HP, Leito JTD, Bagowski CP: **Gene expression profiling of the long-term adaptive response to hypoxia in the gills of adult zebrafish.** *Am J Physiol Regul Integr Comp Physiol* 2005, 289:R1512–R1519.
28. Walther K, Anger K, Pörtner HO: **Effects of ocean acidification and warming on the larval development of the spider crab *Hyas araneus* from different latitudes (54° vs. 79°N).** *Mar Ecol Prog Ser* 2010, 417:159–170.
29. Tomanek L, Somero GN: **Evolutionary and acclimation-induced variation in the heat-shock response of congeneric marine snails (genus *Tegula*) from different thermal habitats: implications for limits of thermotolerance and biogeography.** *J Exp Biol* 1999, 202:2925–2936.
30. Anestis A, Lazou A, Pörtner HO, Michaelidis B: **Behavioral, metabolic, and molecular stress responses of marine bivalve *Mytilus galloprovincialis* during long-term acclimation at increasing ambient temperature.** *Am J Physiol Regul Integr Comp Physiol* 2007, 293:R911–R921.
31. Jost JA, Podolski SM, Frederich M: **Enhancing thermal tolerance by eliminating the pejus range: a comparative study with three decapod crustaceans.** *Mar Ecol Prog Ser* 2012, 444:263–274.
32. Buckley BA, Owen ME, Hofmann GE: **Adjusting the thermostat: the threshold induction temperature for the heatshock response in intertidal mussels (genus *Mytilus*) changes as a function of thermal history.** *J Exp Biol* 2001, 204:3571–3579.
33. Hamdoun AM, Cheney DP, Cherr GN: **Plasticity of HSP70 and HSP70 gene expression in the Pacific Oyster (*Crassostrea gigas*): Implications for thermal limits and induction of thermal tolerance.** *Biol Bull* 2003, 205:160–169.
34. Cummings V, Hewitt J, Van Rooyen A, Currie K, Beard S, Thrush S, Norkko J, Barr N, Heath P, Halliday NJ, Sedcole R, Gomez A, McGraw C, Metcalf V: **Ocean acidification at high latitudes: potential effects on functioning of the antarctic bivalve *Laternula elliptica*.** *PLoS One* 2011, 6(Suppl1):1–11.
35. Melzner F, Gutowska MA, Langenbuch M, Dupont S, Lucassen M, Thorndyke MC, Bleich M, Pörtner HO: **Physiological basis for high CO_2 tolerance in marine ectothermic animals: pre-adaptation through lifestyle and ontogeny.** *Biogeosciences* 2009, 6:2313–2331.
36. Briones P, Vilaseca MA, Ribes A, Vernet A, Lluch M, Cusi V, Huckriede A, Agsteribbe E: **A new case of multiple mitochondrial enzyme deficiencies with decreased amount of heat shock protein 60.** *J Inherit Metab Dis* 1997, 20:569–577.
37. Werner I, Nagel R: **Stress proteins HSP60 and HSP70 in three species of amphipods exposed to cadmium, diazinon, dieldrin and fluoranthene.** *Environ Toxicol Chem* 1997, 16(Suppl11):2393–2403.
38. Zhou J, Wang WN, He WY, Zheng Y, Wang L, Xin Y, Liu Y, Wang AL: **Expression of HSP60 and HSP70 in white shrimp, *Litopenaeus vannamei* in response to bacterial challenge.** *J Invertebr Pathol* 2010, 103(Suppl3):170–178.
39. Al-Fageeh MB, Marchant RJ, Carden MJ, Smales CM: **The cold-shock response in cultured mammalian cells: harnessing the response for the improvement of recombinant protein production.** *Biotechnol Bioeng* 2006, 93(Suppl5):829–835.
40. Colinet H, Lee SF, Hoffmann A: **Temporal expression of heat shock genes during cold stress and recovery from chill coma in adult *Drosophila melanogaster*.** *FEBS J* 2010, 277(Suppl1):174–185.
41. Langenbuch M, Pörtner HO: **Energy budget of hepatocytes from Antarctic fish (*Pachycara brachycephalum* and *Lepidonotothen kempi*) as a function of

ambient CO$_2$: pH-dependent limitations of cellular protein biosynthese? *J Exp Biol* 2003, **206**:3895–3903.

42. Henry RP, Cameron JN: **The role of carbonic-anhydrase in respiration, ion regulation and acid–base-balance in the aquatic crab** *Callinectes-sapidus* **and the terrestrial crab** *Gecarcinus lateralis. J Exp Biol* 1983, **103**:205–223.

43. Edge SE, Morgan MB, Gleason DF, Snell TW: **Development of a coral cDNA array to examine gene expression profiles in** *Montastraea faveolata* **exposed to environmental stress.** *Mar Pollut Bull* 2005, **51**:507–523.

44. Metz JR, van den Burg EH, Wendelaar Bonga SE, Flik G: **Regulation of branchial Na+/K + –ATPase in common carp** *Cyprinus carpio* **L. acclimated to different temperatures.** *J Exp Biol* 2003, **206**:2273–2280.

45. Lannig G, Storch D, Pörtner HO: **Aerobic mitochondrial capacities in Antarctic and temperate eelpout (Zoarcidae) subjected to warm versus cold acclimation.** *Polar Biol* 2005, **28**:575–584.

46. Strobel A, Bennecke S, Leo E, Mintenbeck K, Pörtner HO, Mark FC: **Metabolic shifts in the Antarctic fish** *Notothenia rossii* **in response to rising temperature and PCO$_2$.** *Front Zool* 2012, **9**(Suppl28):1–15.

47. Simpson DP: **Regulation of renal citrate metabolism by bicarbonate ion and pH: observations in tissue slices and mitochondria.** *J Clin Invest* 1967, **46**(Suppl2):225.

48. Acin-Perez R, Salazar E, Kamenetsky M, Buck J, Levin LR, Manfredi G: **Cyclic AMP produced inside mitochondria regulates oxidative phosphorylation.** *Cell Metab* 2009, **9**(Suppl3):265–276.

49. Tresguerres M, Barott KL, Barron ME, Roa JN: **Established and potential physiological roles of bicarbonate-sensing soluble adenylyl cyclase (sAC) in aquatic animals.** *J Exp Biol* 2014, **217**:663–672.

50. Pane EF, Barry JP: **Extracellular acid–base regulation during short-term hypercapnia is effective in a shallow-water crab, but ineffective in a deep-sea crab.** *Mar Ecol Prog Ser* 2007, **334**:1–9.

51. Lewis E, Wallace DWR: *Program Developed for CO$_2$ System Calculations, ORNL/CDIAC-105.* Oak Ridge, Tennessee: Carbon Dioxide Information Analysis Center, Oak Ridge National Laboratory; 1998.

52. Harms L, Frickenhaus S, Schiffer M, Mark FC, Storch D, Pörtner HO, Held C, Lucassen M: **Characterization and analysis of a transcriptome from the boreal spider crab** *Hyas araneus. Comp Biochem Phys D* 2013, **8**:344–351.

53. Noack S: *Statistische Auswertung von Mess- und Versuchsdaten mit Taschenrechner und Tischcomputer.* Berlin: Walter de Gruyter; 1989.

13

Sparing spiders: faeces as a non-invasive source of DNA

Daniela Sint[*], Isabella Thurner, Ruediger Kaufmann and Michael Traugott

Abstract

Introduction: Spiders are important arthropod predators in many terrestrial ecosystems, and molecular tools have boosted our ability to investigate this taxon, which can be difficult to study with conventional methods. Nonetheless, it has typically been necessary to kill spiders to obtain their DNA for molecular applications, especially when studying their diet.

Results: We successfully tested the novel approach of employing spider faeces as a non-invasive source of DNA for species identification and diet analysis. Although the overall concentration of DNA in the samples was very low, consumer DNA, suitable for species identification, was amplified from 84% of the faecal pellets collected from lycosid spiders. Moreover, the most important prey types detected in the gut content of the lycosids were also amplified from the faecal samples.

Conclusion: The ability to amplify DNA from spider faeces with specific and general primers suggests that this sample type can be used for diagnostic PCR and sequence-based species and prey identification such as DNA barcoding and next generation sequencing, respectively. These findings demonstrate that faeces provide a non-invasive alternative to full-body DNA extracts for molecular studies on spiders when killing or injuring the animal is not an option.

Keywords: Lycosidae, Molecular species identification, Molecular prey detection, Multiplex PCR

Introduction

Spiders are a diverse invertebrate group with more than 37,000 described species [1]; they inhabit almost all terrestrial ecosystems, where they are important arthropod predators [2-4]. Identifying spiders based on their morphological traits can be challenging (especially in the case of juveniles and females), but DNA-based identification can simplify this task [5,6]. Similarly, as spiders are liquid feeders, morphological identification of prey remains is of limited use in studying their feeding ecology. Again, molecular tools significantly improved the diet analysis of spiders: solid prey remains (e.g. from webs) are no longer needed to analyse what has been consumed [7-9]. The flip side of the coin is that both molecular identification and prey detection usually requires killing the animals to dissect their gut or using the whole animal for DNA extraction. For species identification, it is at minimum necessary to injure the spider

when inducing autotomy of legs [10]. This is because, with few exceptions, only juveniles will moult and be able to replace missing limbs [2]. This renders these techniques unsuitable for situations in which spider identity and/or prey need to be determined non-invasively, e.g. when investigating threatened species or to identify spiders for subsequent ethological studies. Different types of non-invasively collected samples have been successfully used for molecular analysis, but most of these assays have been developed for vertebrates, which more readily provide suitable sample types such as hairs, feathers, saliva, regurgitates, shed skin, and faeces [11]. Less work has been conducted on invertebrates: slugs were identified based on body swabs [12], high-quality prey DNA was retrieved from carabid regurgitates [13,14], and DNA was successfully analysed from faeces of lobsters [15], millipedes [16], and beetle larvae [17]. In spiders, the complete exuviae of large tarantulas provided enough DNA for molecular species identification [18], but mature spiders usually do not moult [2], making it is impossible to obtain this sample type

* Correspondence: Daniela.Sint@uibk.ac.at
Institute of Ecology, University of Innsbruck, Technikerstraße 25, 6020 Innsbruck, Austria

from adult individuals. Even if an individual will continue moulting, it might be necessary to keep that spider for a lengthier time until the skin is shed. Spiders do not provide regurgitates or mucus, which proved to be a useful source of DNA in other invertebrates. Faeces may provide a valuable source for non-invasively collected DNA samples, but the suitability of spider faeces for molecular analysis has never been tested.

Here we explore whether spider faeces enable non-invasive species identification and diet analysis. Field-collected wolf spiders (Lycosidae) and the corresponding faecal material produced upon collection were used to compare the success in molecular prey detection and species identification between the two sample types (full-body DNA extract and faecal sample) using a series of multiplex PCR assays. This effort included the application of two newly developed PCR systems.

Results

In all 189 full-body DNA extracts of wolf spiders, DNA of *Pardosa* spp. was detected at genus and at species level using the multiplex PCR systems IPC [19] (detects intraguild predation and collembolan DNA) and DUP (duplex PCR system detecting DNA of *P. nigra* and *P. saturatior*), respectively. A total of 185 faecal samples (98%) contained amplifiable DNA: these were compared to the respective full-body DNA extracts regarding their suitability for molecular analysis.

Overall the detection of lycosid DNA was possible in 84% of the 185 faecal samples: 73% of the faecal samples tested positive at genus level (IPC assay) and 72% scored positive at species level for DNA of *P. saturatior* or *P. nigra* (DUP assay). In 95% of the 130 cases in which both the faecal and full-body sample could be assigned to one of the two *Pardosa* species, the identification was congruent between the two sample types. In five out of the six divergent samples, DNA of both *Pardosa* species was found in the full-body DNA extract (see below).

In 22 of the full-body samples and in one faecal sample, DNA of both *Pardosa* species was detected by DUP. This indicates predation among the two species. In all but two of these samples, one of the two amplicons showed at least double the signal strength compared to the other one (mean signal ratio strong : weak was 17:1). This enabled differentiating the consumer (strong signal) from the prey (weak signal). Taking this differentiation into account, 99% of the lycosids could be identified from the full-body DNA extracts as *P. nigra* (146 individuals) and P. *saturatior* (37 individuals).

When testing the full-body extracts for prey DNA using the LIN (PCR system detecting DNA of five linyphiid species) and IPC assays, nine non-*Pardosa* prey taxa were detected: DNA of collembolans and the linyphiid *Erigone tirolensis* were amplified in 123 and 56 samples, respectively. This was significantly more often

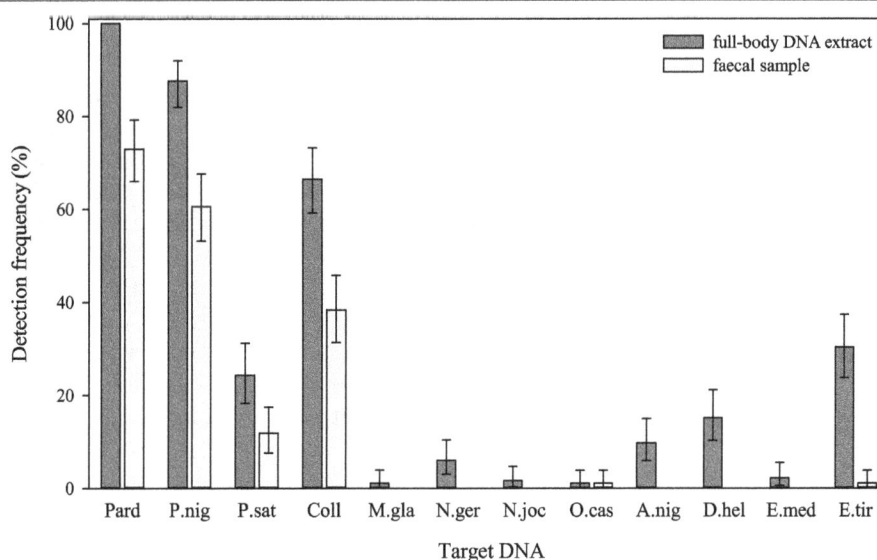

Figure 1 Detection frequencies of consumer- and prey-DNA. Detection frequencies of lycosid and prey DNA in DNA extracts of full-body and faecal samples of lycosid spiders collected in three glacier forelands using three diagnostic multiplex PCR assays. Targeted DNA: *Pardosa* spp. at genus level (Pard), *P. nigra* (P.nig), *P. saturatior* (P.sat), Collembola (Coll), glacier harvestman *Mitopus glacialis* (M.gla), carabid beetles [*Nebria germari* (N.ger), *N. jockischii* (N.joc), *Oreonebria castanea* (O.cas)], and linyphiid spiders [*Agyneta nigripes* (A.nig), *Diplocephalus helleri* (D.hel), *Entelecara media* (E.med), *Erigone tirolensis* (E.tir)]. DNA of the linyphiid *Janetschekia monodon* and of the carabid *N. rufescens* were not detected. Error bars represent 95% tilting confidence intervals from 9999 bootstrap resamples; non-overlapping confidence intervals are interpreted as significant differences.

than the many other prey tested for (Figure 1). The number of individuals where at least one prey type was detected was significantly higher in the full-body (79%) compared to faecal samples (39%; $\chi^2 = 61.32$, p < 0.001). Overall, only three non-*Pardosa* prey taxa could be detected in the faecal samples, whereby DNA of collembolans was detected most frequently (38%) and two faecal samples each tested positive for DNA of *E. tirolensis* and *Oreonebria castanea* (Figure 1).

Discussion

This study demonstrates that spider faeces can be used for both species identification and prey detection, which enables a non-invasive examination of spiders. Although mainly genus- and species-specific primers were applied in this study to identify the consumer and its prey, DNA from spider faeces was also successfully amplified with general primers which are commonly used for DNA barcoding species and for sequence-based prey identification (Additional file 1; [20,21]). This demonstrates that the suite of molecular techniques that can be employed to analyse DNA from spider faeces is not limited to highly-specific diagnostic PCR assays, but that other molecular methods for species and/or prey identification such as DNA barcoding approaches and next generation sequencing could potentially be applicable to this non-invasively derived sample type.

The ability to molecularly identify the consumer and the food from the same sample is important in ecological studies. We found that faeces were especially useful for species identification: consumer DNA was amplified from 84% of the samples. This percentage ranges on the higher end of the scale for the presence of consumer DNA in non-invasive samples reported from other animals (e.g. [16,22-24]). Molecularly differentiating full-body DNA extracts between *P. nigra* and *P. saturatior* by the newly-developed duplex PCR system, which was optimized to amplify both species with approximately the same efficacy, enabled us to identify whether DNA of *P. nigra* or *P. saturatior* dominated the sample [19]. As full-body DNA extracts contain consumer DNA in large excess over prey DNA [25], we could identify predator and prey also in samples where both types of DNA were present. The few discrepancies in species identification between full-body and faecal samples probably reflect the absence of consumer DNA in the faeces, i.e. only the DNA of the consumed wolf spider was present and misidentified as indicating the consumer. This interpretation is supported by the fact that in five out of six questionable samples, DNA of both *Pardosa* species was present in the full-body DNA extracts. In one wolf spider, DNA of *P. nigra* and *P. saturatior* was even

present in both sample types, but the identification of the predator and prey, based on amplicon strength, was not congruent (the difference was clear for the full-body DNA extract, but in the faeces the prey gave a slightly stronger signal). Nevertheless, this ability to identify a spider based on its faeces is an encouraging result for further investigations, as for example when living spiders need to be identified for field or laboratory studies.

Although prey DNA detection success was significantly lower in faeces than in full-body samples, the most common prey items were also detected in the non-invasive samples. The reduced ability to detect prey DNA in faeces is probably due to the high efficiency with which spiders digest their food [2]. This yields a very low DNA content in the faecal samples, which was <1 ng/µl in the current study (see Additional file 1). Nonetheless, using spider faeces as a source of dietary information is useful to get a general idea of the spider diet in situations where the whole spider cannot be killed to obtain a full-body DNA extract to perform molecular gut content analysis. Such situations arise, for example, when prey choice needs to be examined in rare and/or protected spiders, in habitats such as in national parks where killing animals is prohibited, or when the impact of lethal sampling on the studied system needs to be minimized.

We expected spider faeces to have a DNA content comparable to beetle regurgitates, typically containing between 1 and 10 ng ds DNA/µl [26], and which are well suited to track the beetle's diet [13,14]. In the present faecal samples, however, the observed DNA content was much lower (<1 ng ds DNA/µl). We therefore suggest optimizing the DNA extraction process to increase the DNA concentration and, consequently, the robustness to detect and identify prey DNA. One strategy to achieve a higher DNA concentration would be to reduce the volume of buffer when eluting (silica-based extraction protocols) or resolving (e.g. CTAB protocols) the DNA at the end of the DNA extraction process. A concentrating step following DNA extraction could also be performed. Another approach to increase the concentration of prey DNA in faecal samples would be to pool several faecal pellets of one individual into one sample for DNA extraction.

Our work indicates that obtaining a reasonable number of faecal samples is straightforward because more than 50% of the captured lycosid spiders defecated within a few hours upon collection. The number of faecal pellets that can be gathered after spider collection might be increased even further by feeding the animals, as many species defecate stored excrements when provided with fresh prey [2]. Prey DNA can be tracked in spiders for extended times post-feeding

[27-29]: for example, cricket DNA could be detected in full-body DNA extracts of *Pardosa* spp. for a minimum of 84 h after a single meal [30]. It is likely that, along with small amounts of predator DNA, also prey DNA is continuously excreted during these long digestion times. While a single faecal pellet might not contain enough DNA to enable successful detection of prey and/or consumer DNA, the amount contained in several pellets might be sufficient. For prey detection it is important that the spider is not fed during the timespan in which faeces are collected to avoid mixing laboratory- and field-consumed prey. Alternatively, if feeding the spider is desired to enhance faecal production (see above), the spider can be fed with prey which does not occur in the habitat the spider was collected from. These strategies, together with the application of highly sensitive PCR assays or a next generation sequencing approach [31] – where the sequence information of individual molecules is read – might further improve the molecular information that can be obtained from spider faeces beyond the possibilities reported in this study.

Methods

Pardosa nigra and *P. saturatior* (Araneae: Lycosidae) were collected by dry pitfall-trapping between 8 and 23 July 2010 in three neighbouring glacier forelands (Gaisbergtal, Rotmoostal, Langtal) in Tyrol, Austria. Spiders were placed individually in 2 ml reaction tubes, transported to a field station in a cool box, and frozen at –24 °C. In case defecation occurred during transport,

the spider was transferred into a clean tube before freezing, and lysis-buffer was added to the tube containing the faeces. In total, 189 faecal pellets were obtained from 347 sampled spiders. For details on sample processing and DNA extraction see Additional file 1.

Two new multiplex PCR systems (DUP, LIN) were developed (see Additional file 1) and, together with an already existing multiplex PCR system (IPC, [19]), used to screen the DNA extracts of the spiders and their faecal samples for consumer- and prey-DNA. First, the duplex PCR system (DUP, Figure 2) was applied to molecularly identify the samples as either *P. nigra* or *P. saturatior*. Second, all samples were screened for DNA of potential prey taxa common in the three glacier forelands using the LIN and IPC multiplex PCR systems: the LIN system (Figure 2) enables tracking of predation on five linyphiid spider species commonly occurring in the glacier forelands [6], while the IPC system detects predation on four species of carabid beetles (*Nebria germari*, *N. jockischii*, *N. rufescens*, *Oreonebria castanea*), the glacier harvestman *Mitopus glacialis*, and collembolans (IPC, [19]). Additionally, IPC amplifies DNA of *Pardosa* spp. as an internal positive control. PCR products were separated, visualized, and scored on QIAxcel (Qiagen, Hilden, Germany), an automatic capillary electrophoresis system.

To compare the detection frequencies of the different targets, 9999 bootstrap resamples were drawn with replacement from the observed data using the software TIBCO Spotfire S+ 8.1. Non-overlapping 95% tilting confidence intervals were used as a conservative estimation of significant differences.

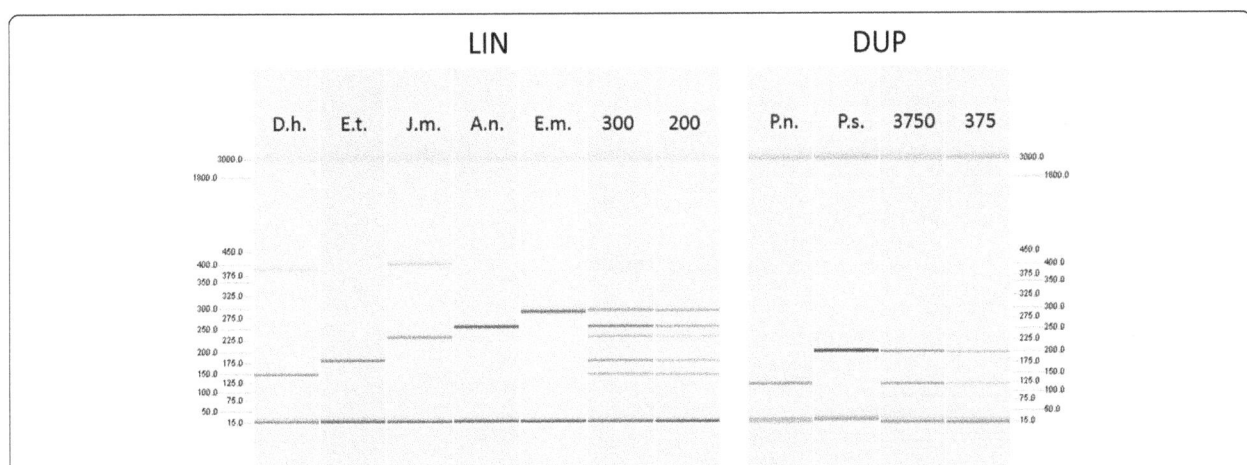

Figure 2 Gel image of the newly developed multiplex PCR systems. QIAxcel gel image of PCR products generated with the multiplex PCR system LIN and the duplex PCR system DUP. LIN: *Diplocephalus helleri* (D.h.; 151 bp), *Erigone tirolensis* (E.t.; 186 bp), *Janetschekia monodon* (J.m.; 240 bp), *Agyneta nigripes* (A.n.; 264 bp), *Entelecara media* (E.m.; 298 bp), artificial mixes containing 300 and 200 double stranded (ds) templates per target, respectively. DUP: *Pardosa nigra* (P.n.; 118 bp), *Pardosa saturatior* (P.s.; 202 bp), artificial mixes containing 3750 and 375 ds templates per target species, respectively. An internal marker is run alongside each sample (15 and 3000 bp), and the scale on the left and right side enables an estimation of fragment length. At higher template DNA concentrations, D.h. and J.m. may produce an additional amplicon of ~390 and ~400 bp, respectively.

Additional file

Additional file 1: Details on sample processing and multiplex PCR systems.

Competing interests
The authors declare that they have no competing interests.

Authors' contributions
DS participated in study design, conducted field work and molecular analysis, developed the multiplex PCR systems and drafted the manuscript. IT conducted molecular analysis. RK participated in study design and conducted field work. MT participated in study design and conducted field work. All authors read and improved the manuscript.

Acknowledgments
We thank L. Raso, who supported this study substantially but forwent co-authorship. This work was funded by the Austrian Science Fund (FWF; "Invertebrate food webs in recently deglaciated alpine areas", P20859-B17), the "Aktion D. Swarovski KG 2013", the Alpine Research Centre Obergurgl, and the Mountain Agriculture Research Unit of the University of Innsbruck.

References
1. Adis J, Harvey MS. How many Arachnida and Myriapoda are there world-wide and in Amazonia? Stud Neotrop Fauna Environ. 2000;35:139–41.
2. Foelix RF. Biologie der Spinnen. 2nd ed. Stuttgart: Thieme Verlag; 1992.
3. Wise DH. Spiders in ecological webs. Cambridge: Cambridge University Press; 1993.
4. Hodkinson ID, Coulson SJ, Harrison J. What a wonderful web they weave: spiders, nutrient capture and early ecosystem development in the high Arctic - some counter-intuitive ideas on cummunity assembly. Oikos. 2001;95:349–52.
5. Barrett RDH, Hebert PDN. Identifying spiders through DNA barcodes. Can J Zool. 2005;83:481–91.
6. Raso L, Sint D, Rief A, Kaufmann R, Traugott M. Molecular identification of adult and juvenile linyphiid and theridiid spiders in Alpine glacier foreland communities. PLoS ONE. 2014;doi: 10.1371/journal.pone.0101755.
7. Öberg S, Cassel-Lundhagen A, Ekbom B. Pollen beetles are consumed by ground- and foliage-dwelling spiders in winter oilseed rape. Entomol Exp Appl. 2011;138:256–62.
8. Traugott M, Bell JR, Raso L, Sint D, Symondson WOC. Generalist predators disrupt parasitoid aphid control by direct and coincidental intraguild predation. Bull Entomol Res. 2012;102:239–47.
9. Welch KD, Schofield MR, Chapman EG, Harwood JD. Comparing rates of springtail predation by web-building spiders using Bayesian inference. Mol Ecol. 2014;23:3814–25.
10. Longhorn SJ, Nicholas M, Chuter J, Vogler AP. The utility of molecular markers from non-lethal DNA samples of the CITES II protected "tarantula" Brachypelma vagans (Araneae, Theraphosidae). J Arachnol. 2007;35:278–92.
11. Waits LP, Paetkau D. Noninvasive genetic sampling tools for wildlife biologists: A review of applications and recommendations for accurate data collection. J Wildl Manage. 2005;69:1419–33.
12. Morinha F, Travassos P, Carvalho D, Magalhaes P, Cabral JA, Bastos E. DNA sampling from body swabs of terrestrial slugs (Gastropoda: Pulmonata): a simple and non-invasive method for molecular genetics approaches. J Molluscan Stud. 2014;80:99–101.
13. Waldner T, Traugott M. DNA-based analysis of regurgitates: a noninvasive approach to examine the diet of invertebrate consumers. Mol Ecol Resour. 2012;12:669–75.
14. Raso L, Sint D, Mayer R, Plangg S, Recheis T, Brunner S, et al. Intraguild predation in pioneer predator communities of alpine glacier forelands. Mol Ecol. 2014;23:3744–54.
15. Redd KS, Jarman SN, Frusher SD, Johnson CR. A molecular approach to identify prey of the southern rock lobster. Bull Entomol Res. 2008;98:233–8.
16. Seeber J, Rief A, Seeber GUH, Meyer E, Traugott M. Molecular identification of detritivorous soil invertebrates from their faecal pellets. Soil Biol Biochem. 2010;42:1263–7.
17. Lefort MC, Boyer S, Worner SP, Armstrong K. Noninvasive molecular methods to identify live scarab larvae: an example of sympatric pest and nonpest species in New Zealand. Mol Ecol Resour. 2011;12:389–95.
18. Petersen SD, Mason T, Akber S, West R, White B, Wilson P. Species Identification of Tarantulas using Exuviae for International Wildlife Law Enforcement. Conserv Genet. 2006;8:497–502.
19. Sint D, Raso L, Traugott M. Advances in multiplex PCR: balancing primer efficiencies and improving detection success. Methods Ecol Evol. 2012;3:898–905.
20. Folmer O, Black M, Hoeh W, Lutz R, Vrijenhoek R. DNA primers for amplification of mitochondrial cytochrome c oxidase subunit I from diverse metazoan invertebrates. Mol Mar Biol Biotechnol. 1994;3:294–9.
21. Simon C, Frati F, Beckenbach A, Crespi B, Liu H, Flook P. Evolution, weighting, and phylogenetic utility of mitochondrial gene-sequences and a compilation of conserved polymerase chain-reaction primers. Ann Entomol Soc Am. 1994;87:651–701.
22. Bosnjak J, Stevanov-Pavlovic M, Vucicevic M, Stevanovic J, Simeunovic P, Resanovic R, et al. Feasibility of Non-Invasive Molecular Method for Sexing of Parrots. Pak J Zool. 2013;45:715–20.
23. Marrero P, Fregel R, Cabrera VM, Nogales M. Extraction of high-quality host DNA from feces and regurgitated seeds: a useful tool for vertebrate ecological studies. Biol Res. 2009;42:147–51.
24. Sastre N, Francino O, Lampreave G, Bologov VV, López-Martín JM, Sánchez A, et al. Sex identification of wolf (Canis lupus) using non-invasive samples. Conserv Genet. 2008;10:555–8.
25. Piñol J, San Andrés V, Clare EL, Mir G, Symondson WOC. A pragmatic approach to the analysis of diets of generalist predators: the use of next-generation sequencing with no blocking probes. Mol Ecol Resour. 2014;14:18–26.
26. Sint D, Raso L, Niederklapfer B, Kaufmann R, Traugott M: Group-specific multiplex PCR detection systems for the identification of flying insect prey. PLoS ONE 2014, 9:doi: 10.1371/journal.pone.0115501
27. Hosseini R, Schmidt O, Keller MA. Factors affecting detectability of prey DNA in the gut contents of invertebrate predators: a polymerase chain reaction-based method. Entomol Exp Appl. 2008;126:194–202.
28. Northam WT, Allison LA, Cristol DA. Using group-specific PCR to detect predation of mayflies (Ephemeroptera) by wolf spiders (Lycosidae) at a mercury-contaminated site. The Science of the total environment. 2011;416:225–31.
29. Traugott M, Symondson WOC. Molecular analysis of predation on parasitized hosts. Bull Entomol Res. 2008;98:223–31.
30. Sint D, Raso L, Kaufmann R, Traugott M. Optimizing methods for PCR-based analysis of predation. Mol Ecol Resour. 2011;11:795–801.
31. Pompanon F, Deagle BE, Symondson WOC, Brown DS, Jarman SN, Taberlet P. Who is eating what: diet assessment using next generation sequencing. Mol Ecol. 2012;21:1931–50.

14

Effect of chromosomal reorganizations on morphological covariation of the mouse mandible: insights from a Robertsonian system of *Mus musculus domesticus*

Jessica Martínez-Vargas[1], Francesc Muñoz-Muñoz[1*], Nuria Medarde[1], María José López-Fuster[2] and Jacint Ventura[1]

Abstract

Introduction: Morphological integration and modularity depend on genetic covariation between traits, which emerges from pleiotropic effects of single loci and genetic linkage between loci. Since chromosomal reorganizations alter meiotic recombination, they might modify groups of linked genes and entail the fixation of new alleles with new pleiotropic effects. As a result, they could contribute to the intraspecific variation of the covariance structure of morphological traits. Although the mouse mandible has long been studied in terms of development and evolution, little is known about how its covariance structure varies in natural populations with chromosomal reorganizations. Consequently, here we analyzed the magnitude and patterns of morphological covariation of mandible shape in groups of mice with different karyotypes from a Robertsonian system of *Mus musculus domesticus*.

Results: The organization of the mouse mandible into two main modules was confirmed in all chromosomal groups, since *RV* coefficients for the corresponding subdivision of landmarks were always significant. However, substantial variation in the magnitude of integration was detected between groups, especially when the effect of allometry was not removed. A significant positive correlation between differences in magnitude of integration of the symmetric component of shape and karyotypic distances between groups was detected when not correcting for size. Moreover, the degree of dependence of symmetric shape variation on size showed a negative association with the chromosome number and a positive association with the magnitude of integration. All groups showed similar patterns of morphological integration of the mandible, especially regarding the symmetric component of shape. However, the display of landmark displacements and the computation of vector angles highlighted some differences. In addition, distances between groups in terms of covariation matrices of the symmetric component were positively correlated with geographic distance.

Conclusions: Robertsonian translocations do not alter the organization of the mouse mandible into two main modules, but do affect the magnitude of integration between them. This effect is mainly due to changes in the allometric relationship. In the 'Barcelona' Robertsonian system, geographically structured sources of variation seem to affect the patterns of integration by producing parallel variation in separate developmental pathways. Overall, our results suggest that Robertsonian translocations could play a role in intraspecific differentiation processes by producing changes in the covariance structure of morphological traits.

Keywords: 'Barcelona' Robertsonian system, Mandible, Modularity, Morphological covariation, Morphological integration, *Mus musculus domesticus*, Robertsonian translocations

* Correspondence: francesc.munozm@uab.cat
[1]Departament de Biologia Animal, de Biologia Vegetal i d'Ecologia, Universitat Autònoma de Barcelona, Facultat de Biociències, Bellaterra, E-08193 Cerdanyola del Vallès, Spain
Full list of author information is available at the end of the article

Introduction

Organisms are composed of several parts that need to be coordinated in order to allow them to function as a whole. In a morphological context, this coordination is known as morphological integration [1], and it is expressed through statistical covariation. However, not all parts within an organism covariate with each other to the same extent. This heterogeneity in the degree of covariation is the basis of the concept of modularity [2]. Accordingly, modules have been defined as complexes of tightly integrated traits, which are relatively independent from other such complexes [3]. Therefore, modularity and integration are complementary concepts that emphasize different aspects of covariation.

Morphological traits originate from the ensemble of molecular and cellular processes taking place during development, which are known as developmental pathways [4]. Consequently, covariation between morphological traits also arises during development, specifically from direct interactions between developmental pathways and parallel variation of separate developmental pathways. Covariation emerging from direct interactions originates in the pathways themselves through different mechanisms, such as the division of a precursor tissue into parts that respectively give rise to a different trait, or the transmission of variation through inductive signaling from one pathway to another. Instead, covariation emerging from parallel variation is due to the simultaneous influence of the same external factor (e.g. environmental conditions) on separate developmental pathways [4].

A substantial component of morphological covariation is genetic covariation, which can arise from genetic linkage between loci and pleiotropic effects of single loci [3]. Presumably, both sources of genetic covariation can be affected by chromosomal reorganizations such as Robertsonian (Rb) translocations, which consist in the fusion of two non-homologous acrocentric chromosomes at their centromeres to originate a metacentric chromosome [5]. Rb translocations cause a decline in chiasma frequency and a more distal distribution of chiasmata during meiotic recombination [6–10], which could affect the linkage between alleles of loci that influence different traits within a structure. Besides, the reduction in recombination entailed by Rb translocations can prompt the fixation of different positively selected alleles [11], which may have different pleiotropic effects [12–15]. In relation to this, the hybridization between chromosomally different populations, differing in their fixed alleles, can entail developmental alterations and hence changes in terms of genetic covariation between traits in the subsequent generations [14].

The Western European house mouse (*Mus musculus domesticus* Schwarz and Schwarz, 1943 [16]) constitutes a model organism for the study of evolutionary processes linked to chromosomal reorganizations, as it shows great karyotypic diversity mainly due to Rb translocations [5]. Within the distributional area of *Mus musculus domesticus*, there are many geographic regions in which populations with different sets of metacentrics hybridize with each other and/or with populations with the standard (St) karyotype (40 acrocentric chromosomes). Jointly, these sets of populations are called Rb systems [5]. One of these, named 'Barcelona' Rb system, is present on the Northeastern Iberian Peninsula, specifically in part of the provinces of Barcelona, Tarragona and Lleida [17-20]. The set of metacentrics that characterizes this Rb system consists of Rb(3.8), Rb(4.14), Rb(5.15), Rb(6.10), Rb(7.17), Rb(9.11) and Rb(12.13), where the pairs of numbers refer to the acrocentric autosomes that gave rise to the metacentric in question. In this system, the frequency of metacentrics is distributed in a staggered way over 5,000 km^2, leading to a progressive reduction in diploid number towards the center of the zone [20].

The mouse mandible has long been useful as a model system to study the development and evolution of complex morphological structures [21,22]. The mandible of the mouse has been divided into two functional modules: a distal one bearing the teeth (alveolar region), and a proximal one that articulates with the skull and constitutes the attachment point for most of the masticatory muscles (ascending ramus) [21]. The study of the genetic basis of the mandible shape has revealed that genetic modularity also occurs in this structure, in the same way as functional modularity does [23–25]. Besides, two concurring developmental modules can be distinguished in the mouse mandible, according to several lines of evidence [26,27]. Furthermore, evolutionary independence between these two mandibular modules has also been detected [28]. Instead, scarce analyses have been performed on the intraspecific variation of the covariance structure of the mouse mandible in natural populations. Similarly, there is a shortage of studies on how changes in the covariance structure prompted by karyotypic variation can contribute to the processes of morphological evolution (but see [29,30]). Covariance structure is a population-level feature whose role in dictating the directions and pace of evolutionary transformations has long been discussed [31,32]. In this regard, strong integration has been considered to constrain evolution because it implies that potentially favorable changes in some traits could entail adverse changes in associated traits [2]. Instead, modularity has been regarded as a driving force of evolution because it enables changes in certain traits to happen without affecting notably the rest of traits, thereby making it easier for evolutionary transformations to occur [2]. Keeping this in mind, it seems plausible that differences between populations in terms of their covariance structure could make them differ in their evolvability, or ability to evolve. If evolvability was itself the object of natural selection, which is actually a controversial hypothesis [33], variation between populations would then be expected to drive evolution within species. In this context, if chromosomal reorganizations turned out to affect the covariance structure, they could play a role in the evolvability of populations. In order to shed light on these topics, the present study aims to analyze the structure of morphological covariation of the mandible in wild populations of *Mus musculus domesticus*, and to assess the effect of Rb translocations on it.

Since the 'Barcelona' Rb system is characterized by a relatively large number of metacentrics and a wide range of diploid numbers, we considered it to be suitable for this study. While conducting it, several specific objectives were approached. In order to determine how Rb translocations affect morphological covariation of the mouse mandible, we assessed the magnitude and patterns of integration of this structure and tested its bimodular organization in groups with different karyotypes. Besides, given that allometry is considered to be a strong integrating factor [27], we evaluated its connection with these chromosomal reorganizations and its effect on the covariance structure of the mandible. As stated above, morphological covariation can result from both direct interactions between developmental pathways and parallel variation of separate developmental pathways. While co-variation between symmetric shape changes can result from both sources, covariation between asymmetric shape changes is only due to direct developmental interactions [4]. In order to know the relative importance of direct interactions and parallel variation of developmental pathways in generating morphological covariation in the mouse mandible, we studied symmetric and asymmetric shape changes separately, which are usually termed symmetric component and asymmetric component of shape variation respectively. Lastly, because chromosomal variation in the 'Barcelona' Rb system is geographically arranged [18,20], we also assessed the effect of geographic distance on the covariance structure of the mandible.

Results

Sources of shape and size variation
Procrustes analyses of variance (ANOVAs) carried out on the replicated subsample revealed a significant effect of the individual and side factors, as well as their interaction, on mandible shape (Table 1). On the other hand, a significant effect of the individual factor and the interaction term, but not of the side factor, was detected on mandible size (Table 2). Since the significant effect of the interaction term indicates that variation in fluctuating asymmetry exceeds variation resulting from measurement error, subsequent analyses were based on a single digitization of landmarks per hemimandible.

Table 1 Procrustes ANOVA conducted on the replicated subsample to evaluate the influence of measurement error on shape data

| Effect | Shape | | | | |
	SS	df	MS	F	P
Individual	1.919	4520	4.245×10^{-4}	6.72	< 0.001
Side	0.024	20	1.173×10^{-3}	18.56	< 0.001
Individual × Side	0.286	4520	6.318×10^{-5}	4.62	< 0.001
Measurement error	0.248	18160	1.367×10^{-5}		

SS, sum of squares; df, degrees of freedom; MS, mean squares; F, F statistic; P, P-value.

Procrustes ANOVAs conducted on each chromosomal group separately (40St, 40Rb, Rb(38–39), Rb(34–37), Rb (31–33), Rb(27–30)) consistently showed significant differences between individuals, regarding both shape and size variation. While directional asymmetry in shape was detected in all groups, directional asymmetry in size was only detected in group Rb(31–33) (see Additional file 1).

Allometry
A significant dependence of the symmetric component of shape on size was detected in all groups ($P < 0.001$). Size accounted for low-to-moderate fractions of this component (40St: 10.78%; 40Rb: 8.30%; Rb(38–39): 12.88%; Rb (34–37): 13.07%; Rb(31–33): 13.63%; Rb(27–30): 23.78%). The asymmetric component of shape variation showed a significant dependence on size asymmetry ($P < 0.05$) only in groups 40St, 40Rb and Rb(27–30). On the whole, fairly low percentages of this component were predicted by size (40St: 2.41%; 40Rb: 4.08%; Rb(38–39): 1.94%; Rb (34–37): 1.76%; Rb(31–33): 1.08%; Rb(27–30): 3.56%). The linear regression revealed a significant negative association between the percentage of symmetric shape variation explained by size and chromosome number (r = −0.86, p < 0.05). No significant association was observed for the asymmetric component (r = 0.15, p = 0.78).

Modularity and magnitude of integration
The *RV* coefficient for the tested partition of landmarks was significant for both components of shape variation in all groups, both for raw and size-corrected data

Table 2 Two-way ANOVA conducted on the replicated subsample to evaluate the influence of measurement error on size data

| Effect | Centroid size | | | | |
	SS	df	MS	F	P
Individual	8.068	226	3.570×10^{-2}	191.33	< 0.001
Side	0.001	1	4.670×10^{-4}	2.51	0.115
Individual × Side	0.042	226	1.870×10^{-4}	2.28	< 0.001
Measurement error	0.074	908	8.200×10^{-5}		

SS, sum of squares; df, degrees of freedom; MS, mean squares; F, F statistic; P, P-value.

(Figure 1). Therefore, the predicted hypothesis of modularity was always confirmed. Overall, the *RV* coefficients corresponding to the symmetric component were higher than those corresponding to the asymmetric component (Figure 1 and Table 3). When using the rarefaction procedure to standardize the *RV* coefficients to a given sample

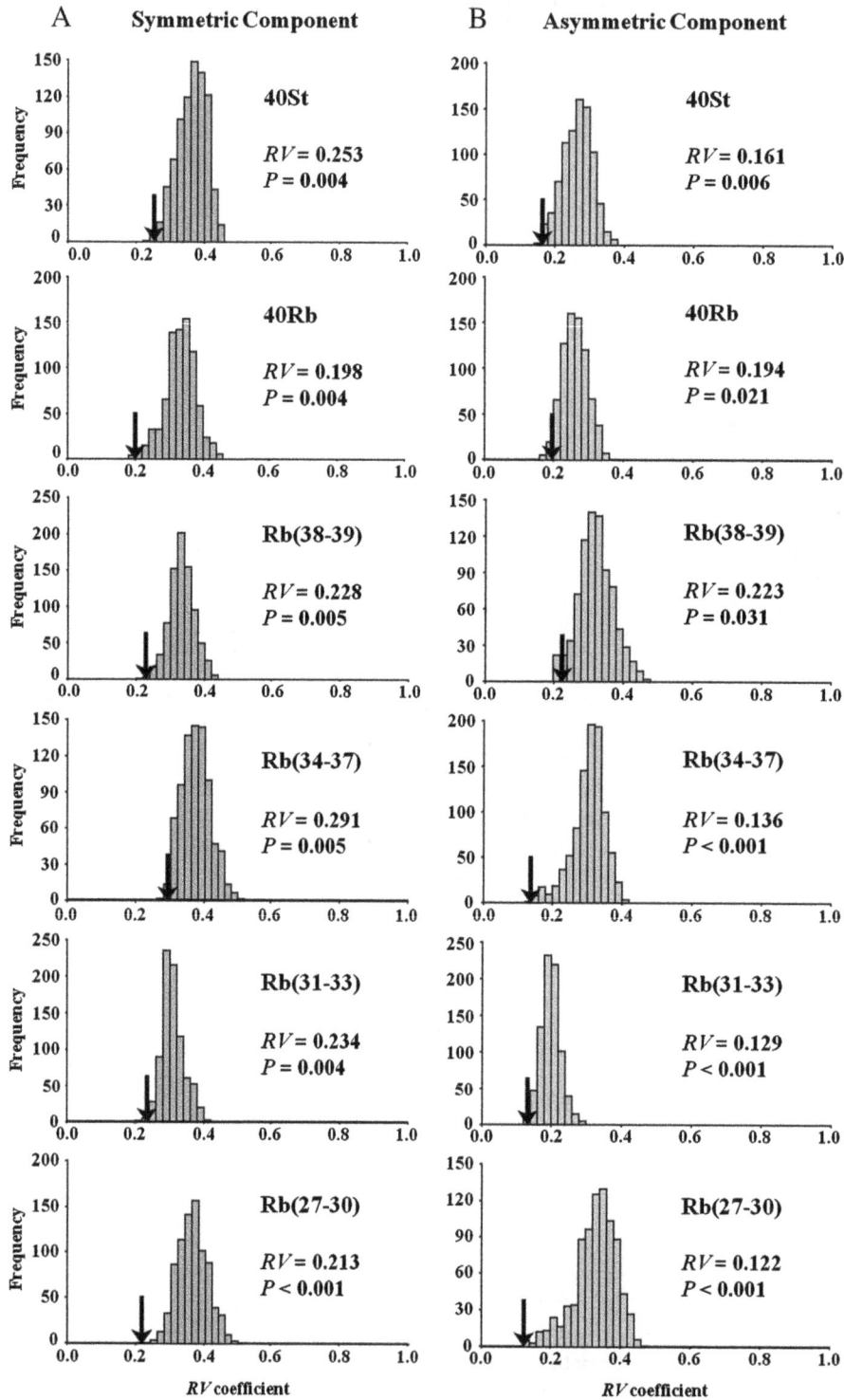

Figure 1 Distributions of *RV* coefficients for the symmetric (A) and asymmetric (B) components for size-corrected data. The values of *RV* coefficients for the partition of landmarks concurrent with the hypothesis of modularity are indicated by arrows and highlighted. *P*-values (*P*) are adjusted through the FDR procedure.

Table 3 RV coefficients non-standardized (NS) and standardized (S) to the same sample size

Group	Symmetric component		Asymmetric component	
	NS	S	NS	S
40St	0.253 (0.357)	0.402	0.161 (0.164)	0.341
40Rb	0.198 (0.300)	0.344	0.194 (0.204)	0.431
Rb(38–39)	0.228 (0.394)	0.416	0.223 (0.227)	0.473
Rb(34–37)	0.291 (0.448)	0.467	0.136 (0.132)	0.389
Rb(31–33)	0.234 (0.372)	0.406	0.129 (0.127)	0.359
Rb(27–30)	0.213 (0.530)	0.378	0.122 (0.117)	0.302

Values between parentheses stand for RVs obtained from raw data.

size (n = 77), they increased for both shape components. However, in both cases the pattern across groups was consistent with that displayed by the non-standardized RV coefficients (Table 3).

The linear regression revealed a significant positive dependence of the magnitude of integration of the symmetric component on allometric percentage (r = 0.92, $P < 0.01$).

Mantel and partial Mantel tests detected a significant positive correlation between distances in magnitude of integration of the symmetric component, calculated from raw data, and karyotypic, but not geographic, distances between groups (Table 4). This association was not obtained with size-corrected data. As for the asymmetric component, analyses with both raw and size-corrected data revealed no significant association between differences in the magnitude of integration and karyotypic or geographic distances.

Patterns of integration

According to the results of the principal component (PC) analyses performed on size-corrected data, mandibular shape variation was mainly concentrated in the first few PCs. In all chromosomal groups, the first two PCs jointly accounted for a substantial fraction of shape variation, for both the symmetric and asymmetric components (Tables 5 and 6). Besides, all groups showed a similar distribution

of the percentages of total variance accounted for by the first ten PCs (Tables 5 and 6). When the shape changes associated with the first two PCs were displayed as diagrams of eigenvectors, most of the variation was concentrated in the ascending ramus, particularly in the coronoid and condylar processes (Figure 2). While the pattern of variation corresponding to PC1 of the symmetric component turned out to be quite similar between groups, conspicuous differences affecting the angular and condylar processes were detected in group 40St (Figure 2A); in fact, vector angles between this group and the others were by far the greatest (Table 7). Regarding the pattern of shape changes associated with PC2 of the symmetric component, noticeable differences were detected between groups (Figure 2A). In this case, groups 40St and Rb(27–30) showed the comparatively most distinct patterns of shape variation (Table 8). As for both PC1 and PC2 of the asymmetric component, some differences in the displacement direction of landmarks were detected between groups (Figure 2B), which was supported by vector angles (Tables 7 and 8).

Significant positive correlations were detected when comparing the covariance matrices of both the symmetric and asymmetric components between groups (Table 9). Correlations between the covariance matrices of both components of each group were also positive and significant, and the correlation coefficient increased as the mean diploid number of the group decreased (Table 9).

According to both Mantel and partial Mantel tests, distances in patterns of integration of the symmetric component, calculated from both raw and size-corrected data, were positively and significantly correlated with geographic distances between groups (Table 4). Although the Mantel test also detected a significant positive correlation between distances in patterns of integration of the symmetric component, computed from size-corrected data, and karyotypic distances, this association was not detected by the partial Mantel test (Table 4). Distances in

Table 4 Mantel and partial Mantel tests correlations between distances in magnitude (RV) and patterns (1-r) of morphological integration, and karyotypic and geographic distances

		Geographic distance		Karyotypic distance	
		Mantel	Partial Mantel	Mantel	Partial Mantel
Raw data	RV_{sym}	0.01	−0.49	0.60*	0.72*
	RV_{asym}	0.17	0.00	0.32	0.27
	$1\text{-}r_{sym}$	0.75*	0.69*	0.39	−0.03
	$1\text{-}r_{asym}$	−0.02	0.14	−0.25	−0.28
Size-corrected data	RV_{sym}	0.00	0.23	−0.32	−0.39
	RV_{asym}	0.11	−0.04	0.26	0.24
	$1\text{-}r_{sym}$	0.77*	0.68*	0.50*	0.14
	$1\text{-}r_{asym}$	−0.06	0.12	−0.31	−0.29

*$P < 0.05$.

Table 5 Percentages of total variance accounted for by the first ten PCs of the symmetric component

	PC1	PC2	PC3	PC4	PC5	PC6	PC7	PC8	PC9	PC10
40St	18.30	13.52	11.15	10.15	6.62	5.96	4.82	3.99	3.27	2.82
40Rb	17.89	16.45	9.22	8.42	7.68	5.43	4.90	4.46	3.78	3.40
Rb(38–39)	17.64	13.07	9.22	8.40	7.69	6.92	5.97	4.30	4.17	3.46
Rb(34–37)	20.58	12.85	9.52	9.02	6.35	5.85	5.37	4.79	3.55	2.97
Rb(31–33)	19.79	11.03	10.63	8.23	7.17	6.07	5.35	4.37	3.53	3.17
Rb(27–30)	19.82	11.53	10.84	9.39	7.20	6.42	4.94	4.33	3.57	3.23

the patterns of integration of the asymmetric component were not correlated with karyotypic nor geographic distances, either with raw or size-corrected data (Table 4).

Discussion

Rb translocations affect the magnitude of integration but not modularity

The results of the tests of modularity lent support to the notion that the mouse mandible consists of two primary modules, the alveolar region and the ascending ramus [25–28]. Besides, the existence of the proximal module is backed by the fact that shape changes within that region were comparatively more coordinated, and that most of the variation in mandible shape was particularly concentrated there [24,34,35]. Despite the confirmation of such modular organization, the fact that only a few PCs accounted for a large part of shape variation indicates that the mouse mandible has a certain degree of overall integration [27,36–39]. As proved, the modular configuration of the mouse mandible is not altered by Rb translocations, regardless of the number of metacentrics. However, the magnitude of integration between the two mandibular modules varied notably among chromosomal groups, especially when allometry was not removed. Because the effects of size simultaneously affect all the parts of a structure, allometry is usually considered to have an integrating influence and thus to obscure modular organizations [27]. In our study, despite significant allometric relationships, modularity was always detected. However, and as expected, magnitudes of integration between the two modules were higher when not correcting for the effect of size. Moreover, magnitudes of integration of the symmetric component of shape increased along with the percentage of allometry, whereas this percentage

was negatively associated with diploid number. Thus, the greater dependence of shape on size in groups with low diploid numbers seems to be the reason why they showed a greater magnitude of integration between the two mandibular modules. At the same time, this would explain why Mantel tests performed with raw data detected an association between differences in karyotype and differences in magnitude of integration between groups. Previous studies have found many quantitative trait loci (QTLs) affecting the size and shape of the mandible [40]. Since Rb translocations reduce meiotic recombination [7], their accumulation could progressively increase the probability of linkage between QTLs affecting size and QTLs affecting shape, which would explain the greater association between size and shape as diploid number decreases.

When the effect of allometry was removed, no linear association was detected between differences in magnitude of integration of the mandible and neither geographic nor karyotypic distances, for none of the two components of shape variation. However, these results do not necessarily preclude an effect of Rb translocations on the integration of this structure. As previously stated (see *Introduction*), these chromosomal reorganizations can affect morphological covariation by modifying the linkage between alleles of loci influencing different mandibular traits, and by entailing the fixation of positively selected alleles. The progressively lower *RV* coefficients obtained in groups Rb(34–37), Rb (31–33) and Rb(27–30) could be due to the modification of the linkage between alleles. Given that metacentrics show the lowest chiasma frequency around the centromere [9], the formation of Rb translocations could progressively imply the emergence of new linkage groups around the centromere of the newly-formed metacentrics. In connection with this, far more QTLs have been assigned to

Table 6 Percentages of total variance accounted for by the first ten PCs of the asymmetric component

	PC1	PC2	PC3	PC4	PC5	PC6	PC7	PC8	PC9	PC10
40St	18.30	11.90	9.87	8.94	7.99	7.01	5.51	4.72	3.90	3.22
40Rb	15.30	12.74	11.41	9.52	6.94	6.42	5.82	4.58	4.20	3.68
Rb(38–39)	20.28	16.17	11.25	7.16	6.20	5.35	5.13	4.29	3.41	2.73
Rb(34–37)	20.78	14.01	10.63	8.63	6.52	5.14	4.72	4.53	3.50	2.80
Rb(31–33)	12.60	12.01	11.24	8.34	6.91	6.51	6.17	5.16	4.49	3.87
Rb(27–30)	22.32	15.22	10.55	8.46	6.72	4.96	4.64	3.76	3.35	2.73

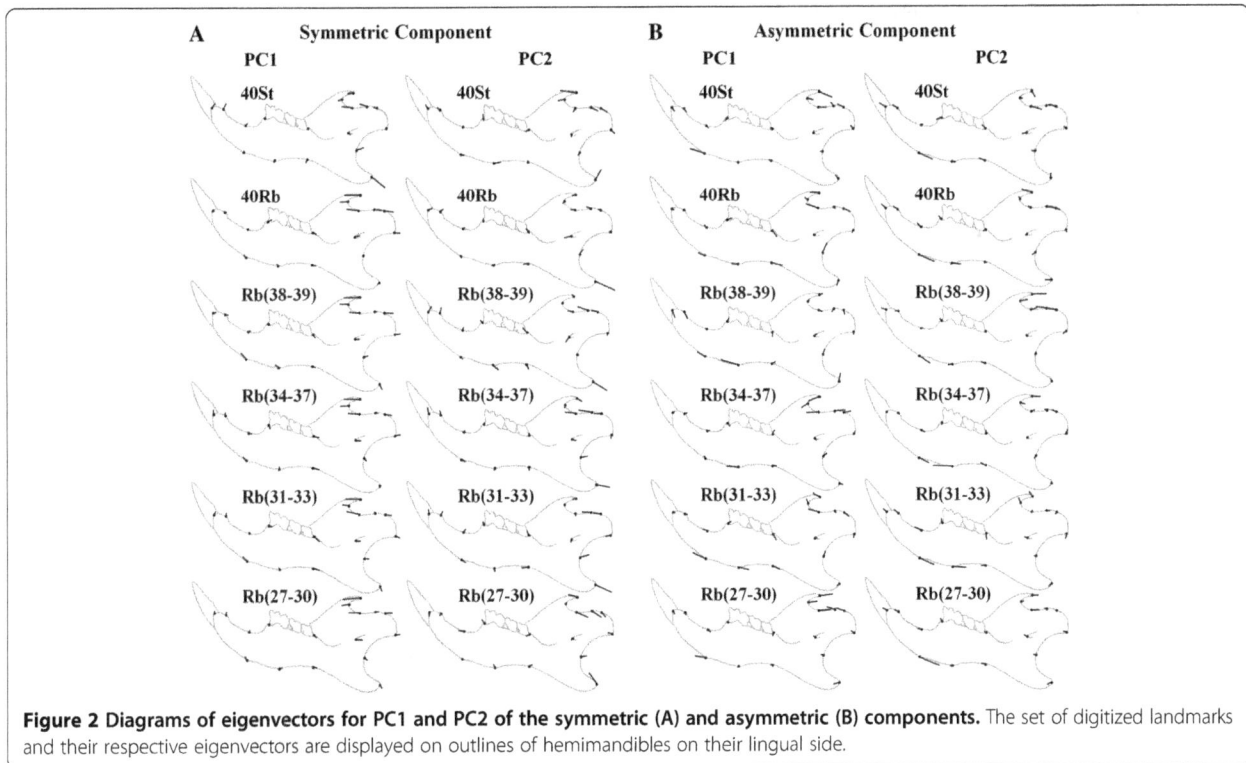

Figure 2 Diagrams of eigenvectors for PC1 and PC2 of the symmetric (A) and asymmetric (B) components. The set of digitized landmarks and their respective eigenvectors are displayed on outlines of hemimandibles on their lingual side.

the ascending ramus than to the alveolar region of the mouse mandible, and several of them are located close to the centromeres of acrocentric chromosomes [23]. Therefore, it is likely that the new linkage groups prompted by Rb translocations would include several QTLs with an effect only on the ascending ramus. If that was the case, covariation, and so integration, within this module would increase along with the amount of Rb translocations, and this would explain the observed decrease in the magnitude of integration of the mandible, in terms of RV, along with the decrease in diploid number. Instead, the increase in the magnitude of integration of the asymmetric component detected in group Rb(38–39) with respect to St populations might not be explained by this mechanism, but by the higher rate of fixation of alleles in metacentrics due to the decrease in meiotic recombination

that they undergo [10,11]. In these metacentrics, the fixed alleles could affect covariation in two different ways. On the one hand, as pleiotropic effects vary together with alleles [12–15], the fixation of new favorable alleles in metacentrics would likely alter genetic covariation of morphological structures. On the other hand, it has been proved that hybrids between populations with different fixed alleles show greater genetic covariation than the parental generations [14]. In particular, when Renaud and collaborators [35] analyzed the strength of covariation between the two main mandibular modules in two subspecies of house mouse and the resulting hybrids, they found that it was higher in the filial 1 (F1) hybrids than in the parental groups and the filial 2 (F2) hybrids. Linking it to our study, given that new alleles are expected to become fixed in Rb translocations, hybridization between St and Rb mouse populations could

Table 7 Vector angles in degrees for PC1

	40St	40Rb	Rb (38–39)	Rb (34–37)	Rb (31–33)	Rb (27–30)
40St	71.29	68.32*	71.68	65.49*	83.50	29.99**
40Rb	76.14	48.23**	73.06	52.06**	43.97**	67.05*
Rb(38–39)	60.87*	26.71**	84.21	81.65	87.77	75.24
Rb(34–37)	67.80*	24.29**	25.80**	44.93**	78.75	47.47**
Rb(31–33)	57.73*	29.90**	22.16**	20.83**	50.01**	80.46
Rb(27–30)	63.36*	22.95**	23.58**	16.93**	22.64**	43.57**

Values between PC1 of symmetric and asymmetric components of each group (on the diagonal) and between PC1s of the symmetric (below) and the asymmetric (above) components among groups. * $P < 0.05$; ** $P < 0.001$.

Table 8 Vector angles in degrees for PC2

	40St	40Rb	Rb (38–39)	Rb (34–37)	Rb (31–33)	Rb (27–30)
40St	86.74	67.70*	56.84*	69.59	62.69*	42.51**
40Rb	88.22	73.35	85.20	76.60	34.91**	62.15*
Rb(38–39)	88.23	42.30**	82.22	49.28**	79.73	43.81**
Rb(34–37)	80.38	53.34**	42.60**	83.06	75.51	57.72*
Rb(31–33)	87.91	37.01**	39.86**	44.39**	84.44	54.20**
Rb(27–30)	64.34*	63.13*	55.74**	49.40**	70.47	85.71

Values between PC2 of symmetric and asymmetric components of each group (on the diagonal) and between PC2s of the symmetric (below) and the asymmetric (above) components among groups. *$P < 0.05$; **$P < 0.001$.

Table 9 Correlation coefficients (r) between covariance matrices

	40St	40Rb	Rb (38–39)	Rb (34–37)	Rb (31–33)	Rb (27–30)
40St	0.473	0.587	0.432	0.414	0.626	0.644
40Rb	0.681	0.504	0.475	0.574	0.704	0.586
Rb(38–39)	0.665	0.742	0.506	0.489	0.538	0.474
Rb(34–37)	0.640	0.702	0.731	0.539	0.623	0.657
Rb(31–33)	0.681	0.719	0.760	0.824	0.520	0.667
Rb(27–30)	0.617	0.683	0.689	0.756	0.780	0.583

Values between covariance matrices of symmetric and asymmetric components of each group (on the diagonal), and of symmetric (below) and asymmetric (above) components among groups. All coefficients are significant ($P < 0.001$).

probably give rise to hybrid populations showing greater co-variation between the two mandibular modules. Bearing in mind the preceding argumentation, we suggest that the differences detected between chromosomal groups in terms of magnitude of integration of the asymmetric component might result from the balance between the divergent effects that the modification of genetic linkage, on the one side, and the differential fixation of alleles, on the other side, may have on morphological covariation in each of them. As it can be noticed by looking back on the results, this argumentation does not fit with the differences between groups in terms of the magnitude of integration of the symmetric component. As stated above, this component of shape variation, unlike the asymmetric one, takes account of variation due to the effect of external stimulus on developmental pathways, which can lead to covariation between traits if those factors affect different pathways simultaneously [4]. Therefore, the impact of certain external sources of variation could be responsible for that discrepancy between results regarding the symmetric and asymmetric components.

Morphological covariation arises from different developmental sources

The degree of congruence between symmetric and asymmetric covariation is said to provide evidence for the relative importance of the two developmental origins of morphological covariation, namely direct interactions between developmental pathways and parallel variation of separate developmental pathways [4,30,34]. When comparing the covariance matrices of the symmetric and asymmetric components of shape variation in each group, intermediate correlation coefficients were generally obtained, which agrees with the results obtained in previous studies [34]. This indicates that, although covariation between symmetric shape changes of the mandible arises from direct developmental interactions to some extent, a considerable amount of this covariation is actually due to the parallel variation of separate developmental pathways. Moreover, the fact that these

correlation values increase as diploid number decreases suggests that the accumulation of new Rb translocations might entail a greater importance of direct interactions over parallel variation in generating morphological covariation in the mouse mandible. However, the mechanisms by which this may happen remain unknown.

Variation in the patterns of integration is geographically structured

Correlation coefficients between the covariance matrices of the chromosomal groups revealed that, in general, they share similar patterns of integration of the mandible. However, diagrams of landmark displacements and vector angles indicated that particular aspects of integration patterns differ between groups. The positive association detected between the differences in the patterns of integration of the symmetric component and the geographic distances between groups suggests that morphological covariation of the mandible is geographically structured. As mentioned above, symmetric covariation takes account of covariation between morphological traits that arises from the simultaneous effect of an external factor on separate developmental pathways. Among these factors, one can distinguish environmental conditions and allelic variability in genes involved in different developmental processes [4], which are sources of variation that can be geographically structured. The positive association detected between differences in the patterns of integration of the symmetric component and geographic distances between groups suggests that such patterns might be influenced by environmental factors and/or genetic differences due to isolation by distance. Several studies state that covariation patterns are remarkably similar between closely related species, as well as between groups belonging to the same species [30,31,41,42], whereas others show that they can vary significantly at small taxonomic scales [43,44]. In reference to this, our results suggest that even though different populations of the same species may show, at a glance, considerably similar patterns of morphological covariation, intraspecific variability can still exist and be detectable. Besides, they highlight that the geographic structure of populations can affect their patterns of morphological integration, and that this effect seems to take place mainly through the simultaneous influence of geographically-structured external factors on separate developmental pathways involved in generating the morphological structure in question.

Conclusions

In the light of our results, we conclude that the organization of the mouse mandible into two modules is a stable attribute that is neither distorted by the presence of Rb translocations nor by the integrative effect of allometry. However, both the accumulation of Rb translocations and allometric relationships affect the magnitude of integration between modules. Notably, the magnitude of integration increases along with the number of Rb translocations due to a parallel increase in the amount of shape variation depending on size. Analyses of size-corrected data also suggest that Rb translocations can affect the magnitude of integration, presumably by linking genes with different effects on mandible morphology and by giving rise to new combinations of alleles. In the first case, because of the genetic constitution of the mouse mandible, a decrease in the magnitude of integration is expected as the number of translocations rises due to an increase in covariation within the ascending ramus. In the second case, a greater magnitude of integration is expected in populations with a mixture of different karyotypes.

Notwithstanding the fact that populations with different sets of Rb translocations share similar patterns of covariation, they actually differ in particular aspects. This variation follows a geographic structure, probably because external sources of variation that affect separate developmental pathways in parallel, such as environmental factors or genetic differences, are geographically structured.

Overall, our study proves that the covariance structure of the mouse mandible shows intraspecific variation between natural populations, not only due to the presence of Rb translocations, but also because of the effect of geographically structured factors. Chromosomal reorganizations are usually thought to take part in speciation processes by acting as barriers to gene flow [11]. Our results support the notion that chromosomal reorganizations can also modify the patterns of morphological integration. Thus, chromosomal reorganizations could also play a role in differentiation processes by producing changes in the covariance structure of morphological traits that might entail evolution of populations in divergent directions.

Materials and methods
Sample and data acquisition
The sample consisted of 1233 right and left hemimandibles from 619 adult wild mice (308 females and 311 males) from the 'Barcelona' Rb system and surrounding St populations. Chromosome preparations were obtained directly from bone marrow [45] and stained using Wright's stain for G-banding [46].

Once their karyotypes were determined (see Additional file 2), the specimens were classified into six groups on the basis of their diploid number: 40St, n = 86; 40Rb, n = 77; Rb(38–39), n = 84; Rb(34–37), n = 107; Rb(31–33), n = 159; Rb(27–30), n = 106. Group 40St included specimens with the standard karyotype (40 acrocentric chromosomes) from localities where there is no evidence of metacentrics. 40Rb stands for the

group including individuals with the standard karyotype from localities where Rb translocations have been reported. Hemimandibles left and right of each individual were detached and laid on a black cardboard. Images of their lingual view, together with a scale in millimeters, were obtained with a Nikon COOLPIX P90 digital camera placed 21.5 cm from the cardboard. Seventeen two-dimensional landmarks were digitized in all the images using the tpsDig2 software (Figure 3A and Additional file 3). In a subsample of 454 hemimandibles (277 right and 277 left; 36.82% of the total), a set of landmarks was digitized three times.

Analyses of integration and modularity
Morphological integration and modularity of the mandible were analyzed by implementing the geometric morphometric methods included in the MorphoJ software, version 1.05e [47]. Size was estimated through centroid size [48]. The landmark configurations of all left hemimandibles were reflected to their mirror images. Then, the configurations obtained via reflection and those of the right hemimandibles were superimposed through a generalized least-squares Procrustes fit and were projected onto the shape tangent space [34,48,49]. As a result of this procedure, variation due to size, position and orientation was eliminated, and so shape information was extracted [24,48].

Coordinates resulting from the Procrustes fit (Procrustes coordinates) were then analyzed by means of multivariate statistics methods. First of all, Procrustes ANOVAs were conducted on the three sets of replicated configurations in order to assess the influence of measurement error on size and shape data [50,51]. Individual and side were entered as random and fixed factors respectively, and Procrustes distances as the dependent variable. The individual factor stands for individual variation, the side factor for directional asymmetry, and the interaction between these two factors represents fluctuating asymmetry [47,49,52]. Measurement error was quantified from the residual variance component between replicates [50]. Since size and shape variation due to measurement error was significantly lower than variation in fluctuating asymmetry, the ensuing analyses (listed below) were based on a single digitization of landmarks per hemimandible. Given that the study was intended to assess the effect of different numbers of Rb translocations on the patterns of morphological integration and modularity, these analyses were conducted for each chromosomal group separately.

To begin with, Procrustes ANOVAs for size and shape were carried out. These analyses separated total variation into its symmetric component, which is the variation between individuals in terms of the averages of the original and reflected landmark configurations, and its asymmetric component, which is the variation within individuals

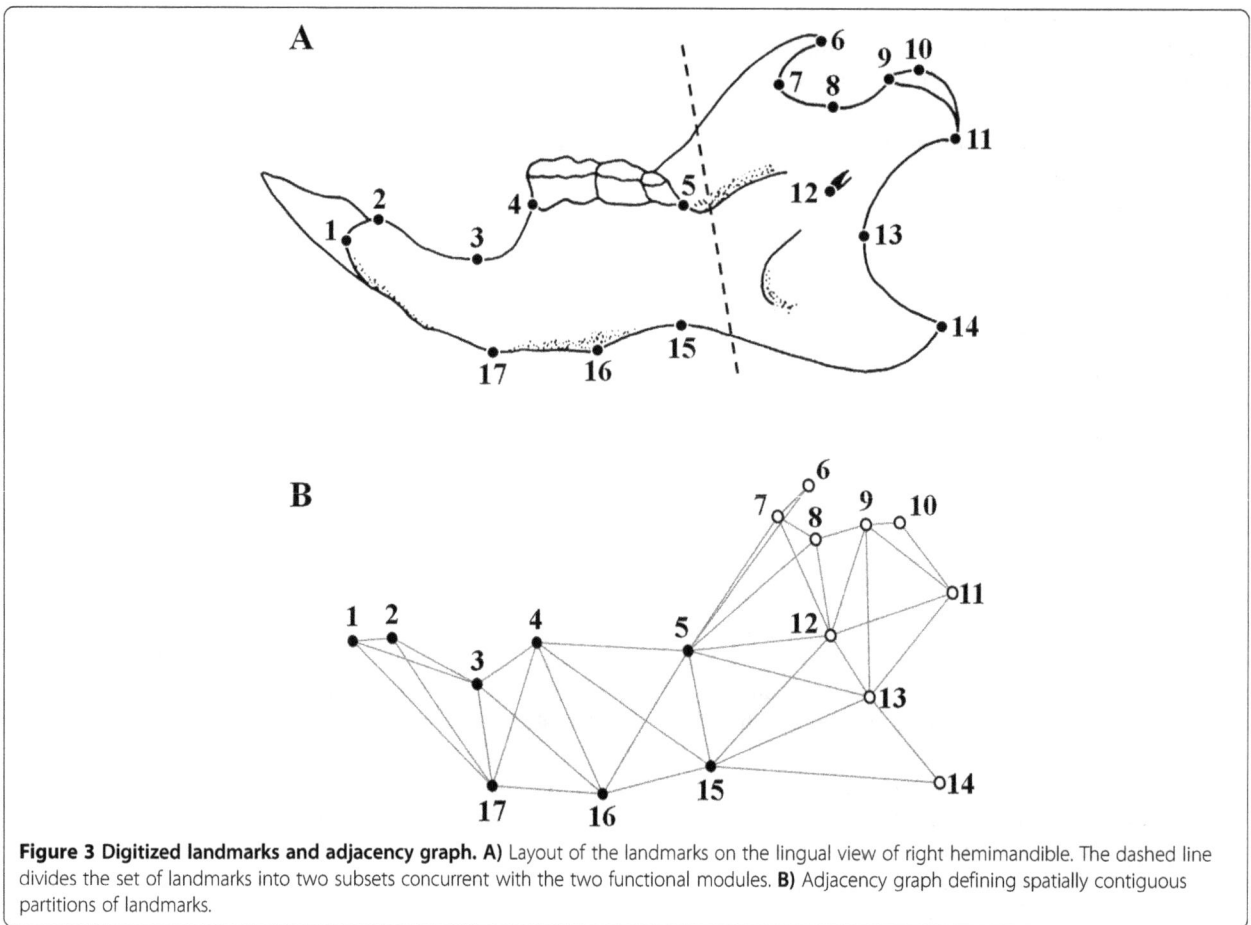

Figure 3 Digitized landmarks and adjacency graph. A) Layout of the landmarks on the lingual view of right hemimandible. The dashed line divides the set of landmarks into two subsets concurrent with the two functional modules. **B)** Adjacency graph defining spatially contiguous partitions of landmarks.

regarding the landmark deviations of the reflected configuration from the original one [30,34,51].

Allometry, that is, the scaling relationship between shape and size, was then evaluated through multivariate regressions of both the symmetric and asymmetric components of shape variation onto symmetric and asymmetric centroid size respectively. Significations were obtained through permutation tests with 10,000 iterations [53,54]. The association between chromosome number and percentage of shape variation explained by size was tested through the regression of the allometric percentages of groups onto their mean diploid number, for both the symmetric and asymmetric components. Since a significant allometric relationship was found in most of the chromosomal groups (see *Results*), subsequent analyses were conducted with the covariance matrices obtained from raw data but also from the regression residuals, in order to assess the role of allometry as an integrating factor [27]. However, and unless the converse is indicated, only the results obtained from size-corrected data are shown.

Evaluations of the hypothesis of bimodular organization of the mandible were conducted for the two components of shape, using MorphoJ. The set of digitized landmarks

was subdivided into two subsets of eight and nine landmarks respectively, corresponding to the two mandibular modules (alveolar region and ascending ramus; Figure 3A). The magnitude of integration between the two subsets was quantified in each group through the computation of the *RV* coefficient [55]. In order to assess the hypothesis of modularity, the resulting *RV* coefficients were compared with the distributions of *RV* coefficients obtained from alternative subsets of landmarks. These subsets were required to include the same number of landmarks as the tested subsets matching the mandibular modules. Since integration cannot occur between spatially separate units [56], comparisons were restricted to subsets whose landmarks were contiguous, that is, connected by the edges of the adjacency graph (Figure 3B). By definition, subsets of landmarks resulting from a subdivision consistent with an actual modular organization are expected to show weaker covariation, and thus lower integration, than subsets not corresponding with actual modules [4,27]. Accordingly, when the *RV* coefficient for the two tested subsets of landmarks was lower than 95% of the distributional values, it was considered to be statistically significant ($P < 0.05$) and the hypothesis of modularity was confirmed [27]. However, it has recently been shown that the *RV* coefficient decreases when sample size increases [57,58]. Therefore, a rarefaction procedure was used in order to obtain sample-size-corrected *RV* values. Through a sampling with replacement, 1,000 random samples of 77 observations (sample size of the smallest group, 40Rb) were drawn from each group. Then, the *RV* coefficient was computed for each dataset, and finally a mean *RV* value was computed for each group [57], which was the actual sample-size-standardized *RV* coefficient. In order to assess the dependence of the magnitude of morphological integration on allometry, a linear regression of the *RV* coefficients from raw data on the percentages of allometry of each group was conducted. Given the low dependence of the asymmetric component of shape variation on size (see *Results*), this analysis was only performed for the symmetric component. Mantel and partial Mantel tests were then conducted with the aim of assessing the relationship between differences in the magnitude of morphological integration (*RV* coefficients) and both the karyotypic and the geographic distances between chromosomal groups (see Additional file 4). These distances were calculated following the procedure used in a previous study [28]. Both raw and size-corrected data, as well as data of the symmetric and asymmetric components, were used.

PC analyses were carried out for the symmetric and asymmetric components of shape in order to extract the patterns of variation across hemimandibles [59]. Shape changes associated with each of the first two PCs (PC1 and PC2) were visualized as diagrams of simultaneous displacements of landmarks [52]. In order to quantify the degree to which these patterns differed, vector angles between normalized PCs were calculated as the arccosine of their vector correlation (for details, see [60,61]). Calculations were conducted independently for PC1 and PC2. For each component of shape, vector angles were calculated between all pairs of groups. Additionally, the angles between corresponding PCs of the two components of shape were calculated in each group. Tests against the null hypothesis of vectors having random directions in the shape tangent space were performed through the "Compare Vector Directions" function on MorphoJ.

Similarity in the patterns of morphological integration was tested computing matrix correlations between covariance matrices. The six groups were compared in pairs for each component of shape variation separately. Besides, the correlation between the covariance matrices of the symmetric and asymmetric components was calculated in each group [34]. Matrix correlations excluded the diagonal blocks of the covariance matrices [62]. Statistical significances were determined through matrix permutation tests, with 10,000 iterations, against the null hypothesis of complete dissimilarity between the covariance matrices concerned [63]. Mantel and partial Mantel tests were conducted with the aim of assessing the possible association

between differences in the patterns of morphological integration and both karyotypic and geographic distances between groups (see Additional file 4). Differences in the patterns of morphological integration between groups were calculated as 1-r, with r being the correlation coefficient between the covariance matrices of the pair of groups under comparison. Both raw and size-corrected data, as well as data of the symmetric and asymmetric components, were used.

In multiple comparisons, *P*-values were adjusted by implementing the false discovery rate (FDR) procedure [64].

Availability of supporting data
The data set supporting the results of this article is provided as an additional file (Additional file 3).

Additional files

> **Additional file 1:** Two-way ANOVA of centroid size and Procrustes ANOVA of shape for the study sample.
>
> **Additional file 2:** Collection sites and individual karyotypes of the study sample, indicating the set of Rb translocations and their structural heterozygosity.
>
> **Additional file 3:** Raw coordinates of the 17 landmarks digitized in the sample analyzed.
>
> **Additional file 4:** Karyotypic distances (below the diagonal) and geographic distances (above the diagonal) between chromosomal groups.

Abbreviations
ANOVA: Analysis of variance; FDR: False discovery rate; F1: Filial 1; F2: Filial 2; PC: Principal component; QTLs: Quantitative trait loci; Rb: Robertsonian; St: Standard.

Competing interests
The authors declare that they have no competing interests.

Authors' contributions
JMV carried out the acquisition of morphometric data, participated in the performance of statistical analyses and drafted the manuscript. FMM conceived the study, participated in its coordination and in the performance of statistical analyses, helped to draft the manuscript and revised it critically. NM carried out the collection of the sample and the acquisition of chromosome preparations and karyotypic data. MJLF and JV revised the manuscript and gave final approval of the version to be published. JV coordinated the research project in which this study is included. All authors participated in the design of the study and the interpretation of data, and read and approved the final manuscript.

Acknowledgments
We are indebted to Christopher Evans (Servei d'Assessorament Lingüístic, University of Barcelona) for revising the English and to Dr. John Abramyan (University of British Columbia) for writing assistance. We also thank two anonymous reviewers for their valuable comments on previous drafts of this manuscript. This work was conducted in the framework of the doctoral program in Biodiversity from Universitat Autònoma de Barcelona, and was funded by the Spanish *Ministerio de Economía y Competitividad* (project reference CGL2010-15243) and by a PIF grant from Universitat Autònoma de Barcelona to JMV.

Author details
[1]Departament de Biologia Animal, de Biologia Vegetal i d'Ecologia, Universitat Autònoma de Barcelona, Facultat de Biociències, Bellaterra, E-08193 Cerdanyola del Vallès, Spain. [2]Departament de Biologia Animal and Institut de Recerca de la Biodiversitat (IRBio), Universitat de Barcelona, Facultat de Biologia, Av. Diagonal 645, E-08028 Barcelona, Spain.

References

1. Olson EC, Miller RL: *Morphological integration*. Chicago: University of Chicago Press; 1958.
2. Schlosser G, Wagner GP: *Modularity in development and evolution*. Chicago: University of Chicago Press; 2004.
3. Klingenberg CP: Evolution and development of shape: integrating quantitative approaches. *Nat Rev Genet* 2010, 11:623–635.
4. Klingenberg CP: Morphological integration and developmental modularity. *Annu Rev Ecol Evol Syst* 2008, 39:115–132.
5. Piálek J, Hauffe HC, Searle JB: Chromosomal variation in the house mouse. *Biol J Linn Soc* 2005, 84:535–563.
6. Bidau CJ: Causes of chiasma repatterning due to centric fusions. *Braz J Genet* 1993, 16:283–296.
7. Bidau CJ, Giménez MD, Palmer CL, Searle JB: The effects of Robertsonian fusions on chiasma frequency and distribution in the house mouse (*Mus musculus domesticus*) from a hybrid zone in northern Scotland. *Heredity* 2001, 87:305–313.
8. Castiglia R, Capanna E: Chiasma repatterning across a chromosomal hybrid zone between chromosomal races of *Mus musculus domesticus*. *Genetica* 2002, 114:35–40.
9. Dumas D, Britton-Davidian J: Chromosomal rearrangements and evolution of recombination: comparison of chiasma distribution patterns in standard and Robertsonian populations of the house mouse. *Genetics* 2002, 162:1355–1366.
10. Capilla L, Medarde N, Alemany-Schmidt A, Oliver-Bonet M, Ventura J, Ruiz-Herrera A: Genetic recombination variation in wild Robertsonian mice: on the role of chromosomal fusions and Prdm9 allelic background. *Proc Biol Sci* 2014. doi:10.1098/rspb.2014.0297.
11. Navarro A, Barton NH: Chromosomal speciation and molecular divergence – accelerated evolution in rearranged chromosomes. *Science* 2003, 300:321–324.
12. Wagner GP, Pavlicev M, Cheverud JM: The road to modularity. *Nat Rev Genet* 2007, 8:921–931.
13. Graham JH: Genomic coadaptation and developmental stability in hybrid zones. *Acta Zool Fennica* 1992, 191:121–131.
14. Grant PR, Grant BR: Phenotypic and genetic effects of hybridization in Darwin's finches. *Evolution* 1994, 48:297–316.
15. Wagner GP, Zhang J: The pleiotropic structure of the genotype-phenotype map: the evolvability of complex organisms. *Nat Rev Genet* 2011, 12:204–213.
16. Schwarz E, Schwarz HK: The wild and commensal stocks of the house mouse, *Mus musculus* Linnaeus. *J Mammal* 1943, 24:59–72.
17. Adolph S, Klein J: Robertsonian variation in *Mus musculus* from Central Europe, Spain, and Scotland. *J Hered* 1981, 72:219–221.
18. Gündüz I, López-Fuster MJ, Ventura J, Searle JB: Clinal analysis of a chromosomal hybrid zone in the house mouse. *Genet Res* 2001, 77:41–51.
19. Sans-Fuentes MA, Muñoz-Muñoz F, Ventura J, López-Fuster MJ: Rb(7.17), a rare Robertsonian fusion in wild populations of the house mouse. *Genet Res* 2007, 89:207–213.
20. Medarde N, López-Fuster MJ, Muñoz-Muñoz F, Ventura J: Spatio-temporal variation in the structure of a chromosomal polymorphism zone in the house mouse. *Heredity* 2012, 109:78–89.
21. Atchley WR, Hall BK: A model for development and evolution of complex morphological structures. *Biol Rev Camb Philos Soc* 1991, 66:101–157.
22. Klingenberg CP, Navarro N: Development of the mouse mandible: a model system for complex morphological structures. In *Evolution of the house mouse*. Edited by Macholán M, Baird SJE, Munclinger P, Piálek J. Cambridge: Cambridge University Press; 2012:135–149.
23. Ehrich TH, Vaughn TT, Koreishi SF, Linsey RB, Pletscher LS, Cheverud JM: Pleiotropic effects on mandibular morphology I. Developmental morphological integration and differential dominance. *J Exp Zool B Mol Dev Evol* 2003, 296:58–79.
24. Klingenberg CP, Leamy LJ, Cheverud JM: Integration and modularity of quantitative trait locus effects on geometric shape in the mouse mandible. *Genetics* 2004, 166:1909–1921.
25. Burgio G, Baylac M, Heyer E, Montagutelli X: Exploration of the genetic organization of morphological modularity on the mouse mandible using

a set of interspecific recombinant congenic strains between C57BL/6 and mice of the *Mus spretus* species. *G3 (Bethesda)* 2012, 2:1257–1268.
26. Hall BK: Unlocking the black box between genotype and phenotype: cell condensations as morphogenetic (modular) units. *Biol Philos* 2003, 18:219–247.
27. Klingenberg CP: Morphometric integration and modularity in configurations of landmarks: tools for evaluating a priori hypotheses. *Evol Dev* 2009, 11:405–421.
28. Muñoz-Muñoz F, Sans-Fuentes MA, López-Fuster MJ, Ventura J: Evolutionary modularity of the mouse mandible: dissecting the effect of chromosomal reorganizations and isolation by distance in a Robertsonian system of *Mus musculus domesticus*. *J Evol Biol* 2011, 24:1763–1776.
29. Jojić V, Blagojević J, Ivanović A, Bugarski-Stanojević V, Vujošević M: Morphological integration of the mandible in yellow-necked field mice: the effects of B chromosomes. *J Mammal* 2007, 88:689–695.
30. Jojić V, Blagojević J, Vujošević M: B chromosomes and cranial variability in yellow-necked field mice (*Apodemus flavicollis*). *J Mammal* 2011, 92:396–406.
31. Marroig G, Cheverud JM: A comparison of phenotypic variation and covariation patterns and the role of phylogeny, ecology, and ontogeny during cranial evolution of New World monkeys. *Evolution* 2001, 55:2576–2600.
32. Merilä J, Björklund M: Phenotypic integration as a constraint and adaptation. In *Phenotypic integration: studying the ecology and evolution of complex phenotypes*. Edited by Pigliucci M, Preston K. New York: Oxford University Press; 2004:107–129.
33. Pigliucci M: Is evolvability evolvable? *Nat Rev Genet* 2008, 9:75–82.
34. Klingenberg CP, Mebus K, Auffray J-C: Developmental integration in a complex morphological structure: how distinct are the modules in the mouse mandible? *Evol Dev* 2003, 5:522–531.
35. Renaud S, Alibert P, Auffray J-C: Modularity as a source of new morphological variation in the mandible of hybrid mice. *BMC Evol Biol* 2012, 12:141–156.
36. Cheverud JM, Rutledge JJ, Atchley WR: Quantitative genetics of development: genetic correlations among age-specific trait values and the evolution of ontogeny. *Evolution* 1983, 37:895–905.
37. Wagner GP: On the eigenvalue distribution of genetic and phenotypic dispersion matrices: evidence for a nonrandom organization of quantitative character variation. *J Math Biol* 1984, 21:77–95.
38. Leamy L: Morphological integration of fluctuating asymmetry in the mouse mandible. *Genetica* 1993, 89:139–153.
39. Cheverud JM, Routman EJ, Irschick DJ: Pleiotropic effects of individual gene loci on mandibular morphology. *Evolution* 1997, 51:2006–2016.
40. Leamy LJ, Klingenberg CP, Sherratt E, Wolf JB, Cheverud JM: A search for quantitative trait loci exhibiting imprinting effects on mouse mandible size and shape. *Heredity* 2008, 101:518–526.
41. González-José R, Van der Molen S, González-Pérez E, Hernández M: Patterns of phenotypic covariation and correlation in modern humans as viewed from morphological integration. *Am J Phys Anthropol* 2004, 123:69–77.
42. Debat V, Alibert P, David P, Paradis E, Auffray J-C: Independence between developmental stability and canalization in the skull of the house mouse. *Proc Biol Sci* 2000, 267:423–430.
43. Ackermann RR: Patterns of covariation in the hominoid craniofacial skeleton: implications for paleoanthropological models. *J Hum Evol* 2002, 43:167–187.
44. Drake AG, Klingenberg CP: Large-scale diversification of skull shape in domestic dogs: disparity and modularity. *Am Nat* 2010, 175:289–301.
45. Ford CE: The use of chromosome markers. In *Tissue Grafting and Radiation*. Edited by Micklem HS, Loutit JF. New York: Academic; 1966:197–206.
46. Mandahl N: Methods in solid tumor cytogenetics. In *Human cytogenetics. A practical approach*. Edited by Rooney DE, Czepulkowski BH. London: IRL Press; 1992:155–187.
47. Klingenberg CP: MorphoJ: an integrated software package for geometric morphometrics. *Mol Ecol Resour* 2011, 11:353–357.
48. Dryden IL, Mardia KV: *Statistical shape analysis*. Chichester: Wiley; 1998.
49. Klingenberg CP, McIntyre GS, Zaklan SD: Left-right asymmetry of fly wings and the evolution of body axes. *Proc Biol Sci* 1998, 265:1255–1259.
50. Klingenberg CP, McIntyre GS: Geometric morphometrics of developmental instability: analyzing patterns of fluctuating asymmetry with Procrustes methods. *Evolution* 1998, 52:1363–1375.
51. Klingenberg CP, Barluenga M, Meyer A: Shape analysis of symmetric structures: quantifying variation among individuals and asymmetry. *Evolution* 2002, 56:1909–1920.
52. Klingenberg CP, Zaklan SD: Morphological integration between developmental compartments in the *Drosophila* wing. *Evolution* 2000, 54:1273–1285.

53. Monteiro LR: **Multivariate regression models and geometric morphometrics: the search for causal factors in the analysis of shape.** *Syst Biol* 1999, **48:**192–199.
54. Good P: *Permutation tests: a practical guide to resampling methods for testing hypotheses.* New York: Springer; 1994.
55. Escoufier Y: **Le traitement des variables vectorielles.** *Biometrics* 1973, **29:**751–760.
56. Martínez-Abadías N, Esparza M, Sjøvold T, González-José R, Santos M, Hernández M, Klingenberg CP: **Pervasive genetic integration directs the evolution of human skull shape.** *Evolution* 2012, **66:**1010–1023.
57. Fruciano C, Franchini P, Meyer A: **Resampling-based approaches to study variation in morphological modularity.** *PLoS One* 2013, **8:**e69376. doi:10.1371/journal.pone.0069376.
58. Smilde AK, Kiers HAL, Bijlsma S, Rubingh CM, van Erk MJ: **Matrix correlations for high-dimensional data: the modified RV-coefficient.** *Bioinformatics* 2009, **25:**401–405.
59. Jolliffe IT: *Principal component analysis.* New York: Springer; 1986.
60. Klingenberg CP: **Multivariate allometry.** In *Advances in morphometrics.* Edited by Marcus LF, Corti M, Loy A, Naylor GJP, Slice DE. New York: Plenum Press; 1996:23–49.
61. Young RL, Badyaev AV: **Evolutionary persistence of phenotypic integration: influence of developmental and functional relationships on complex trait evolution.** *Evolution* 2006, **60:**1291–1299.
62. Klingenberg CP, Badyaev AV, Sowry SM, Beckwith NJ: **Inferring developmental modularity from morphological integration: analysis of individual variation and asymmetry in bumblebee wings.** *Am Nat* 2001, **157:**11–23.
63. Cheverud JM, Wagner GP, Dow MM: **Methods for the comparative analysis of variation patterns.** *Syst Zool* 1989, **38:**201–213.
64. Benjamini Y, Hochberg Y: **Controlling the false discovery rate: a practical and powerful approach to multiple testing.** *J R Stat Soc Series B Stat Methodol* 1995, **57:**289–300.

Permissions

All chapters in this book were first published in FZ, by BioMed Central; hereby published with permission under the Creative Commons Attribution License or equivalent. Every chapter published in this book has been scrutinized by our experts. Their significance has been extensively debated. The topics covered herein carry significant findings which will fuel the growth of the discipline. They may even be implemented as practical applications or may be referred to as a beginning point for another development.

The contributors of this book come from diverse backgrounds, making this book a truly international effort. This book will bring forth new frontiers with its revolutionizing research information and detailed analysis of the nascent developments around the world.

We would like to thank all the contributing authors for lending their expertise to make the book truly unique. They have played a crucial role in the development of this book. Without their invaluable contributions this book wouldn't have been possible. They have made vital efforts to compile up to date information on the varied aspects of this subject to make this book a valuable addition to the collection of many professionals and students.

This book was conceptualized with the vision of imparting up-to-date information and advanced data in this field. To ensure the same, a matchless editorial board was set up. Every individual on the board went through rigorous rounds of assessment to prove their worth. After which they invested a large part of their time researching and compiling the most relevant data for our readers.

The editorial board has been involved in producing this book since its inception. They have spent rigorous hours researching and exploring the diverse topics which have resulted in the successful publishing of this book. They have passed on their knowledge of decades through this book. To expedite this challenging task, the publisher supported the team at every step. A small team of assistant editors was also appointed to further simplify the editing procedure and attain best results for the readers.

Apart from the editorial board, the designing team has also invested a significant amount of their time in understanding the subject and creating the most relevant covers. They scrutinized every image to scout for the most suitable representation of the subject and create an appropriate cover for the book.

The publishing team has been an ardent support to the editorial, designing and production team. Their endless efforts to recruit the best for this project, has resulted in the accomplishment of this book. They are a veteran in the field of academics and their pool of knowledge is as vast as their experience in printing. Their expertise and guidance has proved useful at every step. Their uncompromising quality standards have made this book an exceptional effort. Their encouragement from time to time has been an inspiration for everyone.

The publisher and the editorial board hope that this book will prove to be a valuable piece of knowledge for researchers, students, practitioners and scholars across the globe.

List of Contributors

Stefan M Jahnel
Department of Molecular Evolution and Development, Centre for Organismal Biology, Faculty of Life Sciences, University of Vienna, Althanstrasse 14, 1090 Wien, Austria

Manfred Walzl
Department of Integrative Zoology, Centre for Organismal Biology, Faculty of Life Sciences, University of Vienna, Althanstrasse 14, 1090 Wien, Austria

Ulrich Technau
Department of Molecular Evolution and Development, Centre for Organismal Biology, Faculty of Life Sciences, University of Vienna, Althanstrasse 14, 1090 Wien, Austria

Paul A Bromiley
Centre for Imaging Sciences, University of Manchester, Stopford Building, Oxford Road, Manchester M13 9PT, UK

Anja C Schunke
Department for Evolutionary Genetics, Max Planck Institute for Evolutionary Biology, August-Thienemann-Str. 2, 24306 Plön, Germany

Hossein Ragheb
Centre for Imaging Sciences, University of Manchester, Stopford Building, Oxford Road, Manchester M13 9PT, UK

Neil A Thacker
Centre for Imaging Sciences, University of Manchester, Stopford Building, Oxford Road, Manchester M13 9PT, UK

Diethard Tautz
Department for Evolutionary Genetics, Max Planck Institute for Evolutionary Biology, August-Thienemann-Str. 2, 24306 Plön, Germany

Nicolas Bekkouche
Marine Biological Section, Department of Biology, University of Copenhagen, Universitetsparken 4, 2100 Copenhagen Ø, Denmark

Reinhardt M Kristensen
Natural History Museum of Denmark, Universitetsparken 15, 2100 Copenhagen Ø, Denmark

Andreas Hejnol
Sars International Centre for Marine Molecular Biology, University of Bergen, Thormøhlensgate 55, Bergen N-5008, Norway

Martin V Sørensen
Natural History Museum of Denmark, Øster Voldgade 5-7, 1350 Copenhagen K, Denmark

Katrine Worsaae
Marine Biological Section, Department of Biology, University of Copenhagen, Universitetsparken 4, 2100 Copenhagen Ø, Denmark

Benjamin L Allen
School of Agriculture and Food Sciences, The University of Queensland, Warrego Highway, Gatton, QLD 4343, Australia
Robert Wicks Pest Animal Research Centre, Biosecurity Queensland, Tor Street, Toowoomba, QLD 4350, Australia

Lee R Allen
Robert Wicks Pest Animal Research Centre, Biosecurity Queensland, Tor Street, Toowoomba, QLD 4350, Australia

Richard M Engeman
National Wildlife Research Centre, US Department of Agriculture, LaPorte Avenue, Fort Collins, CO 80521-2154, USA

Luke K-P Leung
School of Agriculture and Food Sciences, The University of Queensland, Warrego Highway, Gatton, QLD 4343, Australia

Mathieu Giraudeau
School of Life Sciences, Arizona State University, Tempe, AZ 85287-4501, USA
School of Biological Sciences A08, University of Sydney, Sydney, NSW 2006, Australia

Paul M Nolan
Department of Biology, The Citadel, Charleston, SC 29409, USA

Caitlin E Black
Department of Biology, The College of Charleston, Charleston, SC 29424, USA

Stevan R Earl
Global Institute of Sustainability & School of Sustainability, Arizona State University, Tempe, AZ 85287-5402, USA

Masaru Hasegawa
Graduate School of Life and Environmental Sciences, University of Tsukuba, 1-1-1 Tennoudai, Tsukuba-shi, Ibaraki 305-8572, Japan

Kevin J McGraw
School of Life Sciences, Arizona State University, Tempe, AZ 85287-4501, USA

Guangjian Liu
Key Laboratory of Animal Ecology and Conservation Biology, Institute of Zoology, Chinese Academy of Sciences, 1-5 Beichen West Road, Chaoyang, Beijing 100101, China

Lutz Walter
Primate Genetics Laboratory, German Primate Center, Leibniz Institute for Primate Research, Kellnerweg 4, 37077 Göttingen, Germany
Gene Bank of Primates, German Primate Center, Leibniz Institute for Primate Research, Kellnerweg 4, 37077 Göttingen, Germany

Suni Tang
Department of Biomedical Sciences, School of Pharmacy, Texas Tech University Health Sciences Center, 1300 S. Coulter St, Amarillo, TX 79106, USA

Xinxin Tan
Key Laboratory of Animal Ecology and Conservation Biology, Institute of Zoology, Chinese Academy of Sciences, 1-5 Beichen West Road, Chaoyang, Beijing 100101, China
Institute of Health Sciences, Anhui University, Hefei, Anhui Province 230601, China

Fanglei Shi
Key Laboratory of Animal Ecology and Conservation Biology, Institute of Zoology, Chinese Academy of Sciences, 1-5 Beichen West Road, Chaoyang, Beijing 100101, China

Huijuan Pan
College of Nature Conservation, Beijing Forestry University, Haidian, Beijing 100083, China

Christian Roos
Primate Genetics Laboratory, German Primate Center, Leibniz Institute for Primate Research, Kellnerweg 4, 37077 Göttingen, Germany
Gene Bank of Primates, German Primate Center, Leibniz Institute for Primate Research, Kellnerweg 4, 37077 Göttingen, Germany

Zhijin Liu
Key Laboratory of Animal Ecology and Conservation Biology, Institute of Zoology, Chinese Academy of Sciences, 1-5 Beichen West Road, Chaoyang, Beijing 100101, China
Primate Genetics Laboratory, German Primate Center, Leibniz Institute for Primate Research, Kellnerweg 4, 37077 Göttingen, Germany

Ming Li
Key Laboratory of Animal Ecology and Conservation Biology, Institute of Zoology, Chinese Academy of Sciences, 1-5 Beichen West Road, Chaoyang, Beijing 100101, China

Urszula Weclawski
Department of Ecology and Parasitology, Zoological Institute, Karlsruhe Institute of Technology, Kornblumenstrasse 13, Karlsruhe, Germany

Emanuel G Heitlinger
Department of Ecology and Parasitology, Zoological Institute, Karlsruhe Institute of Technology, Kornblumenstrasse 13, Karlsruhe, Germany

Tobias Baust
Department of Stochastics, Karlsruhe Institute of Technology, Kaiserstrasse 89, Karlsruhe, Germany

Bernhard Klar
Department of Stochastics, Karlsruhe Institute of Technology, Kaiserstrasse 89, Karlsruhe, Germany

Trevor Petney
Department of Ecology and Parasitology, Zoological Institute, Karlsruhe Institute of Technology, Kornblumenstrasse 13, Karlsruhe, Germany

Yu-San Han
Institute of Fisheries Science, College of Life Science, National Taiwan University, Taipei, Taiwan

Horst Taraschewski
Department of Ecology and Parasitology, Zoological Institute, Karlsruhe Institute of Technology, Kornblumenstrasse 13, Karlsruhe, Germany

Yao-Hua Zhang
State Key Laboratory of Integrated Management of Pest Insects and Rodents in Agriculture, Institute of Zoology, Chinese Academy of Sciences, 1# Bei-Chen West Road, Beijing 100101, China

Jian-Xu Zhang
State Key Laboratory of Integrated Management of Pest Insects and Rodents in Agriculture, Institute of Zoology, Chinese Academy of Sciences, 1# Bei-Chen West Road, Beijing 100101, China

Kim G Mortega
Department of Migration and Immuno-Ecology, Max Planck Institute for Ornithology, 78315 Radolfzell, Germany
Department of Ornithology, University of Konstanz, 78457 Konstanz, Germany
Institute of Biodiversity, Animal Health and Comparative Medicine, University of Glasgow, G12 8QQ Glasgow, UK

Heiner Flinks
Am Kuhm 19, 46325 Borken, Germany

Barbara Helm
Department of Ornithology, University of Konstanz, 78457 Konstanz, Germany
Institute of Biodiversity, Animal Health and Comparative Medicine, University of Glasgow, G12 8QQ Glasgow, UK

Patrick Kück
Zoologisches Forschungsmuseum A. Koenig, Adenauerallee 160-163, 53113 Bonn, Germany

Gary C Longo
Center for Ocean Health, 100 Shaffer Road, 95060 Santa Cruz, CA, USA

Ingo Schiffner
Queensland Brain Institute, University of Queensland, St Lucia QLD 4072, Australia

Hong D Vo
Queensland Brain Institute, University of Queensland, St Lucia QLD 4072, Australia

Partha S Bhagavatula
Queensland Brain Institute, University of Queensland, St Lucia QLD 4072, Australia

Mandyam V Srinivasan
Queensland Brain Institute, University of Queensland, St Lucia QLD 4072, Australia
School of Information Technology and Electrical Engineering, University of Queensland, St Lucia QLD 4072, Australia
ARC Centre of Excellence in Vision Science, University of Queensland, St Lucia QLD 4072, Australia

Melanie Schiffer
Integrative Ecophysiology, Alfred-Wegener-Institute for Polar and Marine Research, Am Handelshafen 12, 27570 Bremerhaven, Germany

Lars Harms
Scientific Computing, Alfred-Wegener-Institute for Polar and Marine Research, Am
Handelshafen 12, 27570 Bremerhaven, Germany

Magnus Lucassen
Integrative Ecophysiology, Alfred-Wegener-Institute for Polar and Marine Research, Am Handelshafen 12, 27570 Bremerhaven, Germany

Felix Christopher Mark
Integrative Ecophysiology, Alfred-Wegener-Institute for Polar and Marine Research, Am Handelshafen 12, 27570 Bremerhaven, Germany

Hans-Otto Pörtner
Integrative Ecophysiology, Alfred-Wegener-Institute for Polar and Marine Research, Am Handelshafen 12, 27570 Bremerhaven, Germany

Daniela Storch
Integrative Ecophysiology, Alfred-Wegener-Institute for Polar and Marine Research, Am Handelshafen 12, 27570 Bremerhaven, Germany

Daniela Sint
Institute of Ecology, University of Innsbruck, Technikerstraße 25, 6020 Innsbruck, Austria

Isabella Thurner
Institute of Ecology, University of Innsbruck, Technikerstraße 25, 6020 Innsbruck, Austria

Ruediger Kaufmann
Institute of Ecology, University of Innsbruck, Technikerstraße 25, 6020 Innsbruck, Austria

Michael Traugott
Institute of Ecology, University of Innsbruck, Technikerstraße 25, 6020 Innsbruck, Austria

Jessica Martínez-Vargas
Departament de Biologia Animal, de Biologia Vegetal i d'Ecologia, Universitat Autònoma de Barcelona, Facultat de Biociències, Bellaterra, E-08193 Cerdanyola del Vallès, Spain

Francesc Muñoz-Muñoz
Departament de Biologia Animal, de Biologia Vegetal i d'Ecologia, Universitat Autònoma de Barcelona, Facultat de Biociències, Bellaterra, E-08193 Cerdanyola del Vallès, Spain

Nuria Medarde
Departament de Biologia Animal, de Biologia Vegetal i d'Ecologia, Universitat Autònoma de Barcelona, Facultat de Biociències, Bellaterra, E-08193 Cerdanyola del Vallès, Spain

María José López-Fuster
Departament de Biologia Animal and Institut de Recerca de la Biodiversitat (IRBio), Universitat de Barcelona, Facultat de Biologia, Av. Diagonal 645, E-08028 Barcelona, Spain

Jacint Ventura
Departament de Biologia Animal, de Biologia Vegetal i d'Ecologia, Universitat Autònoma de Barcelona, Facultat de Biociències, Bellaterra, E-08193 Cerdanyola del Vallès, Spain

www.ingramcontent.com/pod-product-compliance
Lightning Source LLC
Chambersburg PA
CBHW080258230326

41458CB00097B/5121

* 9 7 8 1 6 8 2 8 6 0 5 9 5 *